PRESCRIBING HEMODIALYSIS
A Guide to Urea Modeling

DEVELOPMENTS IN NEPHROLOGY

Cheigh, J.S., Stenzel, K.H. and Rubin, A.L. (eds.): Manual of Clinical Nephrology of the Rogosin Kidney Center. 1981. ISBN 90-247-2397-3.
Nolph, K.D. (ed.): Peritoneal Dialysis. 1981 ed.: out of print. 3rd revised and enlarged ed. 1988 (not in this series). ISBN 0-89838-406-0.
Gruskin, A.B. and Norman, M.E. (eds.): Pediatric Nephrology, 1981. ISBN 90-247-2514-3.
Schück, O.: Examination of the Kidney Function. 1981. ISBN 0-89838-565-2.
Strauss, J. (ed.): Hypertension, Fluid-electrolytes and Tubulopathies in Pediatric Nephrology. 1982. ISBN 90-247-2633-6.
Strauss, J. (ed.): Neonatal Kidney and Fluid-electrolytes. 1983. ISBN 0-89838-575-X.
Strauss, J. (ed.): Acute Renal Disorders and Renal Emergencies. 1984. ISBN 0-89838-663-2.
Didio, L.J.A. and Motta, P.M. (eds.): Basic, Clinical, and Surgical Nephrology. 1985. ISBN 0-89838-698-5.
Friedman, E.A. and Peterson, C.M. (eds.): Diabetic Nephropathy: Strategy for Therapy. 1985. ISBN 0-89838-735-3.
Dzurik, R., Lichardus, B. and Guder, W. (eds.): Kidney Metabolism and Function. 1985. ISBN 0-89838-749-3.
Strauss, J. (ed.): Homeostasis, Nephrotoxicity, and Renal Anomalies in the Newborn. 1986. ISBN 0-89838-766-3.
Oreopoulos, D.G. (ed.): Geriatric Nephrology. 1986. ISBN 0-89838-781-7.
Paganini, E.P. (ed.): Acute Continuous Renal Replacement Therapy. 1986. ISBN 0-89838-793-0.
Cheigh, J.S., Stenzel, K.H. and Rubin, A.L. (eds.): Hypertension in Kidney Disease. 1986. ISBN 0-89838-797-3.
Deane, N., Wineman, R.J. and Benis, G.A. (eds.): Guide to Reprocessing of Hemodialyzers. 1986. ISBN 0-89838-798-1.
Ponticelli, C., Minetti, L. and D'Amico, G. (eds.): Antiglobulins, Cryoglobulins and Glomerulonephritis. 1986. ISBN 0-89838-810-4.
Strauss, J. (ed.), with the assistance of L. Strauss: Persistent Renal-genitourinary Disorders. 1987. ISBN 0-89838-845-7.
Andreucci, V.E. and Dal Canton, A. (eds.): Diuretics: Basic, Pharmacological, and Clinical Aspects. 1987. ISBN 0-89838-885-6.
Bach, P.H. and Lock, E.H. (eds): Nephrotoxicity in the Experimental and Clinical Situation, Part 1. 1987. ISBN 0-89838-977-1.
Bach, P.H. and Lock, E.H. (eds.): Nephrotoxicity in the Experimental and Clinical Situation, Part 2. 1987. ISBN 0-89838-980-2.
Gore, S.M. and Bradley, B.A. (eds.): Renal Transplantation: Sense and Sensitization. 1988. ISBN 0-89838-370-6.
Minetti, L., D'Amico, G. and Ponticelli, C. (eds.): The Kidney in Plasma Cell Dyscrasias. 1988. ISBN 0-89838-385-4.
Lindblad, A.S., Novak, J.W. and Nolph, K.D. (eds.): Continuous Ambulatory Peritoneal Dialysis in the USA. 1989. ISBN 0-7923-0179-X.
Andreucci, V.E. and Dal Canton, A. (eds.): Current Therapy in Nephrology. 1989. ISBN 0-7923-0206-0.
Depner, T.A. (ed.): Prescribing Hemodialysis: A Guide to Urea Modeling. 1990. ISBN 0-7923-0833-6.

PRESCRIBING HEMODIALYSIS

A Guide to Urea Modeling

Thomas A. Depner, M.D.
University of California, Davis

Kluwer Academic Publishers
Boston/Dordrecht/London

Distributors for North America:
Kluwer Academic Publishers
101 Philip Drive
Assinippi Park
Norwell, Massachusetts 02061 USA

Distributors for all other countries:
Kluwer Academic Publishers Group
Distribution Centre
Post Office Box 322
3300 AH Dordrecht, THE NETHERLANDS

Library of Congress Cataloging-in-Publication Data

Depner, Thomas A.
Prescribing hemodialysis : a guide to urea modeling / Thomas A. Depner.
p. cm.
Includes bibliographical references.
ISBN 0-7923-0833-6
1. Hemodialysis—Evaluation—Mathematical models. 2. Urea--Pharmacokinetics—Mathematical models. I. Title.
[DNLM: 1. Blood Urea Nitrogen. 2. Hemodialysis. 3. Kidney Failure, Chronic—therapy. 4. Models, Biological. 5. Urea--metabolism. 6. Urea—pharmacokinetics. WJ 342 D421p]
RC901.7.H45D47 1990
617.4'61059—dc20
DNLM/DLC
for Library of Congress 90-4842
CIP

Seventh Printing 2002.

Printed on acid-free paper.

Printed in the United States of America.

This printing is a digital duplication of the original edition.

DEDICATION

To my patients, with sincere hope for a healthier future, and to my family, Celeste, Charles, Kristine, and Ivy, for their love and understanding.

TABLE OF CONTENTS

FOREWORD

What regulation shall we have for the operation? Shall a man transfuse he knows not what, to correct he knows not what, God knows how (1)?

Dr. Henry Stubbs
Royal College of Physicians
circa 1670

If dialysis therapy were a new pharmaceutical product being evaluated by the FDA now, it might not be approved for marketing. The recommended dose, its potential toxicity, the side effects of under- or over-dialysis as well as its efficacy have been the subject of very few studies. The high mortality rate associated with the treatment may raise a few eyebrows.

That it is a life-saving modality of treatment is undoubtedly true for more than 100,000 patients in the United States and for more than a million patients worldwide. Because dialysis has extended the lives of many people by a variable period of time, most nephrologists have "rested on their laurels" and did not vigorously pursue studies to optimize these treatments. But facts have a way of intruding in all our lives and the facts are that the overall mortality rate of dialysis patients in the United States is rising and stands close to 25% per year and is closer to 33% per year for patients between the ages of 65 and 74 (2). These mortality figures are considerably higher for age-adjusted dialysis populations in Europe and particularly in Japan, and certainly for the age-adjusted normal population.

The reasons for this trend of increasing mortality can and have been debated in several forums particularly in the recent past (3); however, there is increasing evidence that overlaying the multitude of factors that impact on morbidity and mortality of this patient population (4,5) is the prescription of inadequate amounts of dialysis (6,7). It is perhaps no coincidence that the mortality rate is rising at a time when the average dialysis time is decreasing, and as government reimbursements for dialysis procedures have declined (4).

Why a book on prescribing hemodialysis? One can argue that a book dedicated to the potentially single most important issue in the lives of a million people is justification enough. But there are more positive reasons for this book. Dialysis prescription is no longer an art, but a science founded on good theoretical principles that should be relatively simple to apply. A book that explains the fundamentals and shows the "how to" of applying these principles to a rational prescription is sorely needed. Another good reason for such a book is that it allows us to have the courage of our convictions when we are negotiating with patients about their time on dialysis. It is only by understanding the rationale of prescribing 4 or more hours instead of 3 hours of dialysis each time that we can convince our patients of the importance and necessity of this additional burden. It is by understanding the specific consequences that would result from changing blood flow or dialyzer surface area that one can end up with a rational prescription.

The book by Thomas Depner will provide the practicing Nephrologist with both the theoretical and more importantly the practical, step-by-step approach to the prescription of dialysis, the monitoring of its delivery and outcome. Dr. Depner is clearly a scholar who enjoys teaching. He is able to use a combination of text, tables and figures to lucidly present materials that make learning fun while imparting practical information. Dr. Depner also uses a tool that has become indispensable to the practice of nephrology, the computer. His step-by-step approach will make anybody who can type an easy expert on the subject (I know whereof I speak). In having each chapter self-contained, the readers can elect to pick and choose among the chapters depending on their interest, background and goals. The chapters are organized so that the reader can understand the "how to" as well as the "why" and the "why not." Most importantly, as Dr. Depner correctly points out, learning about dialysis prescription can be fun.

This book is of value not only to nephrologists, but also to nurses and quality assurance personnel who are increasingly interested in the appropriate delivery of dialytic care to their patients. It is a book that I recommend highly.

Raymond Hakim, M.D.

References

1. As quoted by Dau PC: Plasmapheresis Therapy in Myathenia Gravis. *Muscle & Nerve* 3:468-482, 1980.
2. Hull AR, Parker TF (eds): Proceedings from the Morbidity, Mortality and Prescription of Dialysis Symposium. Am J Kidney Dis 15:375-383, 1990.
3. Hakim RM: Assessing the adequacy of dialysis (Nephrology Forum). Kidney Int 37:822-832, 1990.
4. Held PJ, Garcia JR, Pauly MV, Cohn MA: Price of dialysis, unit staffing and the length of dialysis treatments. Am J Kidney Dis 15:441-450, 1990.
5. Lowrie EG, Lew NL: Death risk in hemodialysis patients: The predictive value of commonly measured variables and an evaluation of death rate differences between facilities. Am J Kidney Dis 15:458-482, 1990.
6. Sargent JA: Shortfalls in the delivery of dialysis. Am J Kidney Dis 15:500-510, 1990.
7. Gotch FA, Yarian S, Keen M: A kinetic survey of US hemodialysis prescriptions. Am J Kidney Dis 15:511-515, 1990.

PREFACE

Patient: "I had a good dialysis today."
Friend: "How do you know. . . .?"

The hemodialysis industry is a multifaceted spectrum of goods and services designed to confer longevity on patients with end-stage renal disease. The industry had its beginnings in the early 1960s, but not until 1973, when the U.S. Medicare program began to subsidize these life-sustaining programs, did it really begin to take root and grow to the proportions we see today. Recent statistics compiled by the Health Care Finance Administration (table 1) reflect the magnitude of this effort and the extent of the commitment of those who have agreed to fund it (1). Such a commitment cries for feedback from the industry, so that its supporters, often people of nonmedical background, can have some measure of its success, or heaven forbid, failure.

Table 1. Dialysis statistics (U.S., 1988)

Treatment centers	1,819
Patients	105,958
Federal yearly budget	$1,975,600,000

Approximately 85% of the patients shown in table 1 receive in-center hemodialysis funded by Medicare. The very fact that these people are kept alive when they would unquestionably die from lack of a vital organ is an indicator of the program's success. However, since dialysis therapy can have varying intensities, measures of the amount of dialysis delivered and the quality of life achieved seem reasonable as vital statistics. Among the multiplicity of documents required by the bureaucracy from its constituent dialysis centers, one is conspicuously absent: a measure of dialysis adequacy.

Assessment of the completeness and adequacy of dialysis has been an elusive goal of the fledgling dialysis community. Failure to provide an estimate of the dialysis effect is not surprising since the legacy of attempts to quantify uremia is replete with failure and harks back to the middle of the last century. Richard Bright, the first to describe chronic renal failure, a disease that for many decades bore his name, made these observations in 1836:

It is indeed an humiliating confession that, although much attention has been directed to this disease for nearly 10 years, and during that time there has probably been no period in which at least twenty cases might not have been pointed out in each of the large hospitals of the metropolis. . . yet little or nothing has been done towards devising a method of permanent relief, when the disease has been confirmed; and no fixed plan has been laid down, as affording a tolerable certainty of cure in more recent cases (2).

This enlightened scientist and author might turn over in his grave if he could see that today, over 150 years later, the medical community has failed to identify the cause of most of the clinical symptoms and signs of uremia. Despite these failures in the past, efforts largely based on empiricism have succeeded in prolonging life through therapeutic hemodialysis, hemofiltration, and peritoneal dialysis. Ongoing research promises to unlock the secrets of the uremic environment and to devise a method to measure the severity of uremia and the effects of treatment. Today, until a more specific test is available, routine analysis of urea kinetics is regarded as the best method to assess dialysis adequacy and to detect patient or equipment variances requiring adjustments in the dialysis prescription.

The first report of an elevated blood urea concentration after removal of the kidneys appeared over 150 years ago (3). Since then, physicians have periodically measured urea or urea nitrogen concentrations (BUN) in the serum of their patients to learn about the level and adequacy of intrinsic renal function. Traditionally we have been content with the limited information provided regarding the level of kidney function and its medical implications. By making these same measurements in serum samples from patients receiving dialysis therapy, we can gain additional knowledge about the patient's volume of urea distribution, the patient's protein catabolic rate, and ideal duration of dialysis therapy. How is this possible?

One underemphasized tool of the scientific method is perturbation. We poke at or perturb the system under study and then stand back to take measurements of the results. The cyclotron is a typical example of this technique. Atoms are bombarded with high-energy particles that cause them to break apart. During the bombardment investigators can detect, using sensitive measuring devices, previously unseen subatomic particles. Comparing dialysis to the cyclotron may be stretching the imagination somewhat, but the basic approach is analogous, namely, disturbing the urea concentration (lowering it) and then measuring the changes in urea concentration that follow. New information about the patient is gained both from the fall in urea concentration during dialysis and from the later rise between dialyses. This disturbance of the urea concentration, a goal of the dialysis procedure, constitutes good medical treatment for the patient, but it also gives the physician a chance to learn something more about the patient. Such an opportunity should not be wasted. This book is written with the hope of showing dialysis personnel how to take advantage of this opportunity afforded by the disturbance in urea concentration, using quantitative urea modeling techniques. The book begins by reviewing our latest understanding of the significance of urea accumulation and its cyclic behavior in patients with end-stage renal disease.

Over the past eight years at the University of California, Davis, we have devoted a substantial effort to the design and teaching of urea kinetics as a tool for improvement of patient care. This program also serves as a helpful tool for teaching the principles of dialysis. A focus on this single solute, quantitatively the most important solute removed by the dialyzer, helps to establish a foundation of understanding that the student of extracorporeal therapy can use to comprehend a process that at first seems bewildering. Once an understanding of urea dynamics is firmly established, the behavior of other solutes can be compared to it. This provides an efficient mechanism for teaching, a common glossary of terminology, and a forum for discussion that bring the quantitation of dialysis closer to a conscious day-to-day level in our students and trainees.

A key to this teaching technique is removing the burden of mathematical computations by using the computer. Dialysis personnel are delighted to see that solutions to the complex relationships among urea concentration, fluid shifts, patient size, and time on dialysis are instantly available from the computer. Previously frightening mathematical equations become fun to work with as they provide us with valuable patient data and tools for research as well. The manager of dialysis services, often extracted from a nonmedical background, benefits from the opportunity to compare patients within the dialysis center and to compare one center with another. The dietitian is provided with objective data about nitrogen balance that helps with the formulation of a dietary prescription of protein for each patient. The work of third-party providers is eased by the periodic reports that result from routine application of urea modeling.

Routine examination of urea kinetics has become the heart of our quality assurance program. Our efforts have been rewarded in the above disciplines by a better understanding and a sense of confidence and control over the dialysis process itself. We are also provided with an independent measure of each patient's compliance with the dietary prescription and a technique for comparison of dialysis intensity and need among our patients.

We clinicians are generally less adept with mathematics than our colleagues in basic science who contribute to teaching in the medical school curriculum. Perhaps a loss of love for mathematics directed us along the path leading to the application of scientific knowledge rather than along the path leading to development of theory. In any case, it appears that sometimes we go to extremes to avoid use of an equal sign or any other semblance of a mathematical equation. Several clinical textbooks and manuscripts appearing in recent years have openly professed a sense of pride in the authors' ability to avoid complex mathematical formulations. Since mathematics is a kind of language, statements of this sort could be likened to pride in avoidance of French or Italian for an English-speaking audience. Most readers of clinical texts are not familiar with mathematical expressions, so it might be considered inappropriate to include text with too many plus, minus, and equal signs lest the reader become lost or bored. Unfortunately, to quantify dialysis therapy using BUN concentrations, some mathematical expressions are unavoidable. Most of these expressions are concentrated in chapters 4 and 5, where derivations of the single-pool and double-pool models are found. Each of these chapters starts with simple and moves on to more complex expressions, hoping to satisfy the theoretician, but in a relatively small block that can be avoided by the equation-fearing clinician.

For the clinician, the computer is a godsend that permits quick and easy solutions to mathematical problems while avoiding tedium and overindulgence in the details of mathematics. Once the equations describing dialysis urea kinetics have been programmed, no further consideration of mathematics is required; only input and output are examined. The clinician must know how each input variable affects the output variables, but only in a general sense. Indeed, the clinician can use the computer programs to learn how one variable affects another by altering input parameters and observing the results. For example, computer programs for urea kinetics can show that variations in patient weight during dialysis affect urea clearance, generation rate, and volume of urea distribution. This approach bypasses the mathematics, opting instead for an analysis of trends, a valid application of modeling. As a substitute for the diseased patient, the model itself becomes the object of study and a source of learning. If periodic comparisons with actual clinical data are made, modeling techniques can improve the efficiency of learning and provide solutions to problems that would not be approachable by clinical experimentation. The computer allows the clinician to take advantage of the power inherent in mathematical expressions without needing to become involved in the language of mathematics.

This development is not unique to the field of dialysis. There was a time when to listen to the radio, you literally had to build one. Many years ago, building crystal radio sets was a popular activity for children and families. Attempting to foresee the future, some thought that it would be necessary or advantageous to become familiar with sound modulators, oscillating circuits, and reverberators. Today, few people understand how a radio or television receiver works, yet everybody uses them. We can probably use them better because we are not burdened with the "knowledge overhead" of the inner workings of these instruments. Similarly, the computer can be used as an instrument of learning despite a lack of understanding of its interior design and mechanisms. When using data base management software, for instance, it is not necessary to know how the computer found your client's file through string comparisons and phonetic indexing algorithms, only that when you gave it the name, it produced the file. Similarly, mathematical models can be used to study and examine clinical treatments such as urea removal even though the user lacks in-depth understanding of the details behind the model's structure. The model should be tested and retested, perhaps in part by the user, to establish confidence in it. Once it is perfected and confidence is instilled, the model can be used as a tool to guide the physician, who should not be required to explain how it works except in a general sense. In some disciplines, users of complex computer programs, who have no knowledge or experience with the language of programming, master the program's application with more skill and understanding than the original programmer. The complexities of the program are transparent to the user. This discussion is not meant to discourage the programmer, whose special skills will always be required to implement new or modify existing tasks. However, it should be possible for those who do not wish to become familiar with fast Fourier transforms, matrix algebra, or fourth-order Runge-Kutta numerical analysis to reap the benefits granted by those who have developed the model.

This book is intended for physicians, nurses, technicians, and students of nephrology and dialysis. Major emphasis is placed on the hemodialysis prescription, but reference is also made to peritoneal dialysis, most often in comparison to hemodialysis. The text and companion software can be used as a teaching tool for the study of chronic renal failure and its dialytic treatment. Although the software was not conceived for teaching purposes and as such does not strictly fit the definition of Computer Aided Instruction (CAI), it does help the user to see quickly the influence of prescription and patient variables on dialysis outcome while generating a visual image of the BUN/time profile. A little time spent fiddling with the program will help solidify an understanding of the complexities of urea kinetics, which are difficult to comprehend otherwise.

The last few years have seen a proliferation of techniques and software for evaluation of urea kinetics that are often advertised as a simplified approach. These have ranged from hand-held calculators and nomograms for quick bedside calculation of *Kt/V* to desktop computer programs and fragments of larger data base programs or medical information systems designed to track dialysis parameters and laboratory data. Some have appeared on the market without critical review and have serious flaws. Some are based on earlier published versions of algorithms that have since been abandoned in favor of more precise models. At a recent major nephrology meeting this author surveyed several prototype programs that gave radically different results for urea distribution volume, generation rate, and *Kt/V*. The theoretical basis for urea modeling has undergone careful scrutiny and scientific review over the past 25 years and is generally accepted. Unfortunately, the implementation of that theory in the realm of clinical practice is often scrutinized less carefully. There is a need for a central clearinghouse or standard against which new programs or techniques can be measured. One goal of this book is to provide enough information for the readers to develop their own software to evaluate urea kinetics either as a stand-alone program or as a module that can be integrated with existing data base programs. Pitfalls encountered with the mathematics are enumerated and discussed in detail so that the software developer can avoid reinventing the wheel. This approach unfortunately requires inclusion of mathematical details that many readers would prefer to avoid. Examples are given throughout the book; an effort was made to supply parameters used to generate the data for the appropriate kinetic model. Short fragments of computer source code are included in the appendices to help those who would like to develop custom programs. Although absence of flaws in the mathematical logic and algorithms contained within this text cannot be guaranteed, it is hoped that these efforts will serve as a forum for establishing a much-needed standard.

The organization of this book requires some explanation. Chapters 1 and 2 review our understanding of uremic toxins, with focus on urea as a toxin and as a surrogate for other highly dialyzable toxins. Chapter 3 introduces urea modeling, the clinical requirements and advantages of modeling, and various models used in the past. Chapter 4 is an in-depth discussion of the single-compartment model that includes a derivation of the logic and mathematics of this most popular model. Chapter 5 extends the single-compartment model to two compartments, listing the advantages of and the clinical settings that require this more complex model. Extensive comparisons are made between the two models. Chapter 6 is a guide to UREAKIN, a user-friendly program for rapid implementation of urea modeling.

Chapter 7 addresses questions often raised about the corrections for plasma volume, plasma water, and red blood cell water that are required when calculating urea removal rate, volume of distribution, clearance, and urea generation rate. When the dialyzer is perfused with whole blood, urea clearance is lower than the clearance measured when the dialyzer is perfused with water or saline at the same urea concentration. Chapter 7 addresses these questions and the effects of ultrafiltration, saline infusions, and other refinements to the model. Chapter 8 reviews the most controversial aspect of urea modeling, the criteria for establishing ideal or target outcomes. This chapter includes a concise summary of the National Cooperative Dialysis Study (NCDS) and its various interpretations. Chapter 9 provides concrete examples of urea modeling, detailing benefits and pitfalls. Chapter 10 reviews our projections for the future.

With the tools provided here and with techniques and instruments to come, we hope we can at least partially answer the question posed by the dialysis patient's friend.

Listing of variables used throughout this book

To avoid confusing the experts, symbols used here are similar to those already entrenched in the literature. The terms *urea* and *urea nitrogen* are used interchangeably and in almost all cases refer to urea nitrogen (e.g., urea concentration = urea nitrogen concentration; urea generation = urea nitrogen generation; blood urea = blood urea nitrogen). In some cases the distinction does not matter (e.g., urea clearance = urea nitrogen clearance). Solute concentrations are usually expressed as mg/ml, because the formulas require milliters instead of deciliters for calculations. BUN values appear as mg/dl on reports.

Table 2. Meaning of variables

Variable	*Unit of measurement*	*Meaning*
A	cm^2	area
AUC	min • mg/dl	area under the BUN versus time curve
Av pre	mg/ml	average predialysis BUN
av. wt. loss	kg	average weight loss during dialysis
B	ml/min	change in V, during or between dialyses
or,	g/min	
C	mg/ml	solute concentration (usually urea)
C_0	mg/ml	C when time = 0
C_1	mg/ml	patient BUN at the start of dialysis
C_2	mg/ml	patient BUN at the end of dialysis
C_3	mg/ml	patient BUN at the start of second dialysis
C_b	mg/ml	blood concentration
C_d	mg/ml	dialysate concentration
C_e	mg/ml	extracellular concentration
C_{e0}	mg/ml	C_e when time = 0
C_i	mg/ml	intracellular concentration
C_{i0}	mg/ml	C_i when time = 0
C_{in}, C_a	mg/ml	dialyzer inlet BUN
C_o	mg/ml	dialyzer outlet BUN
C_p	mg/ml	peripheral BUN (from opposite arm)
C_v	mg/ml	venous outlet BUN, same as C_o
c		constant of integration
D	cm^2/sec	diffusion coefficient or dialysance
Day	none	day of week, e.g., 3 = Tuesday
ds	mg or g	small change in solute content
dt	min	small interval change in time
dV	ml	small change in urea volume
dW	kg	in reports: average total weight loss during dialysis
	ml/min	in equations: rate of fluid gain
dx	cm	small change in length or distance
e	none	natural logarithm base (2.718)
ECF	none	extracellular fluid
ECV	ml or liters	extracellular fluid volume
f	none	coefficient that converts residual clearance into units equivalent to Kd
F_r	none	venous reflow fraction (Q_r/Q_b)
g	g/day	urea nitrogen generation rate
G	mg/min	urea nitrogen generation rate
GFR	ml/min	glomerular filtration rate

hct	%	blood hematocrit
ht	inches	height of patient, used for surface area
ICF	none	intracellular fluid
ICV	ml or liters	intracellular fluid volume
IC	mg/dl	ideal concentration (BUN)
ID number	none	identification number (alphanumeric)
Ideal hrs	hours	duration of dialysis necessary to achieve ideal TAC
IPREa	mg/dl	target average predialysis BUN
ITAC	mg/ml	target TAC calculated from area under the curve
J_u	mg/min	urea flux through the dialyzer
K	ml/min	short form for K_d
k, k'	per min	elimination constant or other arbitrary constants
KA	ml/min	dialyzer mass transfer area coefficient
KC	ml/min	intercompartment mass transfer area coefficient
K_d	ml/min	dialyzer whole-blood urea clearance adjusted for blood water content and ultrafiltration
K_{d0}	ml/min	dialyzer urea clearance when ultrafiltration rate is zero
Kd_2	ml/min	dialyzer clearance when blood flow is $Q_b - Q_f$
K_{de}	ml/min	expected dialyzer clearance (from KA, Q_b, Q_d)
K_{dp}	ml/min	projected dialyzer urea clearance when V_p is substituted for V
K_r	ml/min	patient residual (native kidney) clearance of urea
Kt/V	per dialysis	dialyzer clearance multiplied by time on dialysis divided by V
K'	ml/min	adjusted clearance to include Kr component in Kt/V expression
ln	none	natural logarithm
Ndays		number of dialyses per week
P	%	percent reduction in predialysis BUN $(C_1 - C_2)/C_1 \cdot 100$
PCR	g/day	net protein catabolic rate
PCRn	g/kg/day	net protein catabolic rate, normalized to V/0.58
PCRp	g/kg/day	PCRn when $V = V_p$
POST	mg/ml	BUN following dialysis
PRE	mg/ml	BUN prior to dialysis
PREa	mg/ml	average predialysis BUN
Q_b, Q_a	ml/min	whole-blood flow (averaged pumped flow rate)
Q_{bi}	ml/min	whole-blood inlet flow into dialyzer
Q_{biw}	ml/min	effective blood water flow through dialyzer
Q_{bo}	ml/min	whole-blood outlet flow from dialyzer
Q_d	ml/min	dialysate flow

Q_f	ml/min	volume change during or between dialyses (sign may change)
Q_p	ml/min	peripheral whole-blood inflow to dialyzer ($Q_b - Q_r$)
Q_r	ml/min	venous reflow of blood into dialyzer
s	none	a solute
Schedule	none	weekly dialysis schedule (e.g., MWF)
T_a	minutes	average time between dialyses
t or T	minutes	time during or between dialyses
TAC	mg/ml	time-averaged BUN from area under curve
Target TAC	mg/ml	calculated ideal TAC
Target time or T_b	minutes or hours	duration of dialysis required to achieve ideal TAC
T_d	minutes or hours	time from start to end of dialysis
T_{da}	minutes or hours	average duration of each dialysis
T_i	minutes or hours	time between dialyses
type	none	code for dialyzer brand name
$t_{1/2}$	minutes	half-life
V	ml or % body weight	urea distribution volume
V_e	ml	immediately equilibrating (extracellular) volume
V_i	ml	remotely equilibrating (intracellular) volume
Vl	liters	urea distribution volume
V_p	ml or % body weight	V projected (estimated) by user
W_1	kg	weight before dialysis
W_2	kg	weight after dialysis
W_3	kg	weight before second dialysis
Wt	kg	postdialysis (dry) weight
x	cm	dialyzer membrane thickness
z	none	temporary variable
#/wk	none	number of dialyses per week

References

1. HCFA report: New facility and patient treatment statistics. Nephrology News & Issues 3:11, 1989 (Oct).
2. Bright R: Cases and observations, illustrative of renal disease accompanied by the secretion of albuminous urine. Guy's Hosp Rep 1:338-380, 1836.
3. Prevost JL, Dumas JA: Examen du sang et de son action dans les divers phénomènes de la vie. (Examination of the blood and its action in the different phenomena of life.) Ann Chim Phys 23:90, 1821.

ACKNOWLEDGEMENTS

I would like to extend sincerest thanks to my colleagues, Paul Gulyassy, M.D. and Ronald Bogusky, M.D., Ph.D., for their help in reviewing selected chapters, and to Angela Cheer, Ph.D for her help with the mathematics in chapter 5 and appendix D.

SOFTWARE AVAILABILITY

The software programs described in Chapter 6 are available from the author at a reduced price to purchasers of this book. A valid receipt is required. Address inquiries to:

Thomas A. Depner, M.D.
University of California, Davis
Nephrology Division
4301 X Street
Sacramento, CA 95817

Please note that the programs are offered directly by the author; the publisher does not provide software services.

PRESCRIBING HEMODIALYSIS
A Guide to Urea Modeling

Chapter 1

UREMIC TOXINS & DIALYSIS

THE UREMIC SYNDROME

The symptoms and signs of renal failure, often called *uremia*, are reversible. Common reversal techniques include manipulation of protein in the diet, transplantation, and various types of therapeutic dialysis. Others include hemofiltration and sorbent therapy. The term *uremia* literally means urine in the blood. It refers to a variety of nonspecific complaints and physical signs that patients inevitably manifest as their renal function falls to below 5% of normal (see tables 1.1 and 1.2). The glomerular filtration rate, measured in the clinical laboratory, is the usual yardstick of renal function and admittedly correlates only crudely with renal damage. The syndrome of uremia expresses itself in different ways in different patients, but there are common typical elements. These serve as a focus for clinicians who would treat uremic patients and for investigators searching for the searching for the elusive toxins responsible for the syndrome. The failure of over 150 years of research to identify a specific toxin or toxins to account for the entire syndrome has led some to propose alternate theories. These are reviewed below. Excellent reviews of uremic toxins are available in standard textbooks (1,2,3). These toxins are briefly summarized here, with special emphasis on how they relate to dialysis, including some of this author's biases.

ROLE OF PROTEIN NITROGEN METABOLISM

For over a century, nephrologists have known that dietary protein restriction will improve the symptoms and signs of uremia (4). Symptoms such as anorexia, nausea, vomiting, lethargy, and somnolence often improve dramatically. Most responsive are gastrointestinal and central nervous system symptoms, including hemorrhagic enteritis and coma. Serositis is not known to respond. From these observations, early investigators concluded that protein metabolites are responsible for the syndrome of uremia.

This effect of protein restriction must be distinguished from the recently popularized low-protein diet to slow the progress of advancing renal failure (5). The latter is proposed as a treatment for mild to moderate renal failure, whereas this discussion refers to treatment of patients with advanced renal failure. In advanced renal failure, the goal of dietary protein restriction is to improve the clinical manifestations of uremia; no improvement in renal function is expected.

Before the arrival of therapeutic dialysis, special diets low in protein content were the only available form of therapy for patients with advanced renal failure (6,7). Consequently, a large experience with dietary treatment accumulated prior to 1970 that is now often ignored. These diets had a high carbohydrate and fat content that caused patients to complain of excessive sweetness and greasiness and often led to caloric deficiency. Despite the best of efforts, as renal function deteriorated, caloric and protein malnutrition became a significant part of the uremic syndrome (8). The clinician and patient were fighting a battle that they would eventually lose, pitting the benefits of lowered protein intake against the risks of malnutrition.

Timely institution of therapeutic dialysis not only prevents the consequences of uremia but also eliminates the necessity for strict low-protein diets. This indirect benefit of dialysis removes the risks of malnutrition that in combination with uremia would often lead to irreversible complications. The patient who died of *uremia* often had a significant element of protein and caloric malnutrition that contributed to the infection and hemorrhage, classically attributed to uremia. Even today we occasionally see a well-dialyzed patient deteriorate from malnutrition who then responds to enteral or parenteral nutritional supplements (9). Thus, the role of nutrition, especially protein nutrition, in renal failure continues to be a two-edged sword. The clinician and dietitian must continue to balance the benefits of a restricted diet against the specter of malnutrition; fortunately, the availability of renal replacement therapy such as therapeutic dialysis makes the job much easier.

Protein in the diet increases the work load of the kidney by increasing glomerular blood flow, glomerular filtration, and the tubular load of sodium (10,11,12). Protein restriction may, because of decreased work load, reduce the metabolic burden of the kidney and at least partially explain the deceleration of renal failure (5,13). This benefit has to be weighed against the potential adverse effect of too much protein restriction. Adequate protein nutrition is essential for immune mechanisms, hormone function, epithelial barrier stability, and maintenance of plasma oncotic pressure. When malnutrition results from dietary insufficiency, refeeding should evoke a prompt recovery. When the patient with end-stage renal disease begins dialysis treatments, dietary restrictions can be relaxed but nutritional recovery is not prompt. So it is best to avoid severe protein restriction such as that practiced in the predialysis era and to institute renal replacement therapy in a timely fashion. This lesson, learned in the past, may have to be learned again with the recent emphasis on severe (less than 0.4 g/kg/day) restriction to salvage renal function (5,13). Once hemodialysis begins, protein restriction is less critical, but excessive protein intake should be avoided because dietary protein is a major determinant of need for dialysis. As explained in chapter 8, the dialysis prescription hinges on accurate assessment of dietary protein or the *protein catabolic rate* (PCR). More oral protein intake demands more dialysis.

Why does protein restriction improve uremic symptoms and decrease the need for dialysis therapy? The presumption has been that protein metabolism generates a toxin, perhaps a polypeptide (14). Other possible toxins include acidic and/or basic metabolites of proteins, such as the indoles or guanidine derivatives. One theory that has substantial support indicts urea as well as the bacteria in the gut that break down urea to the more toxic ammonia (15,16). The finding that the enteritis of uremia can be treated with antibiotics as well as with dialysis supports this theory (17,18,19). Also supporting this theory is the observation that uremic stomatitis responds to antibacterial mouthwashes (20,21). The increase in extracellular urea concentration that stoichiometrically results from protein metabolism provides more substrate for urease-containing gut flora. The superficial, often hemorrhagic ulcerations observed in uremic patients are not unlike ulceration seen after topical exposure to caustic agents such as ammonia. Guanidines and phenols are products of protein metabolism that can cause toxicity in experimental animals (22,23). Their toxicity mimics that of uremia but requires blood and tissue levels far in excess of those measured in patients with uremia. These compounds are discussed below in more detail. Other toxic products of protein metabolism may await discovery.

Clinical measurement of uremia

Since dialysis intervention became widespread in the U.S., the severe uremic state is seen much less frequently. Occasionally a patient will slip through the medical networks and appear, usually in an emergency setting, with advanced uremia. Typically this patient is stuporous or in frank coma with Kussmaul respirations reflecting severe metabolic acidosis with or without pulmonary congestion. The patient usually has signs of malnutrition including weight loss, anorexia, reduced metabolism, hypothermia, and increased susceptibility to infection. Prior to loss of awareness, the patient may have complained of anorexia, loss of taste, nausea, vomiting, diarrhea, pruritus, loss of the ability to concentrate, and somnolence during daylight hours. There may be a typical stocking-glove peripheral neuropathy and a flapping tremor, indistinguishable from that seen with nutritional deficiencies or alcoholism. Hearing loss, "*red eye*," and hoarseness may occur in some patients. The skin is dry and sallow and in the absence of routine bathing, uremic *frost* may be seen in the intertriginous areas. These white deposits represent crystals of urea left behind after the sweat evaporates. The breath has a fishlike odor, often designated *uremic fetor*. Hypertension is the rule, and is sometimes severe. Chest examination discloses a pleuropericardial friction rub and the chest X-ray shows a central infiltrative pattern, resembling a bat wing or butterfly, sparing the periphery. Evidence for bleeding from the upper and/or lower gastrointestinal tract may be seen. This is usually not massive and results from a combined effect of defective clotting mechanisms and local mucosal damage. Medicinal intoxication occurs at this stage if adjustments for renal failure are not made. Almost all patients have extracellular volume expansion at this stage. For many, volume overloading is not severe, probably because the diseased kidney can maintain glomerular-tubular balance, and the patient has poor fluid intake, vomiting, and diarrhea. Typically, those patients who do have problems with excess volume expansion will present earlier, before the other overt signs and symptoms of uremia appear.

Hemoglobin levels in advanced renal failure average about half normal or less. Serum phosphate, magnesium, potassium, chloride, triglyceride, and parathormone concentrations are elevated, while bicarbonate and calcium levels are low. X-rays show bone decalcification with features of hyperparathyroidism and osteomalacia. Many other objective measurements are disturbed in uremia but with less consistency. These include abnormally low or high serum enzyme levels, arrhythmias, glucose intolerance, hyponatremia, tissue calcification, various endocrine organ dysfunctions, and slowing of the electroencephalograph pattern.

The patient described above who presents with full-blown uremia is not difficult to recognize. More pertinent to management of patients with end-stage renal disease in more recent times is the ability to detect subtle symptoms and signs of impending need for dialysis or the ability to recognize inadequate dialysis after treatment begins. If we could answer the question raised previously about the toxic effect of dietary protein by identifying a specific toxin, we could easily provide a measurement of uremia. Unfortunately no single toxin is identified, so we are left with marker solutes that correlate with uremic toxicity but are not themselves very toxic. These include the major nitrogenous end products, urea and creatinine.

Correlation with toxin levels

A major obstacle encountered in the search for uremic toxins has been the necessity to correlate toxic effects of proposed uremic toxins with tissue or blood concentrations in truly uremic patients. Such correlations are necessary to grant credibility to the substance under question as a true contributor to the uremic syndrome. Many substances known to accumulate in renal failure are toxic when given to humans or experimental animals (22,23,24,25,26,27). Many of these solutes produce symptoms and signs similar to that seen in uremia. But their concentrations, measured in the blood of patients suffering from uremia, are usually below that required to produce symptoms or toxic effects in animals. Because of the poor correlation between levels and toxicity, other hypotheses have been proposed. These include possible tissue binding of toxins and toxic synergy, as discussed below.

Marker molecules for uremia

Creatinine is the most popular endogenous marker for intrinsic renal function. In health, the serum creatinine and creatinine clearance correlate reasonably well with glomerular filtration rate (GFR), the commonly accepted measure of overall kidney function (28,29). As renal function deteriorates, tubular secretion accounts for a larger fraction of creatinine excretion, so clearances and serum levels overestimate GFR. This has little clinical consequence because clinical guidelines for intervention are based on the serum creatinine and creatinine clearance, not GFR. Creatinine generation is relatively constant within each patient and is highly correlated with muscle mass. Creatinine is produced in muscle from irreversible dehydration of relatively unstable creatine phosphate, the primary muscle storage repository of high-energy phosphate. Creatinine is an end product anhydride, the most abundant guanidine compound to accumulate in renal failure. Older studies of humans loaded with oral creatinine have shown it to be relatively nontoxic (30,31). Although it is formed from two amino acids, arginine and glycine, there is little or no correlation between creatinine generation and protein intake, so as a marker of uremic liability, it fails. But creatinine can serve as a marker for low-molecular-weight solute accumulation and the effectiveness of dialysis or other treatments designed to remove these solutes. An additional consideration must be the patient's muscle mass. Males with a large muscle mass will have higher serum creatinine levels than females. Unfortunately, there is no simple way to correct for muscle mass. In a patient with no intrinsic (endogenous) renal function, a markedly elevated serum creatinine concentration can mean either a high production rate from increased muscle mass or inadequate dialysis. So serum creatinine levels take second place in the competition for a uremic marker. As renal disease advances in the patient and residual function declines from borderline to dependency on dialysis, the marker of choice changes from creatinine to urea.

Urea concentrations, usually expressed as blood urea nitrogen (BUN) (see *chapter 7*), rise in patients with renal failure, but the levels are erratic and correlate poorly with renal function. In contrast to creatinine concentrations, which depend only on muscle mass and GFR, urea levels depend on additional factors such as protein ingestion, nitrogen balance (anabolic or catabolic), and reabsorption by the renal tubules. Consequently, as a marker for native kidney function in mild to moderate renal failure, urea fails several benchmarks. But as the patient reaches end-stage renal failure, urea levels have several advantages. Single-nephron GFR is high in advanced renal failure, so tubular reabsorption is reduced and is less influenced by states of hydration. Urea levels correlate reasonably well with symptoms and signs of uremia, and average urea levels were found by the National Cooperative Dialysis Study (NCDS) to correlate with dialysis outcome (32). Urea itself is only a "mild" renal toxin (1), but concentrations in the blood correlate with toxicity in advanced renal failure.

Other compounds, such as methylguanidine, guanidinosuccinic acid, hippurate, indoxyl sulfate (indican), aliphatic and aromatic amines, polyamines, various phenols, and myoinositol, have been proposed as markers of uremia. Yet none is as elevated, as easily measured, or correlates as well with clinical illness as urea. Endogenous compounds other than creatinine or urea have also been proposed as markers of renal function (33,34,35), but to date, none offers sufficient additional benefit to supplant these two time-honored markers.

Effect of dialysis on the uremic syndrome

Dialysis is a life-saving therapy that substitutes for normal kidney function and allows patients to survive who would otherwise die. Because the normal kidney is not a dialysis device, we might expect some deficiencies or failure of the hemodialyzer to substitute for the more complex native organ. As any patient treated with maintenance hemodialysis for a few months can document, this treatment, although life-sustaining, does not restore complete health. Table 1.1 lists the symptoms and signs of uremia that respond to dialysis and table 1.2 lists those that respond incompletely or not at all. Fortunately, the more life-threatening complications of uremia such as encephalopathy, serositis, and enteritis do respond to dialysis, but many other complications do not. As a result, many patients kept alive by maintenance hemodialysis or by peritoneal dialysis suffer from an illness that is often difficult to define but is perhaps best viewed as a insidious form of uremia (36). To understand this illness better, investigators have focused on potential toxins that accumulate in uremia and on the evidence for their toxicity (table 1.3). Identification of specific toxins will help to pinpoint the cause(s) of this illness so that dialysis equipment and adjunctive therapy may be designed and adjusted to improve outcome.

Table 1.1 Manifestations of uremia that respond to dialysis

- Uremic encephalopathy
- Peripheral neuropathy
- Pruritus
- Hypertension due to volume expansion
- Electrolyte imbalance
    - Acidosis
    - Hyperkalemia and potassium retention
    - Hypermagnesemia
    - Water retention, osmotic disorders
    - Hypocalcemia
- Uremic fetor
- Uremic enteritis
- Uremic pericarditis, serositis
- Bleeding disorder of uremia

Table 1.2 Manifestations of uremia that do not respond well to dialysis

- Uremic osteodystrophy
- Anemia of renal failure
- Sexual dysfunction
    - Infertility
    - Amenorrhea
    - Impotence
- Altered immune response
- Hormonal imbalance
    - Glucose intolerance
    - Elevated serum levels of:
        - growth hormone
        - gastrin
        - melanocyte-stimulating hormone
        - glucagon
    - Low testosterone levels
- Hyperlipidemia
- Serum enzyme alterations
- Skin pigmentation
- Stunted growth in children
- Altered plasma protein binding of drugs
- Decreased energy levels, fatigability
- Psychological depression

Dialysis is a procedure for separating solutes according to their differential permeability across semipermeable membranes. For the inert membrane in use today, these permeabilities are heavily dependent on molecular size. Fortunately, for patients with renal failure, the unwanted or toxic solutes are much smaller than the more desirable solutes. The difference is of such magnitude that dialysis is effective as a therapeutic toxin remover while causing little of the depletion syndrome that was anticipated in the past (37,38). There are, of course, vital small solutes that are easily removed by dialysis. Fortunately, these solutes are inexpensive and abundantly available, so they can be included in the dialysate at concentrations that cause no net removal. These smaller solutes include the electrolytes, Na^+, K^+, Cl^-, HCO_3^-, Ca^{++}, Mg^{++}, and glucose. It is this fortunate combination of molecular size separation and availability of small vital solutes that accounts for the success of dialysis as a life-sustaining therapy. Removal of toxins by passive diffusion is the major goal of hemodialysis therapy, and it is also the only significant life-sustaining effect (see chapters 3 and 7 for a more detailed review of the physical chemistry of dialysis). Keep in mind the fact that dialysis works, i.e., removal of solutes, the prime accomplishment of therapeutic dialysis, relieves the symptoms and signs of uremia and allows life to proceed in patients otherwise facing certain death. Any theory that proposes to explain the uremic syndrome must account for this observation.

Proposed uremic toxins

Many decades of research have provided us with a list of potential uremic toxins, most which are displayed in table 1.3 with the approximate range of their molecular weights.

It would be convenient to list these toxins in the order of magnitude of their toxicity, but unfortunately this is not possible. In addition, despite gargantuan research efforts spanning the past 150 years, it is impossible to indict any single toxin in table 1.3 as a responsible agent for any single part of the uremic syndrome described above. Skeptics have said that the real toxin is yet to be discovered. This seems hard to imagine, given modern chemical separation techniques, chromatography, analytical spectroscopy, and other tools, all of which have been extensively applied to uremic body fluids. Nonetheless, it is possible that a potent toxin, in trace concentrations, has been overlooked. The low probability of an oversight has spawned other theories to explain uremic toxicity. These include hormonal theories, the trade-off hypothesis, toxic synergy, and tissue binding of toxins. Each of these theories is reviewed below in more detail.

ALTERNATIVES TO THE SINGLE-TOXIN THEORY

Hormonal theory

Hormone deficiency as a contributor to uremic complications is best exemplified by uremic osteodystrophy. This disorder of bone architecture is a potentially devastating and life-threatening complication of the uremic state. Uremic osteo-

Table 1.3 Proposed uremic toxins

Toxins	*M.W.*
Products of protein metabolism	
Urea	60
Creatinine	113
Guanidines	
methylguanidinine	73
guanidinoacetic acid	117
guanidinosuccinic acid	175
others	
Phenols and phenolic acids	100-300
Indoles	200-400
Amines	
aliphatic (e.g., dimethylamine)	46
aromatic	150-300
polyamines (e.g., spermine)	202
Peptides and larger proteins	300-50,000
beta-2 microglobulin	11,800
Products of nucleic acid metabolism	
Uric acid	168
Cyclic AMP	240
Pyridines	100-200
Inorganic elements and compounds	
H^+.	1
H_20	18
Na^+.	23
Al^{+++}.	27
Mg^{++}.	24
K^+ .	39
Ca^{++} .	40
PO_4 .	95
SO_4 .	96
Others	
Hormones	
Parathyroid hormone	9500
"Natriuretic hormone"	?

dystrophy has been studied in great detail, partially because of the discovery in 1969 that the kidney is the site of hydroxylation of 25-OH vitamin D3 to its active 1,25-OH metabolite (39). This active vitamin-hormone promotes calcium absorption across the small intestinal mucosa and in normals is responsive to calcium needs of the body. Lack of the hormone is a major cause of uremic osteodystrophy, a state of calcium imbalance resulting in bone decalcification, pain, fractures, and soft-tissue calcification. In this case we can blame hormone deficiency for part of the uremic syndrome without invoking a retained toxin. The "toxicity" results not from failure to excrete a toxin but from lack of a vital hormone normally supplied or activated by the kidney. A similar scenario could be drawn for the effect of erythropoietin deficiency (another renal hormone) on the anemia of chronic renal failure. Since the introduction of erythropoietin therapy, problems such as lethargy, reduced stamina, and impotence, which had been previously attributed to uremia, have been observed to improve after correcting the anemia (40). Thus, two major complications of chronic renal failure can be partially explained by hormone deficiency. Since dialysis cannot be expected to supply these hormones, neither of these components of chronic renal failure is likely to be a major component of the dialysis-reversible uremic state. Each must be considered a peripheral phenomenon. This conclusion is reached from a logical consideration of the well-demonstrated life-sustaining effect of dialysis. The only other impact of dialysis on 1,25-OH vitamin D3 deficiency might be the provision of calcium in the dialysate that helps to counteract malabsorption of Ca^{++}. But older studies of patients with advanced renal failure in the era before dialysis was available showed that although large doses of calcium and vitamin D helped to alleviate osteomalacia, they did not prevent death from uremia (41). The kidney also produces renin and prostaglandins, hormones that regulate vascular resistance, platelet function, and other phenomena, but these agents have not been implicated in the pathogenesis of uremia. Other unidentified renal hormones could potentially play a role.

Trade-off hypothesis

In 1972, after considering another aspect of the uremic syndrome, secondary hyperparathyroidism, Bricker introduced the *trade-off hypothesis* (42). At that time the pathogenesis of uremic hyperparathyroidism was accepted as the ultimate consequence of phosphate retention (43). Accumulation of phosphate led to borderline or overt hypocalcemia, a known potent stimulus for parathyroid hormone (PTH) secretion and parathyroid hyperplasia. Dietary control of serum phosphorus, both in animals and in humans, was shown to attenuate PTH secretion (43,44). It appeared that phosphate could be considered the "toxin" responsible for one manifestation of uremia, namely secondary hyperparathyroidism leading to uremic osteodystrophy. However, osteodystrophy, hyperparathyroidism, and bone disease are seen during the course of progressive renal failure long before hyperphosphatemia and phosphate retention develop. Also, phosphate infusions causing hyperphosphatemia in animals did not cause secondary hyperparathyroidism or osteodystrophy. Since parathormone is known to regulate extracellular phosphate concentration Bricker reasoned that the uremic animals or patients trade off phosphate retention for hyperparathyroidism. PTH hypersecretion is a physiologic adaptation that effectively prevents phosphate retention, but the consequence of the trade-off is bone disease. No retained uremic toxin, other than PTH itself, need be invoked to account for this aspect of the uremic syndrome. Later studies suggested that PTH may have other toxic effects, possibly by altering intracellular calcium concentrations throughout the body (45). Intracellular calcium is an intracellular messenger that regulates membrane transport and other intracellular events. The trade-off hypothesis is another way to describe a broader category of derangements in physiologic regulatory phenomena that occur in uremia. These regulatory mechanisms that operate by feedback control, perhaps designed to maintain homeostasis under acute stress, are invoked as causes of uremic symptoms and signs. Extrapolated further, this theory suggests that in the state of chronic renal failure, regulatory phenomena may turn against the organism, ultimately resulting in its demise. Bricker envisioned that other trade-offs might be occurring to account for sodium, potassium, or glucose homeostasis (46). Although this theory has enjoyed considerable popularity as an explanation for uremic secondary hyperparathyroidism, it has been challenged by some (47) and has not been convincingly shown to account for other aspects of the uremic state. No regulatory phenomena have proven to cause uremic coma, enteritis, serositis, or other life-threatening components of the uremic state. In addition, this theory cannot explain the life-sustaining effect of dialysis. It is highly unlikely that dialysis maintains life simply by removing phosphate, and conventional cellophane dialyzers do not remove PTH. Thus, secondary hyperparathyroidism must be considered another peripheral phenomenon that contributes to poor quality of life but is not a central component of the uremic syndrome.

Additive toxicity and toxic synergy

In 1983, Gulyassy suggested that the toxicity of uremia may be "not from the action of a single molecule, but from the combined effect of a family of related chemicals" (48). He suggested that the combined additive effect of several solutes, families of compounds each at subtoxic levels (e.g., phenols), act in concert to produce uremic toxicity. No single toxic solute would be identifiable at levels sufficiently high to account for any single uremic lesion or symptom. This theory recalls the principles of combined chemotherapy for cancer. Multiple cytotoxic drugs, each at a slightly subtoxic dose, to avoid side effects, are preferred to a single agent when given to destroy the invading cancer cells. The action of each drug alone cannot account for the overall therapeutic effect, i.e., destruction of the malignant tumor, but when their actions combine, the cytolytic effect is apparent. In some cases the action of each drug is additive, and in other cases there is a synergistic effect. The additive or synergistic theory to explain uremia gathers support from chromatographic spectra of uremic extracts showing many abnormal peaks (figure

1.1). It remains to be shown, however, that the uremic syndrome can be reproduced in animals by simultaneous infusions of several known toxins (e.g., those in table 1.3).

Tissue binding

The tissue-binding theory explains the failure to isolate uremic toxins from blood by suggesting that the search has been carried out in the wrong place. Most efforts to identify uremic toxins have used plasma, serum, or ultrafiltrates of serum for analysis. The tissue-binding theory suggests that significant toxins are lipophilic compounds that preferentially distribute in the intracellular compartment, bound to cell membranes or to regulatory proteins. These toxins would behave like hydrophobic drugs, such as digoxin, cyclosporine, or phenytoin, that have very low serum concentrations (49). Most of the drug resides in the intracellular compartment bound to cellular components. Apparent volumes of distribution for drugs that preferentially distribute intracellularly may be considerably greater than total body water. If uremic toxins behave like these drugs, the concentration of toxins in extracellular fluid may be quite low, low enough to be overlooked by investigators searching for significant peaks on chromatograms. The toxicity of these compounds would be easily explained by their interaction and interference with vital cell membrane structures, such as hormonal receptors or solute transporters. For these toxins to qualify as "uremic" toxins, a mechanism must exist for their removal by the normal kidney. Such a mechanism has been known for over half a century: the renal secretory pathways for excretion of organic acids, bases, and other compounds (50,51).

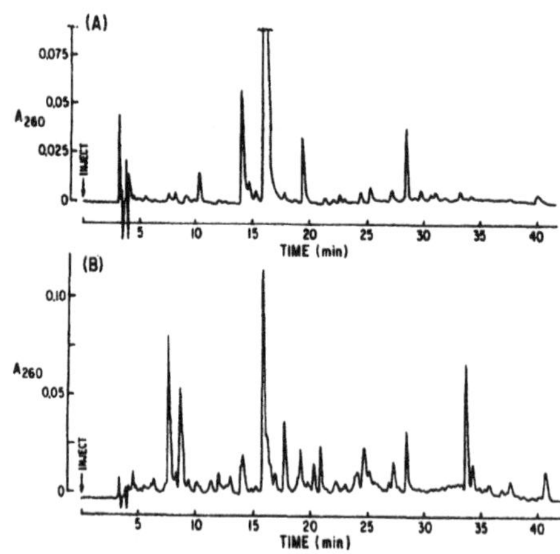

Figure 1.1. High performance liquid chromatography of uremic extracts: **A**, from 1.17 ml normal plasma; **B**, from 0.23 ml uremic serum. Extracts were from serum treated at pH 3.0 with nonionic resin and eluted with methanol. Reprinted with permission from Gulyassy PF et al., Kidney Int 24 (Suppl 16):S239, 1983.

Protein and tissue binding of proposed uremic toxins

Possible tubular secretion of renal toxins

Proximal tubular secretion provides an alternate pathway for excretion of solutes that, due either to their large size or to binding to plasma macromolecules, are poorly filtered. The presence of this pathway in early vertebrate species suggests that secretion is a more fundamental or primitive excretory mechanism than is glomerular filtration (51,52). Its functional role in humans has escaped definition (52,53). After examining the marked efficiency of the secretory pathway for organic acids, Pitts reasoned that tubular secretion must guarantee that certain highly toxic substances will be maintained at very low concentrations in body fluids (53). Gas chromatographic analysis of normal urine reveals over 150 peaks corresponding to organic anions that are normally excreted (54). Many if not most of these arrive in the tubular fluid via the proximal tubular secretory pathway for organic acids.

Serum carriers of tissue-bound toxins

To deliver tissue-bound hydrophobic compounds from the intracellular environment to the kidney, a plasma carrier must be invoked. This transporter must be abundantly available within the blood compartment to enhance delivery of the relatively water-insoluble or protein-bound toxin to the kidney. Figure 1.2 shows a theoretical diagram of the pathway from tissue cells to the kidney. The transporter

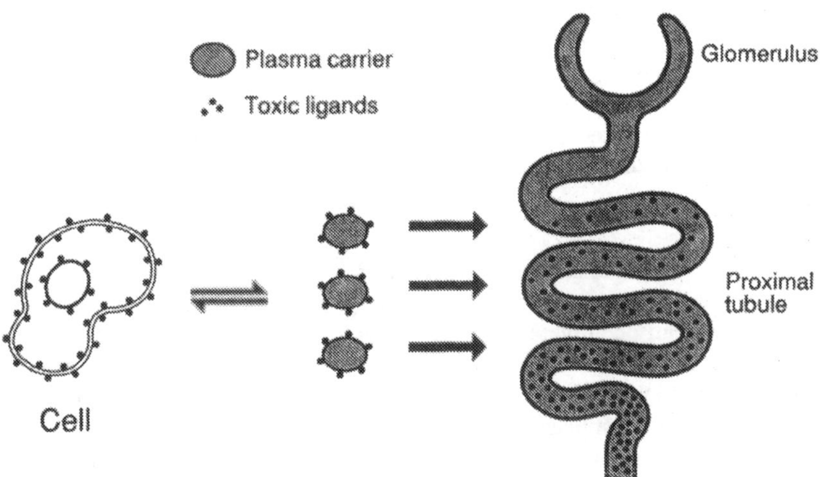

Figure 1.2. Proposed pathway and mechanism for elimination of tissue-bound uremic toxins. A plasma carrier lowers free concentrations in plasma while transporting the toxin to the kidney. Interaction of the carrier with proximal tubular endothelium facilitates transport into the lumen.

would function like albumin, which carries free fatty acids from the liver to peripheral tissues. Albumin has noncovalent binding sites for drugs and endogenous substances including the amino acid tryptophan, fatty acids, and calcium (55). At least three specific binding sites for drugs have been characterized (56,57). Binding to all three of these sites is impaired in plasma from uremic subjects (55,58). This binding inhibition is a reversible phenomenon, i.e., binding can be restored to normal by adsorbing uremic plasma or serum at low pH with charcoal or hydrophobic resins (59,60). The inhibiting compounds have been extracted from the albumin fraction of uremic serum. Unidentified components of the extract inhibit tubular secretion of organic anions both in kidney cortical slices and in the isolated perfused kidney (61,62). Serum ultrafiltrates from the same patients did not have an effect on tubular secretion. Further analysis of the extract revealed multiple chromatographic peaks but only a fraction have been identified (48) (figure 1.1). Hippurate and indoxyl sulfate contribute an average of about 15% to the binding inhibitory action of the extract (63); unidentified compounds contribute the remainder of the inhibitory effect. Some of these may be difficult to isolate and identify because of poor aqueous solubility. Neither hippurate nor indoxyl sulfate have been considered significant uremic toxins, although both accumulate to levels 10 to 20 times normal in uremic serum. Little or no toxic effect can be shown in humans after ingestion of large quantities of hippurate (64), but studies by Dzurik et al. have implicated hippurate as an inhibitor of insulin effect on the isolated rat diaphragm (65,66). Thus, hippurate may partially account for the glucose intolerance of uremia. Van Holder et al. have also implicated hippurate as a toxin (67). An indirect toxic effect might occur if these compounds significantly inhibit binding of more toxic substances to albumin and their subsequent removal by the liver. More recently, the furan carboxylic acids have been identified as new inhibitors of albumin-drug binding (68,69,70,71,72); their role as potential uremic toxins has been suggested (70). The crucial missing element in this theory is the demonstration of a significant tissue-bound or albumin-bound toxin. Such a demonstration is likely to be difficult if the putative toxin is truly hydrophobic and poorly soluble in aqueous solutions.

Facilitated transfer of albumin-bound toxins

Time analysis of injected radiolabeled drugs that bind tightly to serum albumin has shown nearly complete equilibration with intracellular drug within seconds (73). Since diffusion of the unbound species would require more time for complete equilibration to occur, a mechanism for transmembrane diffusion of the bound species has been postulated (74,75). This mechanism requires interaction of the albumin-ligand species with a receptor on the luminal side of the endothelium. Evidence for facilitated transport of bound drugs and other substances has been presented for a variety of organs including the brain, liver, and heart (73,74,76,77). Enhanced transport may be mediated by endothelial albumin receptors, but other interpretations have recently been suggested (78,79). Whether such endothelial receptors exist at the level of the kidney, thereby facilitating tubular secretion of protein-bound compounds, remains to be shown. Besseghir et al. have recently shown that serum albumin facilitates secretion of para-aminohippurate and methotrexate in isolated rabbit proximal tubules (75).

Dialysis of protein-bound toxins

In 1971, Babb et al. introduced the *square meter-hour hypothesis* to explain the beneficial effect of prolonged dialysis in relieving the symptoms and signs of uremia (80). These investigators reasoned that the advantage of additional time on dialysis to reverse uremic peripheral neuropathy could only be explained by removal of higher-molecular-weight species, or *middle molecules*, that exhibited membrane-limited diffusion (see discussion of flow-limited and membrane-limited diffusion in chapter 3). The effectiveness of their Kiil parallel-plate dialyzers correlated more closely with the membrane surface area and time on dialysis than with blood or dialysate flow rate. They also noted that peritoneal dialysis was as effective as hemodialysis if not more effective, despite higher average BUN and creatinine concentrations. This suggested to them that middle molecules are important in the pathogenesis of uremia because peritoneal dialysis can be shown to remove middle molecules more effectively than hemodialysis. The square meter-hour hypothesis evolved to become the middle molecule hypothesis (81).

In their elaboration of the square meter-hour hypothesis, Babb and coworkers considered the possibility of protein binding as an alternate explanation for the above observations (80). Protein binding would cause a low-molecular-weight toxin to behave like a middle molecule, i.e., diffusion across the membrane would be limited because of the low free concentration. Clearance would be minimally affected by blood flow because of the reservoir effect of the bound species. The bound form is usually present at much higher concentrations. Their extensive mathematical analysis of the effect of protein binding included the assumption that binding was constant throughout the length of the dialyzer. This means that the binding to the macromolecule was well below the saturation point and that equilibration was instantaneous or nearly so (80). It would be of interest to consider the possibility that one of these conditions does not hold — that early in dialysis, binding sites are saturated. Another possibility that has not been considered is the interaction of the albumin-ligand complex with dialyzer membranes. It is possible that some dialyzer membranes possess characteristics that can mimic endothelial albumin-ligand receptors.

Toxic contributions of dialysis itself

Besides the toxicity from uremia that dialysis only partially reverses, problems unique to dialyzed patients result from exposure to the dialyzer itself, from inadequacies of the dialyzer, or from therapy given to compensate for the inadequacies of the dialyzer. Table 1.4 lists some of these problems.

Table 1.4 Additional problems related to hemodialysis

Problem	Suspected cause
Hemosiderosis	Transfusions
Disequilibrium syndrome	Osmotic and chemical imbalance
Dialysis osteomalacia	Aluminum intoxication
Dialysis dementia	"
Hemolysis	Chloramine intoxication and red cell trauma
Splenomegaly, pancytopenia	Silicone spallation
Febrile reactions	Endotoxin
Allergic and other reactions to the dialyzer	Surface interactions and released chemicals
Blood access-related problems	Surgical
Hypotension	Acetate intolerance
Dialysis amyloidosis	Poor clearance of amyloid precursors

Many patients require transfusions because dialysis therapy is unable to remove inhibitors of bone marrow erythropoiesis and unable to replace the kidney-derived stimulator, erythropoietin (82). Blood loss, hypersplenism, and dialysis-induced hemolysis also may contribute to the problem (83,84,85,86). Human erythropoietin, recently made available through recombinant DNA research, has greatly relieved this problem (40). Disequilibrium is a necessary adverse effect of a treatment that must, in 2-3 hours, remove excess toxic solutes that accumulate over an interval of 2-3 days. Renal function is normally uninterrupted, so it is not surprising to see intolerance of intermittent dialysis. Aluminum intoxication, causing adynamic osteomalacia and dialysis encephalopathy, has been more recently recognized as a problem resulting in part from contaminated water (87). Because of the large volume of water that comes into close contact with the patient's blood compartment, the source of water must be very pure. Other water contaminants such as chloramines are a source of toxicity, and cause hemolysis and methemoglobinemia in dialyzed patients (88,89). As dialysis membranes become more porous, the contact between water and the patient's blood is more intimate. This increases the probability of adverse effects from contaminants such as endotoxin (90). New dialyzers bring with them potential toxins resulting from the manufacturing process (plasticizers) or sterilization (ethylene oxide) that can cause allergic or toxic reactions (91,92,93,94). The predictable reaction between certain dialysis membranes and inflammatory blood cells has profound systemic consequences (95,96,97). Older methods that used acetate as a base precursor caused problems with vasodilation either in patients intolerant to acetate or when excessive amounts were delivered (98,99,100). More recently, poor removal of a larger-molecular-weight compound (beta-2 microglobulin: 11,800 daltons) has been implicated as a cause of bone and joint disease associated with amyloid deposits (101,102,103). As time progresses, other toxic effects of dialysis may be discovered. These effects of dialysis must be distinguished from true uremia, both by the clinician who attempts to prescribe dialysis therapy and by the investigator who attempts to unravel the complexities of the uremic syndrome apart from dialysis.

Where urea fits into the toxin theory

In 1975, the National Institutes of Health commissioned a cooperative study to evaluate the effects of various dialysis prescriptions (32). The purpose of the study was to decide the relative contributions of small-solute clearance and large-solute clearance to patient well-being. The study involved eight centers and 151 patients. It found that patient outcome correlated with BUN as summarized in table 1.5 (104).

Table 1.5 U.S. National Cooperative Dialysis Study (NCDS)(56)

Dates	1975-80
Number of centers	8
Number of patients	151
Study design	Used various dialysis schedules and techniques to maintain time-averaged BUN at 50 versus 100 mg/dl. Compared patient outcome.
Conclusions	Regardless of the type of dialysis applied, patient outcome correlated inversely with time-averaged BUN.

Those prescriptions that improved urea removal improved patient outcome. Conversely, treatments that resulted in higher BUN values were associated with increased morbidity, regardless of the effect on larger-molecular-weight solutes. The NCDS findings of improved outcome in patients with low average BUN levels served to popularize urea kinetic analysis as a measure of dialysis adequacy and quality assurance. The study results tended to deemphasize the importance of middle molecules while emphasizing the critical role of low-molecular-weight species. This has left investigators of uremic toxins in a quandary, because the low-molecular-weight compounds known to accumulate in uremia listed in table 1.3 do not reach levels high enough in uremic subjects to account for toxicity. Urea itself, the most abundant solute to accumulate in patients with advanced renal failure, has been added to the dialysis bath to prevent a decrease in concentration during hemodialyses. Marked clinical improvement occurred despite lack of net urea removal (24,105). Urea is presumed to serve as a marker or surrogate for more toxic substances that diffuse, like urea, across dialyzer membranes. To date, no candidates for this toxicity have been clearly identified. It is likely that the pathogenesis of uremia is complex. Solutes that diffuse like urea are the most life-threatening, and these are removed by a flow-limited process during conventional dialysis. Other toxins, trade-offs, or yet undescribed toxic mechanisms probably account for the peripheral manifestations of uremia that are not so immediately life-threatening but that contribute to poor quality of life and that may, after many months or years, contribute to morbidity.

There remains a gap in our understanding of uremic toxicity. Fortunately, those of us who are entrusted with the care of these patients can now be confident that the minimum prescribed regimen must keep the average BUN low, regardless of other factors. A good way to accomplish this in a uniform and consistent fashion is through periodic analysis of individual patient urea kinetics using a modeling technique.

References

1. Bergstrom J, Furst P: Uraemic toxins, in *Replacement of Renal Function by Dialysis* (2ed), Drukker W, Parsons FM, Maher JF (eds), Boston, Martinus Nijhoff, pp 354-390, 1983.
2. Powell D, Bergstrom J, Dzurik R, Gulyassy P, Lockwood D, Phillips L: Toxins and inhibitors in chronic renal failure. Am J Kidney Dis 7:292-299, 1986.
3. Vanholder R, Schoots A, Ringoir S: Uremic toxicity, in *Replacement of Renal Function by Dialysis* (3ed), Maher JF (ed), Dordrecht, Kluwer Academic Publishers, pp 4-19, 1989.
4. Schreiner GE, Maher JF: *Uremia: Biochemistry, Pathogenesis and Treatment*, Springfield IL, Charles C Thomas Publishers, 1961.
5. Mitch WE, Walser M, Steinman TI, Hill S, Zeger S, Tungsanga K: The effect of a keto acid-amino acid supplement to a restricted diet on the progression of chronic renal failure. N Engl J Med 311:623-629, 1984.
6. Giovannetti S, Maggiore Q: A low nitrogen diet with proteins of high biological value for severe chronic uremia. Lancet 1:1000, 1964.
7. Giordano C, DePascale C, DeCristofaro D, Capodiscasa G, Balestrieri C, Baczyk K: Protein malnutrition in the treatment of chronic uremia, in *Nutrition in Renal Disease*, Berlyne GM (ed), Baltimore, The Williams & Wilkins Co., pp 23-37, 1968.
8. Gulyassy PF, Aviram A, Peters JH: Evaluation of amino acid and protein

requirements in chronic uremia. Arch Intern Med 126:855-859, 1970.
9. Mitch WE, Klahr S (eds): *Nutrition and the Kidney*, Boston, Little, Brown, 1988.
10. Pullman TN, Alving AS, Dein RJ, Landowne M: The influence of dietary protein intake on specific renal function in normal man. J Lab Clin Med 44:320-332, 1954.
11. Brenner BM, Meyer TW, Hostetter TH: Dietary protein intake and the progressive nature of kidney disease: the role of hemodynamically mediated glomerular injury in the pathogenesis of progressive glomerular sclerosis in aging, renal ablation and intrinsic renal disease. N Engl J Med 307:652-659, 1982.
12. Bosch JP, Saccaggi A, Lauer A, Ronco C, Belledonne M, Glabman S: Renal functional reserve in humans: effect of protein intake on glomerular filtration rate. Am J Med 75:943-50, 1983.
13. Ihle BU, Becker GJ, Whitworth JA, Charlwood RA, Kincaid-Smith PS: The effect of protein restriction on the progression of renal insufficiency. New Engl J Med 321:1773-1777, 1989.
14. Peters JH, Gotch FA, Keen M, Berridge BJ Jr, Chao WR: Investigation of the clearance and generation rate of endogenous peptides in normal subjects and uremic patients. Trans Am Soc Artif Intern Organs 20:417, 1974.
15. Bliss S: The cause of sore mouth in nephritis. J Biol Chem 121:425, 1937.
16. Williams JL, Dick GF: The excretion of nonprotein nitrogen substances by the intestine. JAMA 100:484, 1933.
17. Carter D, Einheber A, Baure H, Rosen H, Burns WF: The role of the microbial flora in uremia. II. Uremic colitis, cardiovascular lesions, and biochemical observations. J Exp Med 123:251, 1966.
18. Simenhoff ML, Burke JF, Saukkonen JJ, Wesson LG, Schaedler RW: Amine metabolism and the small bowel in uraemia. Lancet 2:818, 1976.
19. Simenhoff ML, Saukkonen JJ, Burke JF, Wesson LG, Schaedler RW, Gordon SJ: Bacterial populations of the small intestine in uremia. Nephron 22:63, 1978.
20. Gruskin SE, Tolman DE, Wagoner RD: Oral manifestations of uremia. Minn Med 53:495, 1970.
21. Black DR: Nephritic stomatitis. Urol Cutan Rev 46:75, 1942.
22. Balestri PL, Barsotti G, Camici M, Giovannetti S: High plasma fibrinogen levels and reduced fibrinolytic activity in dogs intoxicated with methylguandidine. Clin Nephrol 2:81, 1974.
23. Balestri PL, Biagini M, Rindi P, Giovanneti S: Uremic toxins. Arch Intern Med 126:843, 1970.
24. Johnson WJ, Hagge WW, Wagoner RD, Dinapoli RP, Rosevear JW: Effects of urea loading in patients with far-advanced renal failure. Mayo Clin Proc 47:21-29, 1972.
25. Mirsky IA, Diengott D, Perisutti G: Insulinase-inhibitory action of metabolic derivatives of L-tryptophan. Proc Soc Exp Biol Med 95:154, 1957

26. Record BN, Princhard JW, Gallagher BB, Seligson D: Phenolic acids in experimental uremia. I. Potential role of phenolic acids in the neurological manifestations of uremia. Arch Neurol 21:387, 1969.
27. Mason MF, Resnik H, Mino AS, Rainey J, Pilcher C, Harrison TR: Mechanism of experimental uremia. Arch Intern Med 60:312, 1937.
28. Bennett WM, Porter GA: Endogenous creatinine clearance as a clinical measure of glomerular filtration rate. Br Med J 4:84-86, 1971.
29. Bauer JH, Brooks CS, Burch RN: Renal function studies in man with advanced renal insufficiency. Am J Kidney Dis 2:30-35, 1982.
30. Rose WC, Dimmit FW: Experimental studies on creatine and creatinine. VII. The fate of creatine and creatinine when administered to man. J Biol Chem 26:345, 1916.
31. Shannon JA: The renal excretion of creatinine in man. J Clin Invest 14:403, 1935.
32. Lowrie EG, Laird NM, Parker TF, Sargent JA: Effect of the hemodialysis prescription on patient morbidity: Report from the National Cooperative Dialysis Study. N Engl J Med 305:1176-1181, 1981.
33. Gulyassy P, Igarashi P: Endogenous solutes as potential indicators of renal function (abstract). Int Soc of Nephrol Abstracts, 1990.
34. Bemert JT, Bell CJ, Guntupalli J, Hannon HW: Pseudouridine as an endogenous renal clearance marker. Clin Chem 34:1011-1017, 1988.
35. Schoots AC, Dijkstra JB, Ringoir SMG, Vanholder R, Cramers CA: Are the classical markers sufficient to describe uremic solute accumulation in dialyzed patients? Hippurates reconsidered. Clin Chem 34:1022-1029, 1988.
36. Lundin PA III: Prolonged survival on hemodialysis, in *Replacement of Renal Function by Dialysis* (3ed), Maher JF (ed), Dordrecht, Kluwer Academic Publishers, 1989.
37. Abel JJ, Rowntree LC, Turner BB: On the removal of diffusible substances from the circulating blood by means of dialysis. Trans Assoc Am Physicians 28:51, 1913.
38. Quadracci LJ, Cambi V, Christopher TG, Harker LA, Striker GE: Assay of serum abnormalities in uremic and dialysis patients: Evidence for depletion of vital substances in hemodialysis. Trans Am Soc Artif Intern Organs 17:96-101, 1971.
39. Fraser DR, Kodicek E: Unique biosynthesis by kidney of a biologically active vitamin D metabolite. Nature 228:764, 1970.
40. Eschbach JW, Egrie JC, Downing MR, Browne JK, Adamson JW: Correction of the anemia of end-stage renal disease with recombinant human erythropoietin: Results of a combined phase I and II clinical trial. N Engl J Med 316:73-78, 1987.
41. Stanbury SW: Azotemic renal osteodystrophy. Br Med Bull 13:57-60, 1957.
42. Bricker NS: On the pathogenesis of the uremic state: An exposition of the "Trade-off hypothesis." N Engl J Med 286:1093-1099, 1972.

43. Slatopolsky E, Caglar S, Gradowska L, Canterbury J, Reiss E, Bricker NS: On the prevention of secondary hyperparathyroidism in experimental chronic renal disease using "proportional reduction" of dietary phosphorus intake. Kidney Int 2:147, 1972.
44. Slatopolsky E, Caglar S, Pennell DB, Taggart JM, Canterbury J, Reiss E, Bricker NS: On the pathogenesis of hyperparathyroidism in chronic experimental renal insufficiency in the dog. J Clin Invest 50:492, 1971.
45. Massry SG, Goldstein DA: Role of parathyroid hormone in uremic toxicity. Kidney Int 13 (Suppl 8):S39-42, 1978.
46. Bricker NS, Fine LG: The trade-off hypothesis: current status. Kidney Int 13 (Suppl 8):S5-8, 1978.
47. Adler AJ, Ferran N, Berlyne GM: Effect of inorganic phosphate on serum ionized calcium concentration in vitro: a reassessment of the "trade-off hypothesis." Kidney Int 28:932-935, 1985.
48. Gulyassy PF, Bottini AT, Jarrard EA, Stanfel LA: Isolation of inhibitors of ligand:albumin binding from uremic body fluids and normal urine. Kidney Int 24 (Suppl 16):S238-242, 1983.
49. Wilkinson GR: Plasma and tissue binding considerations in drug disposition (review). Toxicol Appl Pharmacol 69:427-465, 1983.
50. Marshall EK, Vickers JL: The mechanism of the elimination of phenolsulphonephthalein by the kidney - a proof of secretion by the convoluted tubules. Bull Johns Hopkins Hosp 34:1-7, 1923.
51. Weiner IM: Transport of weak acids and bases, in *Handbook of Physiology*, Orloff J, Berliner RW (eds), Washington, D.C., American Physiological Society, pp 521-554, 1973.
52. Forster RP: Renal transport mechanisms. Fed Proc 26:1008-1019, 1967.
53. Pitts RF: Tubular secretion, in *Physiology of the Kidney and Body Fluids* (3ed), Chicago, Year Book Medical Publishers, pp 140-157, 1974.
54. Tanaka K, Hine DG, West-Dull A, Lynn TB: Gas-chromatographic method of analysis for urinary organic acids. I. Retention indices of 155 metabolically important compounds. Clin Chem 26:1839-1846, 1980.
55. Gulyassy PF, Depner TA: Impaired binding of drugs and endogenous ligands in renal diseases. Am J Kidney Dis, 2:578-601, 1983.
56. Fehske KJ, Muller WE, Wollert U: The location of drug binding sites in human serum albumin. Biochem Pharmacol 30:687-692, 1981.
57. Sudlow G, Birkett DJ, Wade DN: The characterization of two specific drug binding sites on human serum albumin. Mol Pharmacol 11:824-832, 1975.
58. Tavares-Almeida I, Gulyassy PF, Depner TA, Jarrard EA: Aromatic amino acid metabolites as potential protein binding inhibitors in human uremic plasma. Biochem Pharmacol 34:2431-2438, 1985.
59. Craig WA, Evenson MA, Sarver KP, Wagnild JP: Correction of protein binding defect in uremic sera by charcoal treatment. J Lab Clin Med 87:637-647, 1976.

60. Depner TA, Gulyassy PF: Plasma protein binding in uremia: extraction and characterization of an inhibitor. Kidney Int 18:86-94, 1980.
61. Depner TA: Suppression of tubular anion transport by an inhibitor of serum protein binding in uremia. Kidney Int 20:511-518, 1981.
62. Depner TA, Sanaka T, Stanfel LA: Suppression of PAH transport in the isolated perfused kidney by an inhibitor of protein binding in uremia. Am J Kidney Dis 3:280-286, 1984.
63. Gulyassy PF, Jarrard E, Stanfel LA: Contributions of hippurate, indoxyl sulfate, and o-hydroxyhippurate to impaired ligand binding by plasma in azotemic humans. Biochem Pharmacol 36:4215-4220, 1987.
64. Mitch W, Brusilow S: Benzoate-induced changes in glycine and urea metabolism in patients with chronic renal failure. J Pharmacol Exp Ther 222:572-576, 1982.
65. Dzurik R, Spustova V, Gerykova M: Pathogenesis and consequences of the alteration of glucose metabolism in renal insufficiency, in *Uremic toxins*, Ringoir S, Vanholder R, Massry S (eds), New York, Plenum Publishing Co., p 105, 1987.
66. Spustova, V, Dzurik R, Gerykova M: Hippurate participation in the inhibition of glucose utilization in renal failure. Czech Med 10:79-89, 1987.
67. Vanholder R, Schoots A, Cramers C, DeSmet R, Van Landschoot N, Wizeman V, Botella J, Ringoir S: Hippuric acid as a marker, in *Uremic Toxins*, Ringoir S, Vanholder R, Massry S (eds), New York, Plenum Publishing Co., 1987.
68. Liebich HM, Pickert A, Tetschner B: Gas chromatographic and gas-chromatographic-mass spectrometric analysis of organic acids in plasma of patients with chronic renal failure. J Chromatogr 289:259-266, 1984.
69. Lindup WE, Bishop KA, Collier R: Drug binding defect of uraemic plasma: contribution of endogenous binding inhibitors. Hoechst Symposium, Portugal, August 24-September 28, 1985.
70. Mabuchi H, Nakahashi H: A major endogenous ligand substance involved in renal failure. Nephron 49:277-280, 1988.
71. Mabuchi H, Nakahashi H: Inhibition of hepatic glutathion S-transferases by a major endogenous ligand substance present in uremic serum. Nephron 49:281-283, 1988.
72. Mabuchi H, Nakahashi H, Hamajima T, Aikawa I, Oka T: The effect of renal transplantation on a major endogenous ligand retained in uremic serum. Am J Kidney Dis 13:49-54, 1989.
73. Pardridge WM, Oldendorf WH, Cancilla P, Frank HJ: Blood-brain barrier: interface between internal medicine and the brain (clinical conference), Ann Intern Med 105(1):82-95, 1986.
74. Ockner RK, Weisiger RA, Gollan JL: Hepatic uptake of albumin-bound substances: albumin receptor concept. Am J Physiol 245:G13-18, 1983.
75. Besseghir K, Mosig D, Roch-Ramel F: Facilitation by serum albumin of renal tubular secretion of organic anions. Am J Physiol 256:F475-484, 1989.

76. Bassingthwaighte JB, Noodleman L, van der Vusse G, Little SE, Glatz JFC, Reneman RS: Albumin-fatty acid-endothelial membrane interactions and fatty acid transport in the heart. Fed Proc 46:686, 1987.
77. Meijer DKF, Van Der Sluijs P: Covalent and noncovalent protein binding of drugs: implications for hepatic clearance, storage, and cell-specific drug delivery. Pharmaceut Res 6:105-118, 1989.
78. Dubey RK, McAllister CB, Inoue M, Wilkinson GR: Plasma binding and transport of diazepam across the blood-brain barrier: no evidence for in vivo enhanced dissociation. J Clin Invest 84:1155-59, 1989.
79. Smith KR, Borchardt RT: Permeability and mechanism of albumin, cationized albumin, and glycosylated albumin transcellular transport across monolayers of bovine brain capillary endothelial cells. Pharmaceut Res 6:466-473, 1989.
80. Babb AL, Popovich RP, Christopher TG, Scribner BH: The genesis of the square meter-hour hypothesis. Trans Am Soc Artif Intern Organs 17:81-91, 1971.
81. Scribner BH, Farrell PC, Milutinovic J, Babb AL: Evolution of the middle molecule hypothesis, in Proceedings of the Fifth International Congress of Nephrology, Villarreal H (ed), Basel, Karger, pp 190-199, 1974.
82. Solangi K, Lutton JD, Abraham NG, Ascensao JL, Goodman A, Levere RD: Isolation and purification of erythropoiesis inhibitory activity from uremic sera. Nephron 48:22-27, 1988.
83. Bommer J, Ritz E, Waldherr R: Silicone-induced splenomegaly: treatment of pancytopenia by splenectomy in a patient on hemodialysis. N Engl J Med 305:1077-1079, 1981.
84. Hyde SE, Sadler JH: Red cell destruction in hemodialysis. Trans Am Soc Artif Intern Organs 15:50-53, 1969.
85. Leong ASY, Disney APS, Gove DW: Spallation and migration of silicone from blood-pump tubing in patients on hemodialysis. N Engl J Med 306:135-140, 1982.
86. Morley AR, Barron D, Thompson P, Hoenich NA, Harbottle S, Kerr DNS: Surface alterations in dialysis roller pump inserts: a scanning electron microscopy study. J Biomed Eng 8:255-261, 1986.
87. Sherrard DJ: Aluminum toxicity. The Kidney 20:31-35, 1988.
88. Yawata Y, Howe R, Jacob HS: Abnormal red cell metabolism causing hemolysis in uremia. A defect potentiated by tap water hemodialysis. Ann Intern Med 79:362, 1973.
89. Kjellstrand CM, Eaton JW, Yawatta Y, Swofford H, Kolpin CF, Buselmeier TJ, von Hartitzsch B, Jacob HS: Hemolysis in dialyzed patients caused by chloramines. Nephron 13:427-433, 1974.
90. Peterson NJ, Carson JA, Favero MS: Bacterial endotoxin in new and reused hemodialyzers: a potential cause of endotoxemia. Trans Am Soc Artif Intern Organs 27:155-160, 1981.
91. Bommer J, Wilhelms OH, Barth HP, Schindele H: Anaphylactoid reactions in

dialysis patients: role of ethylene-oxide. Lancet 2:1382-1384, 1985.
92. Gibson TP, Briggs WA, Boone BJ: Delivery of di-2-ethylhexyl phthalate to patients during hemodialysis. J Lab Clin Med 87:519-524, 1976.
93. Ono K, Tatsukawa R, Wakimoto T: Migration of plasticizer from hemodialysis blood tubing. JAMA 234:948, 1975.
94. Gutch CF, Eskelson CD, Ziegler E, Ogden DA: 2-Chloroethanol as a toxic residue in dialysis supplies sterilized with ethylene oxide. Dial Transplant 5:21-25, 1976.
95. Craddock PR, Fehr J, Brigham KL, Kronenberg RS, Jacob HS: Complement and leukocyte-mediated pulmonary dysfunction in hemodialysis. New Engl J Med 296:770-774, 1977.
96. Blumenstein M, Schmidt B, Ward RA, Ziegler-Heitbrock HWL, Gurland HJ: Altered interleukin-1 production in patients undergoing hemodialysis. Nephron 50:277-281, 1988.
97. Hakim RM, Fearon DT, Lazarus MJ: Biocompatibility of dialysis membranes: effects of chronic complement activation. Kidney Int 26:194-200, 1984.
98. Mion CM, Hegstrom RM, Boen ST, Scribner BH: Substitution of sodium acetate for sodium bicarbonate in the bath fluid for hemodialysis. Trans Am Soc Artif Intern Organs 10:110-113, 1964.
99. Keshaviah P: The role of acetate in the etiology of symptomatic hypotension. Artif Organs 6:378, 1982.
100. Hakim RM, Pontzer M-A, Tilton D, Lazarus JM, Gottlieb MN: Effects of acetate and bicarbonate dialysate in stable chronic dialysis patients. Kidney Int 28:535-540, 1985.
101. Fenves AZ, Emmett M, White MG, Greenway G, Michaels DB: Carpal tunnel syndrome with cystic bone lesions secondary to amyloidosis in chronic hemodialysis patients. Am J Kidney Dis 7:130-134, 1986.
102. Ullian ME, Hammond WS, Alfrey AC, Schultz A, Molitoris BA: Beta-2-microglobulin-associated amyloidosis in chronic hemodialysis patients with carpal tunnel syndrome. Medicine 68:107-115, 1989.
103. Gejyo F, Odani S, Yamada T, Honma N, Saito H, Suzuki Y, Nakagawa Y, Kobayashi H, Maruyama Y, Hirasawa Y, Suzuki M, Arakawa M: B2-microglobulin: a new form of amyloid protein associated with chronic hemodialysis. Kidney Int 30:385-390, 1986.
104. Laird NM, Berkey CS, Lowrie EG: Modeling success or failure of dialysis therapy: the National Cooperative Dialysis Study. Kidney Int 23 (Suppl 13):S101-106, 1983.
105. Merrill JP, Legrain M, Hoigne R: Observations on the role of urea in uremia. Am J Med 14:519, 1953.

Chapter 2

UREA METABOLISM: CLINICAL CHEMISTRY OF UREA

Urea is a simple compound of low molecular weight that has appeared at the crossroads of scientific discovery. It was first synthesized by Friedrich Wöhler in 1828, a landmark achievement that helped to erase the mystical barriers between organic and inorganic chemistry. Wöhler showed that urea, an organic compound, could be synthesized from ammonium cyanate, an inorganic compound. The synthesis of urea, representing a link between organic and inorganic chemistry was perhaps more significant than the synthesis of DNA. Although technically more difficult, the latter was accepted more readily by the scientific community than was the synthesis of urea.

Apart from its philosophical significance, the synthesis of urea has been expanded and applied commercially during the past two centuries. Synthetic urea has many uses; a prominent one is as a soil fertilizer. Urea is an excellent fertilizer because of its high nitrogen content (47%) in a solid, water-soluble form. Urea nitrogen, in contrast to atmospheric nitrogen, is more active and available to plants for structural synthesis. Animals synthesize urea as a waste product of nitrogen metabolism. In animals, urea cannot be utilized for structure or for energy metabolism and is excreted. Studies of urea excretion by the normal kidney have contributed significantly to our understanding of nitrogen metabolism and renal function.

In mammals, urea appears to have a function only in the kidney, where its accumulation in the renal medulla helps to improve performance of the countercurrent multiplier, the renal mechanism for conserving water. In certain fish and amphibia, urea functions as an osmotic buffer in times of osmotic stress. The common toad, for example, actively transports urea through its skin to prevent solute and water loss in times of drought (1). Humans and other mammals seem to have an obligatory synthesis of urea that goes on even in starvation. Whether this is a spill-over phenomenon or serves a useful purpose is not certain. The major purpose of urea synthesis in mammals is apparently the detoxification and irreversible elimination of nitrogenous waste.

Excretion of nitrogenous waste products: comparative physiology

Primitive microorganisms absorb and use amino acids with absolute efficiency. As they grow and multiply, no protein degradation occurs as they incorporate needed amino acids found in their environment into protein and other nitrogen-containing cell components. No nitrogen returns to the growth medium, and amino acids that are not required do not enter the cell. Conditions are quite different in mammalian tissues. Throughout mammalian life, protein degradation occurs, even during growth phases. In addition, surplus amino acids enter the body and are used for energy production. This creates a continual turnover of amino acid nitrogen, some from exogenous supply and the remainder from endogenous protein catabolism. Constant turnover of body protein probably affords an advantage to the organism. Repair and reconstruction, shedding old damaged structures and replacing them with new, allows the organism to better adapt to the environment. Estimates of protein degradation in normal adults are approximately 4-5 grams/kilogram/day or about 300 grams of protein/day in a 70 kg adult (2). Dietary intake is 80-150 grams/day, so the total protein catabolic rate (PCR) always exceeds dietary intake.

Amino acids are small-molecular-weight compounds that could be eliminated from the organism intact, but this would represent a significant loss of energy because the carbon skeleton of amino acids can be used as fuel. The simplest pathway for elimination of amino acid nitrogen yields ammonium, a highly toxic compound. Various mechanisms have evolved in different animals to deal with ammonium. The term *ammonotelic* describes aquatic invertebrates and some vertebrates that excrete most of their nitrogenous waste as unaltered ammonium. This is the most primitive excretory mechanism; it consists of simple diffusion from the circulatory compartment to the outside environment. Ammonotelic animals take advantage of their proximity to the aquatic environment and eliminate wastes much like the dialyzer, by simply allowing the ammonium to diffuse down its concentration gradient. This mechanism cannot be used by terrestrial animals. Some animals, such as reptiles and birds that have adapted to life on dry land, are *uricotelic* — that is, they excrete their nitrogenous waste as solid uric acid and allantoin, both of which are poorly soluble in water. This allows maximum conservation of water. Mammals are *ureotelic* — they transform ammonium to urea, a highly water soluble-compound that is excreted through the kidney. The production of urea allows the animal to conserve water, but it consumes energy and requires a complex organ that uses the principles of countercurrent multiplication to concentrate the urine. Amphibians excrete ammonium both via direct diffusion and via urea formation. Before metamorphosis, the aquatic tadpole excretes ammonium directly through the skin into the environment, while the adult frog synthesizes and excretes urea. More terrestrial amphibians synthesize urea, while the aquatic frogs revert to ammonium excretion, a less energy-requiring process.

Some animal species (e.g., certain fish and amphibia) can produce urea by degradation of purines via the purine-urea cycle. This pathway involves conversion of purines to uric acid, then to allantoin, and further degradation of allantoin to urea. The enzymes necessary to do this are absent in primates, birds, and reptiles.

BIOCHEMISTRY OF UREA

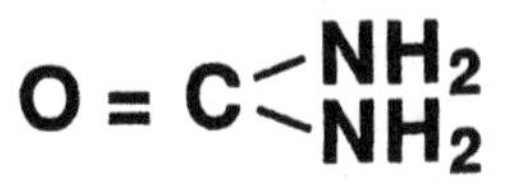

$$O = C\begin{matrix} \diagup NH_2 \\ \diagdown NH_2 \end{matrix}$$

Figure 2.1. Chemical structure of urea.

Urea is a colorless, odorless, crystalline substance with a molecular weight of 60 daltons. It is nonionic but nonvolatile and highly polar. Its polarity results from the doubly bonded oxygen and two amino nitrogen groups shown in figure 2.1. As a chemical structure (figure 2.1), it can be considered the amide of carbamic acid. Urea can be synthesized by direct reaction of carbon dioxide with ammonia at high temperature and pressure, a process used to make fertilizer:

$$2\ NH_3 + CO_2 \longrightarrow N_2H_4CO + H_2O$$

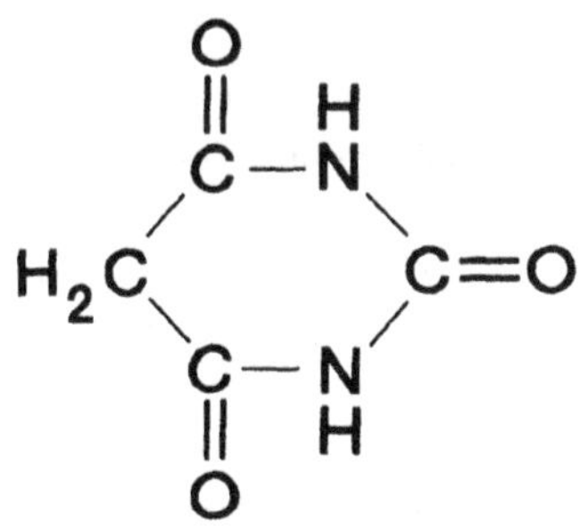

```
          O
          ||  H
          C — N
         /     \
    H2C         C=O
         \     /
          C — N
          ||  H
          O
```

Figure 2.2. Barbituric acid (malonyl urea).

Conversely, it can be broken down at high temperature or enzymatically (e.g., with bacterial urease) to ammonia and CO_2.

Urea condenses with malonic acid to form barbituric acid (malonyl urea), used in the pharmaceutical industry (figure 2.2).

Urea is used in agriculture as a fertilizer because its nitrogen is fixed. Fixed nitrogen is nitrogen combined chemically with hydrogen or other atoms, breaking the tight triple bond that exists between nitrogen atoms in the gaseous state. Plants can incorporate fixed nitrogen into protein but not atmospheric nitrogen. Certain microorganisms found in the roots of leguminous plants can fix atmospheric nitrogen and hence add fertility to the soil.

Urea is the major end product of nitrogen metabolism in mammals, appearing in normal adult urine in amounts of 20 to 40 grams/day or 330 to 670 mmoles/day. The most important biochemical pathway that leads to urea production in humans and other vertebrates is the Krebs urea cycle.

The urea cycle

The urea cycle, also known as the ornithine-urea cycle or the Krebs-Henseleit cycle, is a biochemical pathway used to dispose of surplus amino nitrogen and ammonia as urea. The primary source of nitrogen input to the urea cycle is from the amino acid pool. The cycle is found in primitive plants and animals and may have evolved from the biochemical pathways designed for arginine synthesis. In mammals, urea is produced from hydrolysis of arginine, an irreversible step during the urea cycle. This reaction is catalyzed by arginase, found in many tissues including hepatic cytoplasm:

$$\text{arginine} \xrightarrow{\text{arginase}} \text{ornithine} + \text{urea}$$

Arginase is a widely distributed enzyme in living organisms, and arginine is ubiquitous because it is needed for protein synthesis. This means that urea synthesis should be a universal biochemical phenomenon. However, only organisms that have a functioning ornithine urea cycle quantitatively convert arginine to urea. The urea cycle is shown in figure 2.3. The critical step is capture of nitrogen as carbamyl phosphate. This is the unique step that makes urea.

The synthesis of arginine from ornithine and ammonia becomes a cyclic process after addition of arginase. The cyclic nature of this metabolic pathway allows reutilization of the carbon chain of ornithine, producing one urea molecule per turn of the cycle. The two nitrogen atoms in urea come from two sources: one from ammonium (as carbamyl phosphate) and the other from aspartate. In each case, entry of nitrogen as ammonium or as aspartate requires the consumption of three ATP molecules.

The liver is the major site of urea synthesis since it contains high concentrations of carbamyl phosphate synthetase. The remaining enzymes required for operation of the urea cycle can be found in other organs that lack this critical enzyme.

Hepatectomy in the dog prevents the rise in urea concentration after intravenous administration of amino acids (3). In patients with chronic liver disease, urea synthesis is apparently adequate to detoxify ammonia under ordinary protein loading. But if protein intake is high or when liver damage becomes severe, ammonium intoxication occurs (4). Serum levels of ammonium and amino acids rise under these circumstances. Acute hepatotoxicity induced by mercury or carbon tetrachloride causes a rapid rise in serum ammonium and amino acids, suggesting that urea synthesis may be selectively inhibited (5).

Hepatic transamination of oxaloacetate/glutamate to form aspartate allows entry of amino nitrogen into the urea cycle without going through the formation of ammonia (right side of figure 2.3). This is an energy-conserving pathway to urea formation that consumes only one ATP. After loss of the amino group, the alpha-ketoacid carbon skeleton is rapidly oxidized for further energy production. Transamination can also occur in the reverse direction in the presence of an abundance

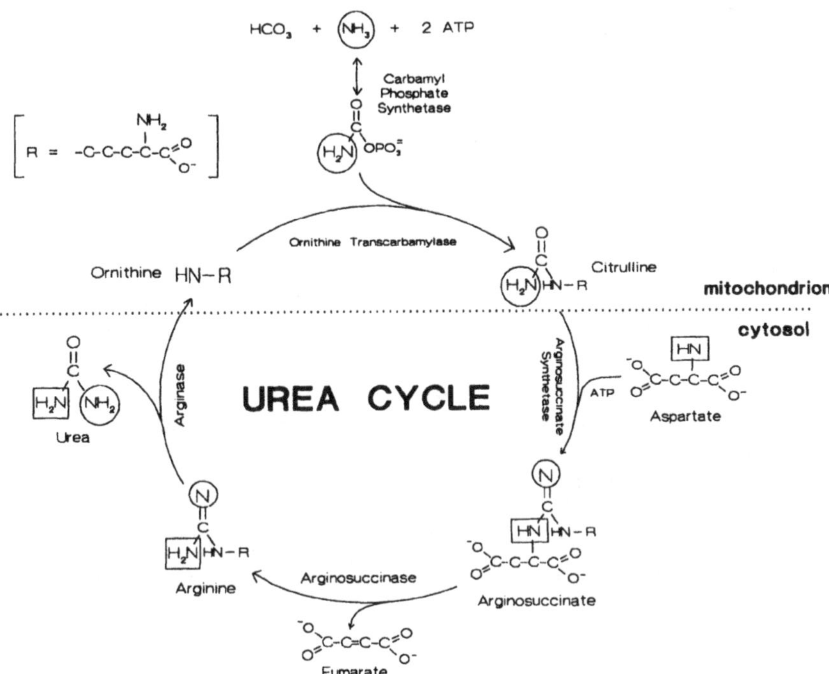

Figure 2.3. The Krebs ornithine-urea cycle. Nitrogen enters the cycle from two sources: 1) from ammonia (circles) and 2) from aspartate (squares). After synthesis of arginine, the cyclic pathway is made possible by the hydrolytic enzyme arginase that releases urea. The resulting ornithine carbon skeleton is recycled to citrulline to arginosuccinate and back to arginine.

of alpha-ketoacid precursors (6). Dietary intake of alpha-keto analogues of the essential amino acids can reduce ammonia and urea production and may have beneficial effects in patients with renal excretory failure, as discussed below.

Nitrogen balance

Gluconeogenesis

The alternate use for amino acids, other than as subunits of structural protein, is as a source of fuel for energy needs. *Gluconeogenesis* is the term given to the set of biochemical reactions that results in formation of glucose from noncarbohydrate sources such as protein. This energy-releasing process starts with protein hydrolysis to form amino acids and is followed by loss of nitrogen from the amino acids. Amino nitrogen is removed from the amino acid via transamination with alpha-ketoglutaric acid to form glutamate. Glutamate can then be deaminated to form ammonia for the urea cycle (top of figure 2.3) or undergo transamination to aspartate (right of figure 2.3). The remaining carbon skeleton can be utilized in the Krebs citric acid cycle, leading to production of CO_2, water, and high-energy phosphate bonds or for gluconeogenesis. This process of amino acid breakdown for fuel is accelerated under conditions of stress, trauma, sepsis, and fever, called *catabolic states* (7).

Minimum nitrogen requirement

The amino acid pool is limited in size and capacity (about 30 grams in an adult). Surplus amino acids from dietary protein intake cannot be stored, so they stimulate the urea cycle and cause increased urea synthesis. Conversely, a decrease in protein intake causes a decrease in urea synthesis until a minimum level of intake is reached. Below this level of protein intake (the minimal nitrogen requirement), the daily amino acid supply from endogenous protein catabolism and exogenous supply is inadequate to supply protein synthetic needs (8). The protein synthetic rate falls below the catabolic rate, and the patient has a negative nitrogen balance. The protein synthetic needs stem primarily from turnover of structural and other body proteins and provision of glutamine for ammonium excretion by the kidney. The standard minimum nitrogen requirement is about 3.5 g amino nitrogen/day for healthy resting medical students (8,9) corresponding to ~20 g protein/day. Table 2.1 lists several factors that impact on this minimal requirement.

Table 2.1. Factors that determine minimum nitrogen requirements

1. Age
2. Physical activity
3. Stress (a catabolic state), e.g., fever, trauma, surgery, infection, etc.
4. Hormones: glucocorticoids, catecholamines, thyroxine
5. Quality of protein intake: essential/nonessential amino acid ratio
6. Acid/base balance
7. Drugs, e.g., tetracycline

A major goal of dietary therapy for patients with advanced renal failure is to decrease ureagenesis and production of other toxins by reducing protein intake (10). The ideal minimum protein intake of ~20 g/day for normal adults, slightly more in patients with advanced renal failure, is rarely achieved (11). The reasons for this include the list of modulating factors in table 2.1, but equally important is the poor palatability of a low-protein diet. Patients either stray beyond the limits of the imposed diet or elect not to eat unappetizing foods, causing a reduction in their total calorie intake.

The ratio of essential to nonessential amino acids is a measure of the protein's biological value. Certain foodstuffs, such as milk protein, egg albumin, and meat, have high biological value whereas, certain vegetable proteins have less of the essential amino acids and are rated low in biological value. A protein with zero biological value would be one altogether lacking one essential amino acid. Without this essential ingredient, the organism is unable to synthesize protein, so the remainder of the amino acids in the diet are surplus. As such, they are metabolized for energy production or stored as glycogen and lipid. Before metabolism or storage, the nitrogen moiety must be removed and eliminated as urea, since there is no storage form of nitrogen other than as protein. Under these circumstances, the organism would suffer a negative nitrogen balance because of the lack of protein synthesis and the constant obligatory protein degradation that characterizes mammalian tissues. The negative nitrogen balance would be essentially the same as in a patient who has no amino acid intake. This is obviously an undesirable state, but especially in patients with renal failure, where urea and other protein-derived toxins will accumulate. Maintaining the proper balance of amino acids in the diet, using proteins of high biological value, is an important goal of dietary therapy (11).

INBORN ERRORS OF UREA METABOLISM: UREA CYCLE ENZYMOPATHIES

In the absence of a functional urea cycle (figure 2.3), ammonium builds up to toxic levels. An experiment of nature occurring in children with inborn errors of urea synthesis has taught us the importance of the urea cycle. Inherited defects in each of the five urea cycle enzymes depicted in figure 2.3 have been described (12,13,14,15). Children who are born with complete loss of one of these enzymes develop hyperammonemia within one week of age that usually leads to coma and death. Manipulation of the diet to reduce ammonia production may prolong survival but is usually accompanied by impairment of intellectual function due to episodic hyperammonemia (15,16). Efforts to combat accumulation of ammonium in these children include stimulation of alternate pathways of waste nitrogen metabolism. Nearly all nitrogen metabolites are excreted as urea in ureotelic animals. The only other quantitatively significant nitrogenous waste products are creatinine, uric acid, and hippurate. The total of creatinine, uric acid, and hippurate excretion in normals is less than 10% of total nitrogen excretion. Hippurate production can be augmented by feeding sodium benzoate, and urea production can be augmented in some of these patients by feeding arginine. Intravenous sodium phenylacetate stimulates production of phenylacetyl glutamine, a nitrogenous compound similar to hippurate. This intravenous supplement is useful during hyperammonemic episodes (16). Ketoacid analogues of the amino acids have also been used to treat these disorders (see below) (17,18).

Urea transport

Urea is a small-molecular-weight (60 daltons) nonionic but polarized compound that is almost infinitely soluble in water (figure 2.1). Despite its affinity for water, it finds its way through lipid biological membranes with ease, much like water. The reason for its high permeability is not completely understood, but in some tissues a pathway for facilitated diffusion has been identified (19). In humans, most tissues that are permeable to water are also permeable to urea, with one exception, the kidney. In the descending limb of Henle's loop, water permeability is high while urea permeability is limited. This differential permeability allows urea to concentrate in the medulla, supporting the kidney's efforts to conserve water through the countercurrent multiplier principle. Urea moves amongst aqueous compartments within the body with ease by simple or facilitated diffusion. The kidney is the only organ that can establish and maintain a urea concentration gradient. It concentrates urea by manipulating water and urea permeabilities without expending energy specifically for urea transport (20). No active transport of urea has been demonstrated in humans but there is considerable evidence for facilitated diffusion in the red cell (see chapter 7). Active transport of urea occurs in other animals (1). Diffusion within human body compartments is rapid but not infinite. The rebound in urea concentration after dialysis supports a finite rate of diffusion from the intracellular and interstitial space (see chapter 5) (21). Reequilibration is rapid, dissipating the gradient after approximately 30-40 minutes.

Nitrogen recycling

In patients with renal failure, there is much interest in reducing protein catabolism and reincorporating the lost amino nitrogen back into protein (22,23,24). Either of these accomplishments would reduce the production of urea and toxic nitrogenous wastes (see chapter 1). Catabolism of protein and amino acids can be controlled by reducing intake of low-biological-value proteins, by provision of adequate but not excessive amounts of high-biological-value proteins in the diet, and by controlling catabolic processes. The latter include the clinical states listed in table 2.1 that accelerate endogenous protein breakdown. Other attempts to control the recycling of nitrogen have focused on the biochemical pathways of nitrogen disposal. The first addresses reutilization of urea nitrogen and the second addresses reutilization of amino nitrogen.

Urea recycling

The concept of urea as an end product of protein metabolism was challenged 30 years ago when it was suggested that ammonia resulting from intestinal urea hydrolysis could be reincorporated into protein (22,25,26). Subsequent studies showed that intestinal degradation of urea is only slightly increased even when urea concentrations are markedly elevated in patients with renal failure (23). Reincorporation of urea nitrogen into protein is usually minimal in patients on ordinary diets.

Ketoacid analogues of the essential amino acids

Reincorporation of amino nitrogen can be stimulated by manipulating the diet. The keto analogues of essential amino acids have been used to supplement the diet in patients with advanced renal failure in an attempt to improve symptoms and to slow the progression of the kidney damage (24,27). Amination of ketoacids to their amino acid counterparts reduces the obligatory production of urea, presumably by recycling amino nitrogen through transamination in the liver (27,28). Figure 2.4 shows the transamination reaction that reclaims amino nitrogen. Keto analogues also may reduce protein degradation and increased protein synthesis, an independent effect similar to hibernation in bears (29). Although expensive, these compounds may reverse symptoms and reduce the rate of progression of renal failure (24,30).

Methods of Urea Measurement

Urea has been measured in whole blood, plasma, deproteinated plasma, urine, and other body fluids. The newer methods are rapid, specific, and relatively inexpensive (31). Almost all current techniques employ the enzyme urease:

$$\text{urea} + 2\ H_2O \xrightarrow{\text{urease}} \text{ammonium} + \text{bicarbonate}$$

The ammonium ion is the target for assay. Ammonium is usually present in very low concentrations in body fluids because of its toxicity, so little or no interference is expected except in urine. Because of the high ammonium content of urine (often 1000 times that of serum), a blank must be subtracted from the urease-treated

$$\text{R-}\overset{\text{O}}{\overset{\|}{\text{C}}}\text{-}\overset{\text{O}}{\overset{\|}{\text{C}}}\text{-O-H} + \text{R'-}\overset{\text{H}_2\text{N}}{\underset{\text{H}}{\text{C}}}\text{-}\overset{\text{O}}{\overset{\|}{\text{C}}}\text{-O-H} \underset{}{\overset{\textit{transaminase}}{\rightleftharpoons}} \text{R-}\overset{\text{H}_2\text{N}}{\underset{\text{H}}{\text{C}}}\text{-}\overset{\text{O}}{\overset{\|}{\text{C}}}\text{-O-H} + \text{R'-}\overset{\text{O}}{\overset{\|}{\text{C}}}\text{-}\overset{\text{O}}{\overset{\|}{\text{C}}}\text{-O-H}$$

Figure 2.4. Transamination. The alpha-ketoacid (R-COCOOH) on the left receives an amino group from an amino acid (R'-CNH_2COOH). Transamination converts the ketoacid to its amino acid counterpart (R-CNH_2COOH) and the amino acid to its ketoacid form (R'-COCOOH).

specimen. Urease has a very high specificity for urea, so interference from ketones, amides, or similar compounds does not occur.

Ammonia gas released after alkalinization can be titrated with dilute acid to complete the assay. Colorimetric methods to measure ammonium directly have also been used. Newer methods use kinetic methods to measure the rate of ammonium ion formed. This eliminates interference from previously formed ammonium that may contaminate either the specimen or the reagents. Two popular rate methods are the coupled urease/glutamate dehydrogenase reaction and the conductimetric method. The former adds a second reaction that consumes NADH, forming NAD that can be monitored by its absorbance at 340 nm:

$$NH_4^+ + \text{alpha-ketoglutarate} + NADH \xrightarrow[\text{dehydrogenase}]{\text{glutamate}} NAD + \text{glutamate}$$

The conductimetric assay measures the rate of increase in electrical conductance of the reaction mixture after urease is added and ammonium ion begins to form. This technique is simple but requires precise timing and instant mixing of reagents. It can be performed with whole serum. Both techniques are rapid and specific. More recent methods measure the change in color of a pH indicator dye after formation of NH_4^+. These are especially suited for dry reagent technology used in newer automated multi-analyzers.

In general, urea analyses are very reliable, relatively inexpensive, and provide results with an accuracy of 3%-4% (coefficient of variation) (31). Dialysis centers that care for large numbers of patients may wish to consider addition of a dedicated urea analyzer to provide better control and more rapid return of results (see chapter 7 for a discussion of the units of measurement commonly used for urea).

The toxicity of urea

Correlation of urea concentrations with disease has become so routine in clinical medicine that terminology has arisen to describe (or sometimes to obscure) typical clinical problems. The term *azotemia* describes an elevation of the blood urea or blood urea nitrogen (BUN) without symptoms. The term *uremia* refers to the symptoms and signs of advanced renal failure, which is always accompanied by an elevated urea concentration; uremia implies azotemia, but not vice versa. We also refer to *prerenal failure* and *intrinsic renal failure* to distinguish causes of an elevated blood urea concentration. The term *prerenal failure* or more properly *prerenal azotemia* absolves the kidney from wrongdoing by implying that either urea appearance has increased or extrarenal hemodynamic factors account for decreased renal excretion. The term *postrenal failure* describes the consequences of urinary obstruction. Intrinsic renal failure is a consequence of disease within the kidney and is more likely to be accompanied by symptoms and signs of uremia at any given level of azotemia. The latter observation is consistent with our understanding of urea as a mild renal toxin. In prerenal failure or pure azotemia, urea levels usually reach higher levels before symptoms and signs of uremia appear.

As emphasized in chapter 1, the toxins responsible for uremic symptoms are not identified. Urea constitutes the bulk of nitrogen metabolites retained in renal failure, so it has come under intense scrutiny as a suspect toxin. Intravenous infusion of urea in animals and humans causes a myriad of symptoms and signs mostly attributable to the osmotic effect of urea causing extracellular fluid loss via the kidney. When urea is added to the dialysate, preventing its loss during dialysis treatments, symptomatic recovery occurs in patients with acute renal failure (32). During similar studies in patients with chronic renal failure, Johnson showed that symptoms appeared only when the concentration was allowed to reach very high levels, in excess of 300 mg/dl (33). Postulated indirect effects of urea include carbamylation of protein (34,35). After prolonged exposure to cyanate, which is derived from plasma urea, serum albumin and other proteins are chemically altered by the nonenzymatic but covalent addition of cyanide (carbamyl) groups. This spontaneous chemical reaction is similar to the nonenzymatic glycosylation of hemoglobin after exposure to glucose in diabetics (36). Carbamylation has the potential to interfere with protein enzyme functions, transport functions, and plasma binding of toxic ligands, but this mechanism could not be confirmed in uremic patients (37). After lengthy review of the potential toxicity of urea, Bergstrom concludes that urea is a "mild uremic toxin" (38).

References

1. Rapoport J, Chaimovitz C, Hays RM: Active urea transport in toad skin is coupled to H^+ gradients. Am J Physiol 256:F830-835, 1989.
2. Mitch WE, Klahr S (eds): *Nutrition and the Kidney*, Boston, Little, Brown, 1988.
3. Mann FC: The effects of complete and of partial removal of the liver. Medicine 6:419-511, 1927.
4. Rudman D, Difulco TJ, Galambos JT, Smith RB, Salam AA, Warren WD: Maximal rates of excretion and synthesis of urea in normal and cirrhotic subjects. J Clin Invest 52:2241-2249, 1973.
5. McLean P, Rossi F: Changes in the activities of urea-cycle enzymes after administration of carbon tetrachloride. Biochem J 81:261, 1964.
6. Walser M, Coulter AW, Dighe S, Crantz FR: The effect of keto-analogues of essential amino acids in severe chronic uremia. J Clin Invest 52:678-690, 1973.
7. Pittiruti M, Siegel JH, Sganga G, Coleman B, Wiles CE, Placko R: Determinants of urea nitrogen production in sepsis. Arch Surg 124:362-372, 1989.
8. Rose WC: The amino acid requirements of adult man. Nutr Abstr Rev 27:631-647, 1957.
9. Rose WC, Wixom RL, Lockhart HB, Lamberth GF: The amino acid requirements of man. XV. The valine requirement: summary and final observations. J Biol Chem 217:987-995, 1955.
10. Gulyassy PF, Aviram A, Peters JH: Evaluation of amino acid and protein requirements in chronic uremia. Arch Intern Med 126:855-859, 1970.

11. Wassner SJ, Bergstrom J, Brusilow SW, Harper A, Mitch W: Protein metabolism in renal failure: abnormalities and possible mechanisms. Am J Kidney Dis 7:285-291, 1986.
12. Russell A, Levin B, Oberholzer VG, Sinclair L: Hyperammonemia: a new instance of an inborn enzymatic defect of the biosynthesis of urea. Lancet 2:699, 1962.
13, McMurray WC, Rathburn JC, Mohyuddin F, Koegler SJ: Citrullinuria. Pediatrics 32:347, 1963.
14. Westall RG: Arginosuccinicaciduria: identification and reactions of the abnormal metabolites in a newly described form of mental disease, with some preliminary metabolic studies. Biochem J 77:135, 1960.
15. Msall M, Batshaw ML, Suss R, Brusilow SW, Mellits ED: Neurologic outcome in children with inborn errors of urea synthesis: outcome of urea-cycle enzymopathies. N Engl J Med 310:1500-1505, 1984.
16. Brusilow SW, Danney M, Waber LJ, Batshaw M, Burton B, Levitsky L, Roth K, McKeethren C, Ward J: Treatment of episodic hyperammonemia in children with inborn errors of urea synthesis. N Engl J Med 310:1630-1634, 1984.
17. Bohles H, Harms D, Heid H, Schmid D, Fekl W: Treatment of argininosuccinic aciduria with keto analogues of essential amino acids. Am J Clin Nutr 31:1808-1811, 1978.
18. Brusilow S, Batshaw M, Walser M: Use of keto acids in inborn errors of urea synthesis. Curr Concepts Nutr 8:65-75, 1979.
19. Kaplan MA, Hays L, Hays RM: Evolution of a facilitated diffusion pathway for amides in the erythrocyte. Am J Physiol 226:1327-1332, 1974.
20. Rabinowitz L, Gunther RA: Excretion of urea in sheep during urea, mannitol, and methylurea osmotic diuresis. Am J Physiol 222:807-809, 1972.
21. Pedrini LA, Zereik S, Rasmy S: Causes, kinetics and clinical implications of post-hemodialysis urea rebound. Kidney Int 34:817-824, 1988.
22. Walser M, Bodenlos LJ: Urea metabolism in man. J Clin Invest 38: 1617, 1959.
23. Walser M: Urea metabolism in chronic renal failure. J Clin Invest 53:1385, 1974.
24. Mitch WE, Walser M, Steinman TI, Hill S, Zeger S, Tungsanga K: The effect of a keto acid-amino acid supplement to a restricted diet on the progression of chronic renal failure. N Engl J Med 311:623-629, 1984.
25. Levenson SM, Crowley LV, Horowitz RE, Malm OJ: The metabolism of carbon-labeled urea in the germfree rat. J Biol Chem 234:2061, 1959.
26. Richards P: Nutritional potential of nitrogen recycling in man. Am J Clin Nutr 25:615-625, 1972.
27. Walser M: Ketoacids in the treatment of uremia. Clin Nephrol 3:180-186, 1975.
28. Richards P: The metabolism and clinical relevance of the keto acid analogues of essential amino acids. Clin Sci and Mol Med 54:589-593, 1978.
29. Cahill GF Jr: Nitrogen versatility in bats, bears and man. N Engl J Med 290:686-687, 1974.

30. Barsotti G, Guiducci A, Ciardella F, Giovannetti S: Effects on renal function of a low-nitrogen diet supplemented with essential amino acids and ketoanalogues and of hemodialysis and free protein supply in patients with chronic renal failure. Nephron 27:113-117, 1981.
31. Kaplan LA: Urea, in *Methods in Clinical Chemistry*, Pesce AJ, Kaplan LA (eds), St. Louis, C.V. Mosby Co, pp 22-26, 1987.
32. Merrill JP, Legrain M, Hoigne R: Observations on the role of urea in uremia. Am J Med 14:519, 1953.
33. Johnson WJ, Hagge WW, Wagoner RD, Dinapoli RP, Rosevear JW: Effects of urea loading in patients with far-advanced renal failure. Mayo Clin Proc 47:21-29, 1972.
34. Bachmann K, Valentovic M, Shapiro R: A possible role for cyanate in the albumin binding defect of uraemia. Biochem Pharmacol 29:1598, 1980.
35. Flückiger R, Harmon W, Meier W, Loo S, Gabbay KH: Hemoglobin carbamylation in uremia. N Engl J Med 304:823, 1981.
36. Bunn HF, Haney DN, Kamin S, Gabbay KH, Gallop PM: The biosynthesis of human hemoglobin A1: slow glycosylation of hemoglobin in vivo. J Clin Invest 57:1652-1659, 1976.
37. Clegg LS, Lindup WE: Drug binding defect of uraemic plasma: unlikely involvement of carbamoylated albumin. Biochem Pharmacol 31:2791-2794, 1982.
38. Bergstrom J, Furst P: Uraemic toxins, in Replacement of Renal Function by Dialysis (2ed), Drukker W, Parsons FM, Maher JF (eds), Boston, Martinus Nijhoff, pp 354-390, 1983.

Chapter 3

UREA MODELING: INTRODUCTION

DEFINITION OF MODELING

Modeling in general means simulating reality. A model can be physical, like a construction model, which is a miniature facsimile of the real structure, or conceptual like a mathematical model. The test of a model is its ability to reproduce or conform to the object that inspired it, in all details. For example, a well-constructed model airplane or model train, except for its size, is often difficult to distinguish from the authentic original.

Urea models used for analysis of urea kinetics are mathematical models that attempt to simulate the movement of urea from, to, and within the body during and between hemodialyses. As a practical goal, urea models attempt to predict urea concentrations at any time during or following dialysis. Mathematical models are more abstract than physical models, but they seek the same objective, i.e., simulation of reality. Urea models are used as clinical tools to measure the extent or effectiveness of dialysis and to examine certain patient variables that are difficult to measure by other means.

A patient's need for dialysis creates an advantageous environment for the physician to learn more about his/her patient by taking blood samples and making simple measurements during dialysis. By sampling the blood urea concentration at opportune times during and between dialyses, physicians can learn more about the patient with end-stage renal disease than they can by measuring the blood urea concentration in the patient who is not undergoing dialysis therapy. The perturbations in urea concentrations induced by the dialyzer provide an opportunity to gain clinical insight that the modeling technique exploits.

Figure 3.1 shows a diagram of countercurrent blood and dialysate flow in modern artificial kidneys. The dialyzer membrane has multiple, fairly uniform pores that allow diffusion of smaller-molecular-weight solutes and exclude macromolecules. For highly diffusible solutes, like urea, the concentrations are highest in both blood and dialysate compartments at the arterial (blood inflow) end of the device.

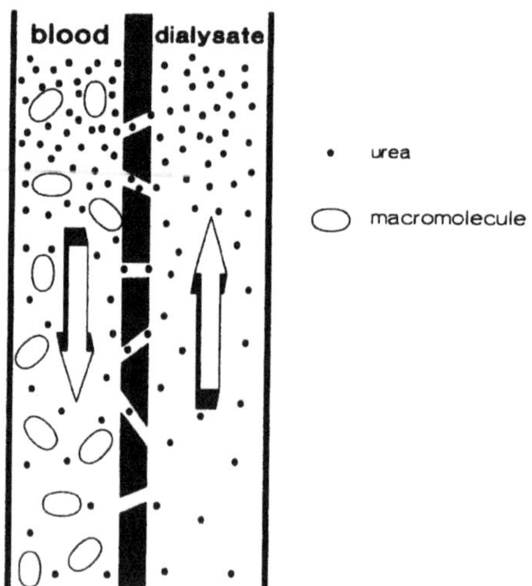

Figure 3.1. Diagram of diffusion pathway through a semipermeable membrane when flow is countercurrent. Macromolecules are excluded while small molecules such as urea easily move from the blood compartment into the dialysate compartment.

EVOLUTION OF UREA MODELING

Dialysis is a treatment founded on empiricism. Little or no theory was developed prior to the early application of hemodialysis. Mathematical description of solute mass transport during dialysis followed the establishment of dialysis as a bona fide life-sustaining treatment. Wolf first described the basic kinetics of hemodialysis nearly 40 years ago (1). He took the idea of clearance developed for the native living kidney and applied it to the artificial kidney, adding dialysate concentration to the denominator of the familiar mathematical expression of clearance:

$$D = \frac{\text{urea removal rate}}{C_b - C_d} \qquad 3.1$$

C_b is blood urea concentration
C_d is dialysate urea concentration

The term D on the left of equation 3.1 was called *dialysance* by Wolf, the first to use this expression. The concept of dialysance is exactly analogous to the concept of clearance. The essence of both is straightforward: the rate of urea removal is directly proportional to the force driving its movement. Clearance or dialysance is the constant ratio between the removal rate and the driving force. For the dialyzer, the driving force is the concentration gradient across the semipermeable dialyzing membrane. This gradient is the difference between inlet dialysate and blood urea concentrations ($C_b - C_d$). The constant nature of the ratio between the removal rate and the gradient means that dialyzer urea transport (diffusion) behaves according to the laws of first-order kinetics. First-order processes are discussed in more detail in chapter 4.

Further extensions to the equations describing urea removal by the dialyzer were provided by Renkin (2), who added membrane permeability, surface area, and path length and helped to develop our now widely applied concept of the dialyzer mass transfer coefficient.

Major contributions to our current understanding and application of urea modeling were made by two investigators, Gotch and Sargent (3,4,5,6,7,8). The unique combination of clinical experience and mathematical skills of these scientists helped to bring urea kinetic analysis to the forefront of clinical practice. Using the hollow fiber kidney, they developed expressions for the volume of urea distribution and urea generation rate that could be translated into meaningful data for the clinician: estimators of effective dialyzer clearance, predicted ideal time on dialysis, and protein catabolic rate (PCR) (3). The latter has been widely used to guide dietary counseling for each patient (9,10,11). Over a period of nearly two decades, these investigators and others have refined the equations by adding parameters that allow a more rigorous definition of mass balance for urea during and between dialyses (8,12,13,14,15,16). Many other contributions have been made by scientists and clinicians over the past 30 years that have helped to refine the models of urea removal and accumulation that are now in use. These are cited individually in the following chapters.

WHY UREA INSTEAD OF OTHER SOLUTES?

Urea is a unique solute that lends itself to the study of hemodialysis in several important ways as shown in table 3.1. Examining the kinetics of other solutes may provide information about dialyzer performance, but only urea modeling can offer the additional crucial information about the patient's nitrogen balance. Since nearly all nitrogen-containing metabolites are eventually converted to urea, and nitrogen intake is almost entirely composed of protein (see chapter 2), urea modeling gives immediate insight into protein metabolism. In the decades preceding the arrival of hemodialysis therapy, protein intake was well known to dictate the need for renal excretory function. In patients with borderline renal function, switching to a low-protein diet was sufficient to reverse symptoms of uremia (17,18). The origin of urea as an end product of protein metabolism is a preeminent feature that lifts urea modeling to a level of practicality well above that of other solute models. It means that the physician can apply the modeling results immediately, without waiting for additional measurements of protein intake. A study of urea kinetics in each patient brings together the two critical elements of each dialysis prescription by giving the physician a measure of dialyzer performance as well as dietary nitrogen intake, the major determinant of the patient's need for dialysis. These two parameters, as illustrated in figure 3.2, must be balanced in each patient to achieve an optimum outcome.

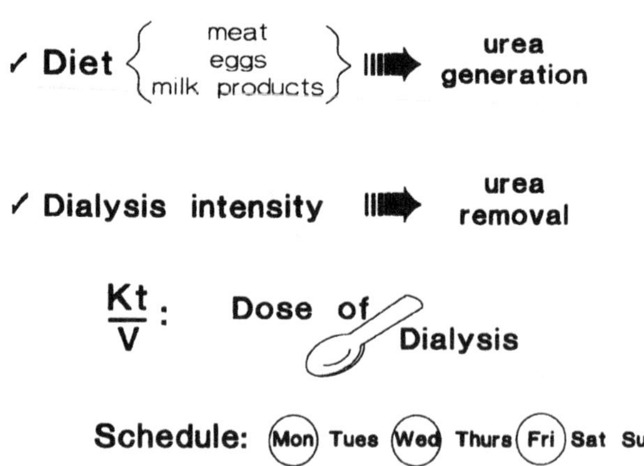

Figure 3.2. The dialysis prescription consists of patient-controlled dietary variables and therapist-controlled dialysis variables. The patient controls urea generation, while the dialysis center controls urea removal.

A study of creatinine kinetics, potassium kinetics, or urate kinetics could be used to gather information about the balance of each of these particular solutes and the dialyzer performance. But the concentration of these solutes in the blood has not correlated with the clinical state of uremia or dialysis need. To prescribe a dialysis regimen adequately, ultimately some measure of protein intake must be available. The evidence linking protein catabolism and dialysis need is examined in detail in chapter 8. It is important to add that the modeling technique used to assess protein intake is totally independent of dietary analysis.

Table 3.1 Qualifications of urea as a model solute

Urea is the most abundant solute removed by the kidney
Urea is an end product of nitrogen balance that quantitatively accounts for nearly all nitrogen metabolism
Urea concentrations correlate with uremic toxicity
Urea distributes in all body water compartments yet moves easily and quickly among compartments
Urea is transported rapidly across erythrocyte cell membranes

Urea has unique chemical and physiologic properties that simplify modeling. Because urea distributes in all body water compartments, measurement of its distribution volume provides information about total body water. The rapidity of equilibration among compartments usually allows a simple single-compartment model to satisfy most clinical needs. Exceptions to this rule are examined in chapter 5. Facilitated transport across red cell membranes simplifies simulation of urea kinetics in the blood compartment. There is little or no need to consider red cell volume separate from plasma volume. Except for adjusting water content, whole-blood volume, whole-blood concentrations, and whole-blood flow can be used for urea modeling. Adjustments for unequal or delayed partitioning between red cells and plasma are not necessary. This unique transport property of urea is discussed further in chapter 7.

Quantifying Hemodialysis Therapy

The adequacy of dialysis and its measurement have been debated over the past 20 years by authorities concerned about how much of this life-sustaining treatment is appropriate for patients with end-stage renal disease (19). Most of the publications that address dialysis adequacy focus on the lack of reliable tests to quantify dialysis and the failure of medical science to identify crucial toxins that the dialyzer removes. Research on toxin disposal by the dialyzer has been directed toward improved removal of middle molecules for nearly two decades (20).

Middle molecules are solutes with molecular weights in the 350-5000 dalton range that are poorly dialyzable across conventional cellulosic membranes (21).

Dialysis of these solutes is membrane-limited, that is, their removal is critically dependent on membrane area and permeability and less so on blood and dialysate flow. Figure 3.3 shows the relationship between blood flow and clearance for flow-limited and membrane-limited dialysis. The appearance of these curves is a function of both the solute and the dialyzing membrane. However, as molecular size increases, the dialysis becomes more membrane-limited. It should be noted that as blood flow increases, a plateau phase is reached even for flow-limited solutes beyond which solute removal becomes membrane-limited. Standard blood and dialysate pumps provide flow rates that approach the plateau phase for urea removal by cellulosic dialyzers.

Interest in middle molecules has waned in recent years because of the report by the National Cooperative Dialysis Study (NCDS) that found blood urea concentrations to correlate strongly with the outcome of dialysis (22). The NCDS is discussed more thoroughly in chapter 8. Because there is little or no demonstrated toxicity associated with high blood urea concentrations (23,24) the NCDS reviewers reasoned that urea serves as a marker for other easily dialyzed but unidentified solutes that have molecular weights below the middle-molecule range. Efforts to improve clearance of middle molecules across dialysis membranes have shifted to an emphasis on clearance of urea by the same membranes that were designed to

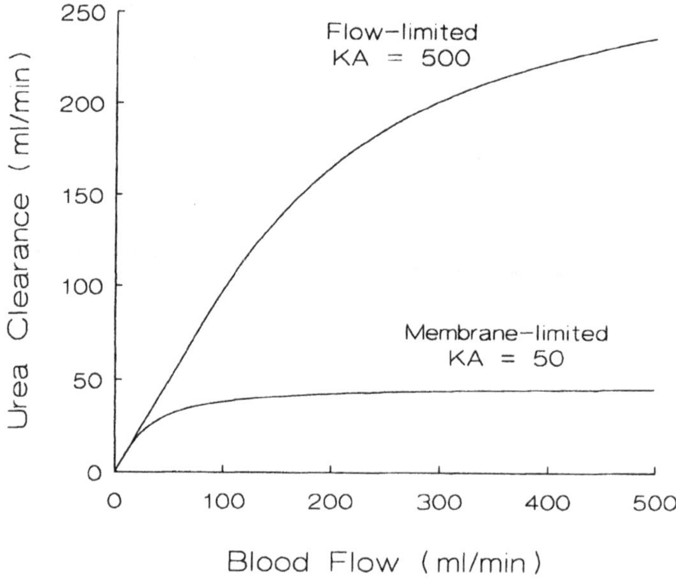

Figure 3.3. Urea clearance usually increases with increasing blood flow. The highly permeable membrane shown by the upper curve is more responsive to increasing blood flow. Solute transport across the membrane shown in the lower curve is fixed above blood flow rates of 100 ml/min, where membrane permeability becomes the limiting factor.

allow larger solutes to pass. This means that the dialyzing physician must use higher blood and dialysate flow rates to take advantage of membranes that are no longer limiting for urea transport at conventional flow rates. For what it is worth, these techniques also improve middle-molecule clearance, but their clearance is less critically dependent on flow rates. Urea clearance has returned to the forefront as a measure of dialyzer membrane effectiveness, and assessment of urea kinetics in the individual patient has been labeled the best measure of dialysis adequacy (25).

As dialyzers and dialysate delivery systems have become more and more reliable, both patients and dialysis staff have become more comfortable with the dialysis procedure. Hemodialysis is a simple process, but it requires a complicated delivery system, studded with safety features and electronic gadgetry to ease the burden of the operator nurse or technician. Modern dialysis equipment provides digital displays of fluid balance, blood compartment pressures, transmembrane pressures, and time, as well as automated adjustments of sodium and bicarbonate concentrations, and precise control of fluid removal. Paradoxically, the measurement of the primary effect, removal of solute by diffusion (dialysis), has not been successfully monitored. A hypothetical case illustrates this point:

> We might propose to "dialyze" a patient with a membrane impermeable to urea or any other solute. Even with the best of modern equipment using multiple fail-safe detection devices, no alarm would be raised to signal that dialysis is not taking place! At the end of the procedure, both the patient and the clinical staff would leave with a false sense of satisfaction, having watched the patient's blood flowing through the dialyzer for several hours, unaware that nothing had been accomplished. In fact, the patient might feel better than he or she usually feels following an effective dialysis.

Closer to reality is the potential for malfunction of the blood access device that would allow dialyzer venous blood to recirculate and cause a drastic reduction in urea clearance. The reduction in solute clearance from recirculation and other causes may not be detected even after several dialyses unless measurement of solute removal is done periodically. Partial clotting of the dialyzer or loss of effective surface area from reuse also may affect clearance during the dialysis procedure. If clotting occurs, dialyzer clearance may decline toward the end of dialysis. Other causes of dialysis variance are listed in table 3.2.

One cause of underdialysis that has been recognized in recent years is partial collapse of the pump segment when prepump pressure falls too low (table 3.2) (26). This effect is illustrated in figure 3.4 as a decrease in cross-sectional area across the roller pump segment. Routine monitoring of prepump pressure with appropriate lower limits will prevent this type of variance.

Urea modeling provides a sensitive test of actual dialysis outcome compared with expected dialysis outcome. Its routine application will allow the physician to detect outcome variance in a timely manner. This will avoid unnecessary compromise of patient health as well as embarrassment to the dialysis treatment center.

Table 3.2 Causes of dialysis variance

Faulty dialyzers or wrong dialyzer
Faulty blood roller pumps
Failure of occlusion
Calibration error
Low prepump pressure causing pump segment collapse
Loss of surface area for diffusion during dialysis
Clotting
Other membrane alterations
Access device failure causing low blood flow and reflow
Outlet (venous) obstruction or stenosis
Inlet (arterial) stenosis
Recirculation of venous blood for other reasons
Dialysate flow error
Minor causes
Temperature changes
Hematocrit variation

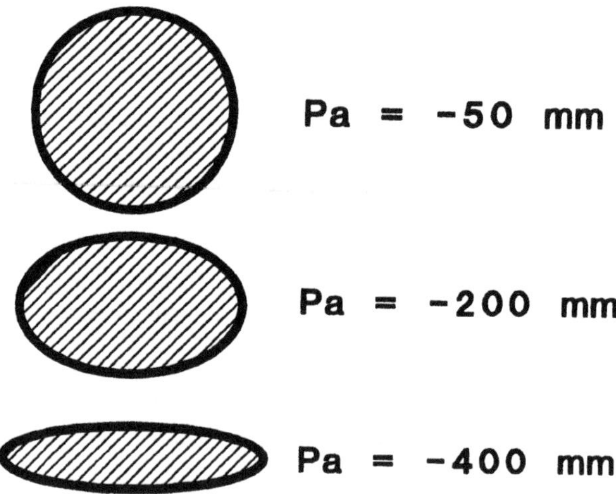

Figure 3.4. Blood flow rates recorded by roller pumps are affected by low prepump pressures (Pa). Low pressure here impairs the elastic recoil of the pump segment, partially collapsing the tubing. This collapse decreases the cross sectional area (shown diagrammatically here) and volume of the pump segment, dissociating RPM meter readings from actual blood flow rates.

Quality assurance programs and urea modeling

Quality assurance (QA) programs, a major concern of industrial management, guarantee the reliability of goods that roll off the assembly line. Implementation of QA programs for the provision of services is more difficult. For medical services, the concern for privacy, litigation, and allowance for practice of the "art" of medicine have hampered development of standard QA programs in the hospital or clinic environment. In years past, QA programs were superficial at best because outcome measurements for illnesses and procedures were not available. QA programs concentrated only on objectively measurable parameters such as blood pressure, patient weight, and laboratory tests. The dialysis industry has followed a similar route. In a recent survey of dialysis centers throughout the Transpacific End Stage Renal Disease (ESRD) Network, the following items were listed as major components of each center's QA program:

Audits of: Medical records
Infection control
Social services
Nutrition
Nursing/technical care
Reuse

Emergency equipment
Patient satisfaction surveys
Mortality data
Incident reports
Hospital admission data
Hepatitis surveillance
Culture reports
Electrical safety testing
Air testing

Conspicuously absent from the above list is a measure of the service itself, i.e., a measure of dialysis or solute removal. This does not mean that the above list is invalid; each quality indicator mentioned is important for patient safety and comfort. But foremost should be a measure of dialysis adequacy, just as a measure of tensile strength is the primary quality measurement for newly manufactured rope.

Urea modeling allows the clinician to examine critically the prescription for small-molecule solute removal in each patient and to compare the expected outcome with the real outcome. The strong correlation between favorable outcome and low blood urea nitrogen (BUN) reported by the NCDS serves as the basis for linking quality assurance to urea removal. The amount of dialysis prescribed is often expressed in simple terms as Kt/V (see chapter 8). Outcome is often measured as midweek predialysis BUN, time-averaged BUN, or modeled Kt/V. If the patient's urea volume (V) is known, urea modeling also provides a means of checking dialysis equipment by comparing expected clearance to modeled clearance. Urea modeling provides a measure of ideal outcome linked to the patient's PCR, a vital determinant of need for dialysis. Finally, the modeling algorithm allows prediction of required changes in the prescription necessary to achieve ideal outcome. The application of time-averaged BUN measurements, *Kt/V*, expected versus measured parameters, and ideal outcome are discussed more thoroughly in chapter 8.

UREA MODELING DEVELOPMENT AND TECHNIQUES

Mathematical modeling of almost any process, including dialysis urea kinetics, goes through two components or phases. The first is the development phase, where the quantitative mathematical expressions are derived, compared to real dialyses, and then honed to match real events as closely as possible. After a satisfactory model is developed, the second phase begins. This is the application phase, where the model is used in individual cases to measure the adequacy of dialysis and/or the size of the urea pool and the rate of urea appearance. Using this information, the model can predict future events when the dialysis prescription changes. When the model predicts future events, these predictions can then be compared with actual events to examine the accuracy of the model. If inaccuracies are found, the modeler can return to phase one and modify the model. This process can be repeated initially or later as conditions change (e.g., high-flux dialysis) until the model is adequate. Those who use the model for patient care need not be concerned with the development phase, but some familiarity with the origin of the mathematics will help clinicians interpret results and troubleshoot problems.

The clinical application of urea modeling involves the *fitting* of measured blood urea concentrations to an *expression* that defines these urea concentrations at any given time. The term *fitting* means that variable parameters in the expression of urea concentration are modified until the measured urea concentrations are adequately described by the expression, i.e., they fit the expression. The *expression* in the above definition is a mathematical equation that has urea concentration or change in urea concentration on the left side of the equal sign and several unknown parameters on the right that can be modified, e.g.,

$$C = f\,(V, G, \text{removal rate}) \qquad 3.2$$

C is measured urea concentration (known)
V is volume of urea distribution (unknown)
G is urea generation rate (unknown)
Removal rate includes dialyzer clearance, residual kidney clearance, and time on dialysis, all measurable constants in a single patient.

Equation 3.2 indicates that *C is a function of* or *is determined by V* and *G* and measurable constants. This is the fundamental expression, derived from mass balance considerations, upon which all urea kinetic modeling is based. Solution of the equation requires reverse reasoning. We are given two or more values of C and are asked to compute values for V and G. This is often done without actually solving the equations for V and G because these equations are often too complex to solve. Calculator or computer programs that give solutions to these equations use a trial-and-error process called *iteration*. This simply means that the programs systematically insert different values for V and G into the equation until all measured values of C fit. An example of this fitting technique is shown graphically in figure 4.7. A series of repeated attempts (iterations) at varying G are made until the measured data fit the graph that describes the weekly changes in BUN. This is the essence of the process. If one understands this process of fitting then the only other conceptual task is to understand the starting parameters. These include urea clearance, ultrafiltration rate, urea distribution volume, and urea appearance or generation rate, all familiar to the clinician.

An additional requirement for successful modeling is that the clinician accepts the mathematical model. Most physicians responsible for treatment of patients want to know something more about the derivation of equations and a description of the technique for fitting to be better prepared to interpret the results and anticipate problems. To satisfy this need, detailed explanations of single- and double-compartment mathematical models are provided in chapters 4 and 5. Use of a computer program is addressed in chapter 6. To proceed with urea modeling, dialysis personnel need only feed the urea concentrations and dialysis conditions shown in figure 3.5 into the computer and it will provide the modeled parameters, including urea volume (V), generation rate (G), and ideal time on dialysis. Chapter

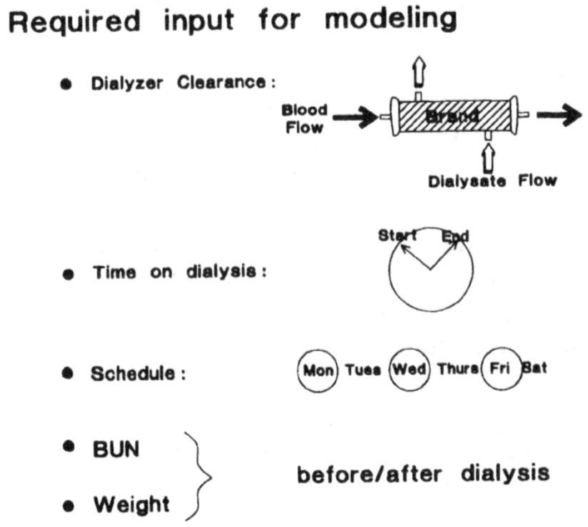

Required input for modeling

- Dialyzer Clearance: Blood Flow, Dialysate Flow
- Time on dialysis: Start, End
- Schedule: Mon Tues Wed Thurs Fri Sat
- BUN
- Weight

before/after dialysis

Figure 3.5. Urea modeling demands a minimum of data input. Most of the information is routinely recorded for each dialysis, leaving BUN measurement as the only additional requirement.

7 addresses further interpretation of the results and the pitfalls of modeling.

What is the purpose of urea modeling?

Measurement of BUN is the traditional approach to evaluate the severity of renal failure. Urea is the end product of protein nitrogen metabolism, and protein metabolism correlates with uremic toxicity. Nearly all concerned with measuring the extent of uremia or the adequacy of dialysis agree that measuring the BUN is very important. All would agree that a high BUN is undesirable and a low BUN is associated with fewer symptoms of uremia. Yet we have all seen patients with uremic symptoms when the BUN is only modestly elevated, and we have seen other patients with no uremic symptoms when the BUN is very high. Studies of uremic toxicity have concluded that urea is not very toxic but serves as a marker for the uremic state (23,24). A consensus is rarely reached on what level to expect in symptomatic patients, when to measure the BUN in dialyzed patients, or what level is best for hemodialyzed patients. Gotch notes a kind of "schizophrenic attitude" toward the BUN amongst nephrologists who agree to measure the BUN but then ignore it because "it doesn't matter" (8). Part of the reason for this troubling behavior is that an isolated BUN is difficult to interpret. Several other parameters must be considered with the BUN in each patient for a rational assessment of the severity of uremia or adequacy of treatment. What is considered a low BUN in one patient may be too high in another. Factors that modify the interpretation of BUN include the urea nitrogen generation rate, timing of the sample, patient size, fluid gain between dialyses, rate of change in BUN, dialysis schedule, and native kidney (residual) urea clearance. Urea modeling supplies the tool to explore these parameters simultaneously. Besides producing measures of the quality of dialysis and the extent of uremia, modeling also gives us a proper understanding of the BUN in hemodialyzed patients.

The routines of hemodialysis quickly consume the attention of the dialysis staff, leaving little time for consideration of the dialysis process itself. Patients dialyzed side by side in the dialysis center may have similar treatment regimens but may undergo markedly different chemical changes. Nurses or technicians who examine the results of urea modeling are often astonished at the variability in response to dialysis which was not appreciated after reviewing routine laboratory data. Figure 3.6 illustrates two examples of this variability. Patient A has a higher protein intake and a smaller urea volume compared to patient B.

Urea modeling gives the physician information about the patient's dialysis, diet, and equipment function. Dialysis information includes measurements of dialysis outcome, estimates of expected outcome, and expressions that describe both the actual intensity and the potential effectiveness of the dialysis. These are illustrated in figure 3.7.

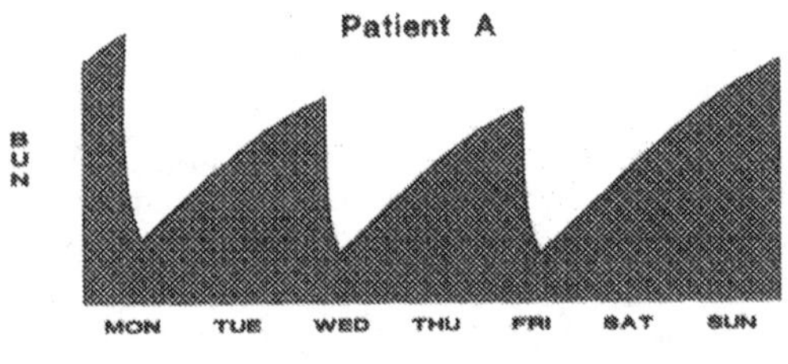

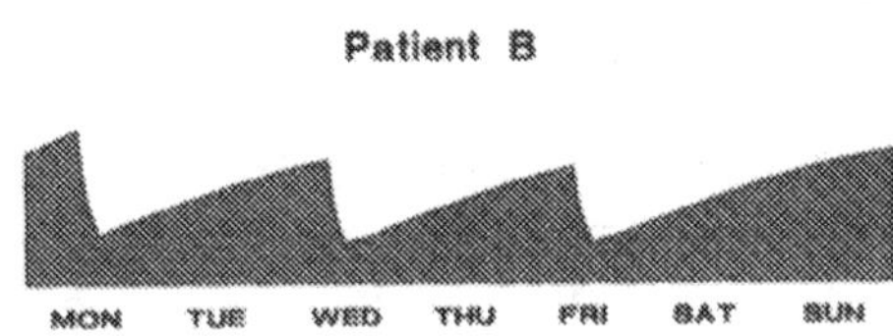

Figure 3.6. These patient profiles are from two real patients maintained in our center. Patient A has a high protein catabolic rate (PCRn) due to dietary overindulgence. Patient B has a flatter profile due to less protein intake, a small residual clearance, and larger size.

Results of Modeling

Patient variables

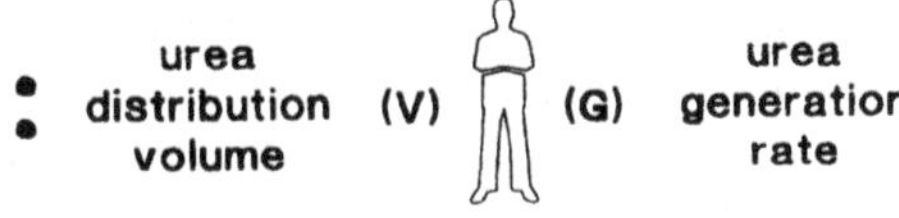

- protein catabolic rate

Dialysis variables

- time-averaged BUN { actual / ideal
- effective dialyzer clearance
- $\frac{Kt}{V}$ (modeled)
- ideal dialysis time

Figure 3.7. Data obtained from urea modeling are of two types, patient variables and dialysis variables. The latter depend on the former. Dialysis variables include actual and ideal measures of outcome, such as time-averaged BUN and modeled *Kt/V*. Effective dialyzer clearance is a whole-body clearance based on a previously established urea distribution volume.

Urea modeling for high-flux, short-duration dialysis

In recent years, attention has been drawn to the benefits of shortened dialysis time using dialyzers that have high urea clearances (high flux). This approach has several advantages for the dialysis center and for the patient. Shortened dialysis time per patient allows more patients to be treated in any given day with minimal increase in staffing. When reimbursement is on a per capita basis, the increased revenue gained from such efficiency helps to offset the added expense of the high-flux dialysis equipment and the additional work required to process more patients within the same period. The major advantages of high flux dialysis are realized by the patient. The synthetic membranes incorporated in high-flux dialyzers are more biocompatible and more permeable to middle molecules. Patients tolerate the dialysis itself better and experience an improved sense of well-being, perhaps from more efficient removal of toxins. Patients also benefit psychologically from the shortened time on dialysis. More time is freed up for the patient during waking hours, which can lead to increased productivity and a sense of reduced dependence on the artificial treatment. For some patients, the psychological benefits may overshadow all others.

Because dialysis time is shorter, fluid accumulated between treatments must be removed faster during dialysis. This would appear to be a disadvantage, because we are all aware of the intolerance some patients have to rapid ultrafiltration. It is somewhat surprising to find that patients usually tolerate the higher ultrafiltration rates better than the lower filtration rates with conventional dialysis. The improved tolerance may be related to decreased acetate exposure or to reduction of the inflammatory effect associated with cellulosic membranes (27,28). Tolerance to ultrafiltration can be further improved by increasing the dialysate sodium concentration (29). The consequences of increasing dialysate sodium concentration from 140 to 145 meq/L is an increase in interdialysis thirst and weight gain (29). But the weight gain is not entirely disadvantageous because predialysis BUN levels are lower (dilutional effect) and the added ultrafiltration slightly improves urea clearance during dialysis (see chapter 7).

Kinetic modeling is especially important for patients who will benefit from high-performance dialyzers and shortened dialysis time. When treatments last four hours and longer, variation in actual dialysis time is less critical. An error of 30 minutes between actual and ideal dialysis times will cause much less concern than a similar error in patients dialyzed for two hours, because the error constitutes a larger fraction of total dialysis time. Ideal time on dialysis becomes a more critical issue for patients using high-performance dialyzers. For the physician managing such patients, especially when a change is made from standard to high-performance dialysis, some measure of dialysis effectiveness is desired. An estimator of ideal dialysis time, usually linked to PCR, is also necessary. Both of these are provided by urea modeling (see chapter 8). Nonspecific symptoms such as nausea, pruritus, weakness, or laboratory changes such as a fall in hematocrit or elevation of potassium concentration may occur after the switch is made to a shortened dialysis schedule. Since these symptoms and signs can be attributed to inadequate dialysis, decisions made by the attending physician are subject to error unless some provision is made to control the mean BUN. This is best done by a well-implemented urea modeling program.

What clinical data are provided by urea modeling?

Three important constants that are difficult to determine by other techniques can be measured using urea modeling. These are the patient's volume of urea distribution (V), effective dialyzer urea clearance (K_d), and urea nitrogen generation rate (G). From these, other patient and dialysis parameters describing actual and ideal dialysis conditions can be derived. As detailed more thoroughly in chapter 4, there are two phases to modeling hemodialysis urea kinetics: 1) an *intra*dialysis phase in which dialyzer action predominates and 2) an *inter*dialysis phase in which the dialyzer is removed and the urea nitrogen generation rate largely determines changes in urea concentration. V and K_d are resolved primarily during the first phase. During dialysis, these two variables have an overwhelming role in determining the decline in urea concentration. G makes a relatively small contribution in the opposite direction. During the second phase, between dialyses, the major determinants of urea concentration are urea generation (G) and the patient's urea pool size (V). If V has already been estimated during the first phase, G can be resolved. The two phases of dialysis can be linked mathematically. The variables K_d and V are resolved during the first phase and used to calculate G during the second phase. Because G plays a small role during the first phase, values for K_d and V can be further improved by using the new value for G during phase 1 calculations. These new values for K_d and V are then used for phase 2 calculations, etc. This is the essence of the iterative approach to urea modeling, for which the computer is especially suited.

Significance of the urea distribution volume (V)

How urea clearance and volume are measured

Urea volume is expressed in milliliters or liters and is equivalent to total body water. For the single-compartment model, V is assumed to be one pool, the sum of extracellular plus intracellular water space. For the two-compartment models described in chapter 5, V is divided into two pools between which urea diffuses passively at a finite rate. Other multicompartment models divide the extracellular space further. For the two-compartment model, the ratio of extracellular to intracellular water space can be determined by modeling with larger numbers of BUN measurements or it can be simplified to a normal 1:2 ratio.

As shown more extensively in chapter 4, after selecting a model type (table 4.1), we can determine the rate constant for elimination of urea simply by measuring predialysis and postdialysis BUN levels. From the rate constant, the *ratio* K_d/V is derived. For the simple one-compartment, fixed-volume model, if we ignore the effects of residual function, weight changes, and urea generation (G) during dialysis, the parameter K_dt/V varies logarithmically with the fractional reduction in BUN as shown by equation 4.14 (30):

$$K_dt/V = \ln(C_1/C_2) \qquad 3.3$$

C_1 is predialysis BUN
C_2 is postdialysis BUN
ln is the natural logarithm

Although the assumptions made for this simplified model usually cause significant errors, the model serves to demonstrate the origin of K_d/V. Since time (t) can be determined precisely, measurements of predialysis and postdialysis BUN allow direct calculation of the ratio K_d/V. Without additional data, the model can go no further to resolve K_d or V. To resolve these individually, either one or the other must be provided by the user. To find V, K_d must be accurately measured. Conversely, to find K_d, V must be determined accurately. Fortunately, to measure dialysis outcome, neither must be determined explicitly; the ratio is enough. If the clinician's sole purpose for urea modeling is to estimate dialysis adequacy, only the ratio of K_d/V is required. But additional insight is gained if the value of K_d can be provided and V calculated or vice versa.

A better way to resolve V and K_d

To avoid the errors inherent in the simplified equation 3.3, V can be calculated formally using the complete modeling equations and a value for K_d. The value for K_d inserted here is usually the expected value, derived from calculations based on the dialyzer mass transfer coefficient, blood flow, and dialysate flow (see chapter 7, equations 7.4 and 7.5). The modeled value for V can be compared to expected values for V, determined as described below from the average of previous values or from formulas based on body size, etc. Comparison of expected with modeled values is a contribution to quality assurance provided by urea modeling.

Similarly, to avoid the errors in equation 3.3, a modeled value for K_d can be obtained using the complete modeling equations and a value for V. Where does one obtain V? This relatively fixed patient constant can be derived from standard formulas (see appendix G) but is best obtained by averaging V from previous kinetic analyses. The modeled value for K_d is compared with the expected value of K_d determined as described above. These two comparative processes, V with expected V and K_d with expected K_d, are equivalent. Either one or the other may be used for quality assurance purposes.

The value of estimated urea volume

The user-provided value for V helps to answer an important "what if" question. If there is a significant difference between the user estimate of V and the program's calculated value for V, then the program alters dialyzer clearance (K_d) to fit a new expression for urea concentration in which estimated V is substituted for V. In other words, we make the assumption that the calculated value for V is wrong and force the model to use another value, estimated V, in which we have more confidence. When V is fixed in this manner, K_d must vary to allow the measured values for urea concentration to continue to fit the model. If estimated V is more accurate than modeled V, then the estimate of K_d is a better measure of urea clearance achieved during the study dialysis. The resolution of K_d and V using this technique of formal urea modeling is analogous to but more rigorous than the simple ratio between K_d and V depicted in equation 3.3. Comparison of the expected urea clearance with the actual urea clearance helps dialysis personnel spot trouble with equipment or problems with the patient's blood access device. The latter is a frequent cause of reduction in K_d (31).

Calculation of urea distribution volume is a necessary byproduct of urea modeling, not a major goal. Either V or the ratio K_d/V is needed to calculate urea generation rate, time-averaged concentration and other patient variables that are helpful to the clinician.

Why not use other methods to estimate urea volume?

It is possible to estimate urea volume using standard formulas for total body water based on patient weight, height, age and sex (see appendix G: anthropometric estimates of total body water) (32,33). When calculated this way, V may not be as accurate compared to the modeled value of V. It is better for two reasons to let the program calculate (model) V than to estimate its value from formulas. First, the modeled value for V is based on kinetic analysis of real data for the individual patient and is likely to be more accurate, provided that dialyzer clearance can be measured accurately. Second, standard formulas for calculation of V rely on a normal body habitus in an edema-free state. Errors of 30% to 50% are not uncommon when these formulas are applied to the population of patients with end-stage renal disease who have a high incidence of edema and malnutrition.

Other methods such as computerized tomography (34), magnetic resonance imaging (35), infrared interactance (36), ultrasonography (36), and whole-body impedance measurements (37) have been used to measure lean body mass and total body water content. These techniques, although successfully deployed in various research efforts, are impractical for routine clinical use in the current dialysis setting.

SIGNIFICANCE OF THE UREA NITROGEN GENERATION RATE (G)

Importance of average values

Urea nitrogen generation (G) is the third fundamental variable derived from urea modeling. G is defined as the amount of urea, usually expressed as urea nitrogen, that appears within the body per unit of time. Although individual patients have relatively constant average nitrogen balance, urea generation may vary each day depending on the protein content of the patient's diet. For the purposes of urea modeling, G must be considered constant from day to day and from week to week. Urea kinetics are usually modeled during a single dialysis or during a single dialysis and the following interdialysis interval. When applying the results of modeling, we assume that BUN values measured before and after the modeled dialyses are representative, i.e., that generation rate and distribution volume are unchanged from dialysis to dialysis. Urea volume, aside from the modeled weight changes between and during dialyses, satisfies this requirement because it is fixed and not likely to change from week to week. Urea generation rate is subject to more variability. If the modeled value for G is not representative of the average, the clinician may be falsely led to make changes in the prescription that may inappropriately increase or decrease the intensity of dialysis. As detailed in chapter 4, both the two-BUN method and the three-BUN models can give errors in G of this type. Paradoxically, the two-BUN method is less subject to error due to non-representative BUN values.

Expressions for urea generation and their units of measurement

In the noncatabolic state, virtually all urea is derived from liver metabolism of the amino acid subunits of dietary protein (see chapter 2). The urea nitrogen generation rate in patients who are in nitrogen balance can be used to estimate protein nitrogen intake. To allow for the possibility of anabolic or catabolic states, the latter is usually expressed as net *protein catabolic rate* (PCR). Net PCR is calculated from the urea appearance rate (see below), which correlates with catabolism of both exogenous (dietary) and endogenous protein. Recall from chapter 2 that total protein catabolism far exceeds this net value. Net protein catabolism of 70 g protein/day in a normal adult represents the difference between a total anabolic rate of about 300 g protein/day and a total catabolic rate of 370 g protein/day (38). For the clinician, the protein catabolic rate is a more practical expression than the urea generation rate. It is especially beneficial to the dietitian, who can use it to assess dietary prescriptions and patient compliance for patients who are in nitrogen balance (9,11).

Net protein catabolism can be expressed in absolute terms as grams/day of protein breakdown or it can be adjusted for body size and expressed as g/kg body weight/day. The latter is more useful and has become the traditional unit of measurement of protein intake. The normal target value in average adult patients with end-stage renal disease is 1.1 g/kg/day, range 0.8 to 1.4 g/kg/day (11). Protein intake is usually adjusted to body weight to allow comparison among patients. Weight instead of body surface area is used because weight is often the only measure of patient size that is available, and protein intake is usually only a rough estimate anyway.

As discussed above, urea modeling provides a measure of total body water, as reflected in the urea distribution volume. Body water volume at patient dry weight is probably a better denominator for PCR than either patient weight or height, since it varies only with lean body mass. Lean body mass is a measure of nonadipose cell mass, the principal site of protein anabolism and catabolism. Urea volume, as a correlate of lean body mass, should determine the patient's protein turnover. If PCR is divided by urea volume instead of body weight, the target value would be 1.9 g/liter of water/day instead of 1.1 g/kg body weight/day. Because this target range strays from convention and is not familiar to nutritionists and nephrologists, the denominator is usually *normalized* to an average body composition. This is done by assuming an average body water content of 58%. Dividing urea volume in liters by 0.58 liters/kg gives a normalized or *ideal* body weight that is equivalent to lean body mass plus a normal fat distribution. The denominator (Vl/0.58) allows expression of PCR in the more familiar units of grams/kg body weight/day, but it must be understood that body weight is derived from urea volume, not measured weight.

Figure 3.8 shows the relationship between urea nitrogen appearance (*G*) and net PCR. These data were obtained from a detailed study of a small group of dialyzed patients who were judged to be in nitrogen balance (39). The results agreed very closely with another independently gathered set of data in patients with renal failure before initiating maintenance hemodialysis (40). The regression formula depicted in figure 3.8 is inverted to obtain PCR from urea generation:

$$g = 0.154 \cdot PCR - 1.7 \quad (3.4)$$

$$PCR = 6.49 \cdot g + 11 \quad (3.5)$$

g is urea nitrogen generation rate (g/day)
PCR is protein catabolic rate (g/day)

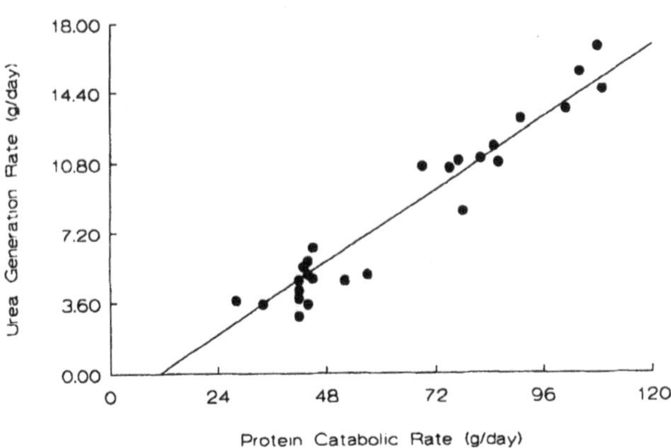

Figure 3.8. Relationship between urea generation rate (*G*) on the *y*-axis and protein catabolic rate (PCRn) on the *x*-axis in hemodialyzed patients at different levels of protein intake. Regression line is $G = 0.154(PCR) - 1.7$. Reproduced with permission from Borah MF et al., Kidney Int 14:496, 1978.

The slope of the line (0.154) is close to the average nitrogen fraction of protein (0.16). If g is expressed as mg/min instead of g/day:

$$PCR = 9.35 \cdot G + 11 \qquad 3.6$$

G is urea nitrogen generation rate (mg/min)

Equation 3.6 indicates that nitrogen released from excess catabolism of dietary and endogenous protein is mostly converted to urea ($9.35 \cdot G$). A small component is converted to other nitrogenous compounds that are excreted or dialyzed. The additional 11 grams/day of protein is converted to nonurea nitrogenous compounds, mainly creatinine, uric acid, hippurate, and, for dialyzed patients, amino acids. Generation of these compounds varies with patient size but not with protein intake. As protein intake decreases, these compounds become a larger fraction of total nitrogen appearance. For an average adult, if net protein catabolism is 100 g/day, then 11% of excess nitrogen appears as nonurea nitrogen; if net protein catabolism is 40 g/day, then 28% is nonurea nitrogen.

Since the average volume of the patients depicted in figure 3.8 was 38 liters, the second term is usually normalized to patient volume by dividing it by 38 liters. This allows the equation to apply to pediatric patients and patients of varying size:

$$PCR = 9.35 \cdot G + 0.29 \cdot Vl \qquad 3.7$$

Vl is urea volume (liters)

Finally, the entire expression is normalized to body size as explained above, using the factor $Vl/0.58$:

$$PCRn = (9.35 \cdot G + 0.29 \cdot Vl) \cdot 0.58/Vl \qquad 3.8$$

$$= 5.42 \cdot G/Vl + 0.17 \qquad 3.9$$

PCRn is net protein catabolic rate, normalized to body size (g/kg/day)
The expression $Vl/0.58$ is normalized body weight or lean body mass plus normal fat content as explained above

Like V, urea nitrogen generation (G) and its derivative, PCRn, are also byproducts of urea modeling, but PCRn, in contrast to V, has practical importance as a measure of protein intake. The latter is a major determinant of the need for dialysis.

Components of the dialysis prescription

Dietary protein prescription

Figure 3.2 shows that the dialysis prescription sets limits for dietary protein in addition to specifying the intensity of dialysis. These two components of the prescription, one managed by the patient, the other regulated by the dialysis center, must be coordinated. The dialysis center should pay attention to what the patient is doing with the diet, and the patient should receive feedback from the center about the success of dialysis (average dialysis outcome). Using this approach, the dietary prescription and dialysis intensity can be adjusted to control optimally the uremic state. If, for example, the patient's protein intake doubles for an extended period, then the intensity of dialysis must be stepped up to avoid uremic complications. Dialysis intensity must be tailored to dialysis needs. The patient and the physician must interact in this way to mimic the effect of normal physiologic controls. Where normal feedback controls have been lost, interaction between patient and physician are vital to the success of therapy. A similar relationship exists for the treatment of hypertension, where the patient's salt intake partially determines the requirement for antihypertensive medications.

Measurement of protein catabolism (PCRn), through urea modeling, gives dialysis personnel the information needed to adjust the prescription for optimal outcome. The dietary prescription is usually obtained through consultation with a trained dietitian. It should be noted that PCR and PCRn calculated by the model are obtained independent of retrospective dietary food analysis. The urea model calculates PCRn from the rate of appearance of urea nitrogen, an end product of protein metabolism. Retrospective dietary analysis, an equally valid technique for estimation of PCRn, attempts to determine protein intake from client history and analysis of foodstuffs, a process that is considerably more tedious and prone to subjective errors. Dietary analysis also suffers from the disadvantage of placing the patient in control of the dialysis prescription. Because the dialysis prescription is tied directly to protein catabolism (see chapter 8), the patient's misrepresentation of the dietary history can cause the physician to alter the dialysis regimen inappropriately. Urea modeling avoids this pitfall by giving the dietitian more objective data to follow when advising the patient or analyzing the dietary history.

Dialysis intensity

Prescription adjustment is most often made by a change in the intensity of dialysis. Dialysis intensity or vigor can be altered by changing the dose of each dialysis or the schedule of dialyses, as shown in figure 3.2. The dose of dialysis is best embodied in the expression Kt/V. This parameter is discussed in more detail in chapter 8; in brief, Kt/V is a measure of the amount of dialysis administered factored for the patient's size much like the dose of a medication expressed as mg/kg body weight/day. Its value is altered by changing the dialyzer urea clearance (K_d), abbreviated as K, or by changing the timed duration of each dialysis (t). The dialyzer clearance is altered by changing the type of dialyzer, the blood flow rate, or the dialysate flow rate.

Measures of prescription effectiveness

Once a prescription has been formulated and applied to the patient, a measure of its effectiveness is needed. Outcome parameters that have been used in the past include average predialysis BUN, midweek predialysis BUN, modeled *Kt/V*, and time-averaged BUN, as illustrated in figure 3.9. Predialysis BUN as a sole measure of dialysis effectiveness has several weakness. No consideration is given to dialysis schedule, postdialysis BUN, patient size, or protein intake. For example, two patients with the same mean BUN but with differing urea generation rates may have different predialysis values (see figure 8.2). *Kt/V*, a measure of the amount of dialysis prescribed, is a reasonable measure of effectiveness, provided it is measured by fitting its value to a model of dialysis urea kinetics (refer to chapter 8 for a more detailed discussion of measured and modeled *Kt/V*). As a gauge of dialysis effectiveness, *Kt/V* suffers from lack of allowances for dialysis schedule, residual clearance, volume shifts, and protein intake.

We consider the best measure of total prescription effectiveness to be the time-averaged BUN, usually abbreviated TAC for time-averaged concentration. TAC is the average or mean urea concentration assessed by measuring the area under the urea-versus-time curve. Clinicians will recognize the analogy between this measurement and the mean blood pressure. TAC has been judged by many to be the best measure of exposure to urea, a solute that admittedly serves only as a marker for uremic toxicity (30,41,42). An indicator of dialysis effectiveness, TAC was used by the National Cooperative Dialysis Study, a multicentered effort to discover the best criteria for establishing ideal dialysis dosage (see chapter 8). TAC as a sole measure of dialysis effectiveness suffers the weakness that no consideration is given to protein intake. So to define completely the patient's risk of uremic toxicity from inadequate dialysis, both TAC and protein catabolic rate must be considered. From

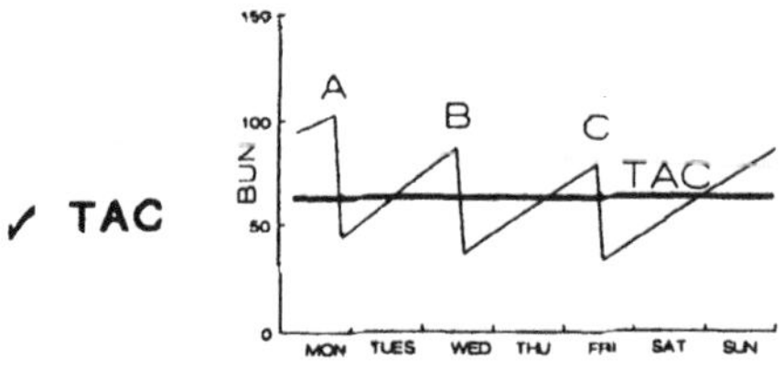

✓ Average predialysis BUN
= (A+B+C)/3

✓ Midweek predialysis BUN
= B

Figure 3.9. Time-averaged BUN (TAC) is the average weekly BUN. TAC is obtained by dividing the area under the curve of BUN concentration during the week by a week's time interval (10,080 minutes). Average predialysis BUN is the average of all peak BUN values during the week. The midweek predialysis BUN closely approximates average predialysis BUN. The latter can vary independently of TAC.

the protein catabolic rate, a value for expected TAC can be derived that allows an individualized interpretation of TAC. Comparison of TAC with expected TAC is one of the best methods we have available today for assessment of dialysis adequacy. Both are individualized parameters, that is, they are unique values for each patient and must be calculated separately for each patient.

References

1. Wolf AV, Remp DG, Kiley JE, Currie GD: Artificial kidney function: kinetics of hemodialysis. J Clin Invest 30:1062-1070, 1951.
2. Renkin EW: The relation between dialysance, membrane area, permeability and blood flow in the artificial kidney. Trans Am Soc Artif Intern Organs 2:102-105, 1956.
3. Sargent JA, Gotch FA: The study of uremia by manipulation of blood concentrations using combinations of hollow fiber devices. Trans Am Soc Artif Intern Organs 20:395-401, 1974.
4. Sargent JA, Gotch FA: The analysis of concentration dependence of uremic lesions in clinical studies. Kidney Int 7 (Suppl 2):S35-44, 1975.
5. Sargent JA, Gotch FA: Mathematic modeling of dialysis therapy. Kidney Int 18:S2-10, 1980.
6. Sargent JA, Gotch FA: Principles and biophysics of dialysis, in *Replacement of Renal Function by Dialysis* (3ed), Maher JF (ed), Dordrecht, Kluwer Academic Publishers, pp 87-143, 1989.
7. Gotch FA, Sargent JA: A mechanistic analysis of the National Cooperative Dialysis Study (NCDS). Kidney Int 28:526-534, 1985.
8. Gotch FA: Kinetic modeling in hemodialysis, in *Clinical Dialysis* (2ed), Nissensen AR, Gentile DE, Fine RN (eds), Norwalk CT, Appleton and Lange, pp 118-146, 1989.
9. Kosanovich JM, Dumler F, Horst M, Quandt C, Sargent JA, Levin NW: Use of urea kinetics in the nutritional care of the acutely ill patient. J Parenteral and Enteral Nutr 9:165-169, 1985.
10. Goldstein DJ, Frederico CB: The effect of urea kinetic modeling on the nutrition management of hemodialysis patients. J Am Diet Assoc 87:474-479, 1987.
11. Sargent J, Gotch F, Borah M, Piercy L, Spinozzi N, Schoenfeld P, Humphreys M: Urea kinetics: a guide to nutritional management of renal failure. Am J Clin Nutr 31:1696-1702, 1978.
12. Popovich RP, Hlavinka DJ, Bomar JB, Moncrief JW, Decherd JF: The consequences of physiological resistance on metabolite removal from the patient-artificial kidney system. Trans Am Soc Artif Intern Organs 21:108-115, 1975.
13. Nolph KD, Bass OE, Maher JF: Acute effects of hemodialysis on removal of intracellular solutes. Trans Am Soc Artif Intern Organs 20:622-627, 1974.
14. Heineken FG, Evans MC, Keen ML, Gotch FA: Intercompartmental fluid shifts in hemodialysis patients. Biotechnol Progr 3:69-73, 1987.

15. Dombeck DH, Klein E, Wendt RP: Evaluation of two pool model for predicting serum creatinine levels during intra- and inter-dialytic periods. Trans Am Soc Artif Intern Organs 21:117-124, 1975.
16. Dedrick RL: Pharmacokinetic and pharmacodynamic considerations for chronic hemodialysis. Kidney Int 7:S2-S15, 1975.
17. Giovannetti S, Maggiore Q: A low nitrogen diet with proteins of high biological value for severe chronic uremia. Lancet 1:1000, 1964.
18. Giordano C, DePascale C, DeCristofaro D, Capodiscasa G, Balestrieri C, Baczyk K: Protein malnutrition in the treatment of chronic uremia, in *Nutrition in Renal Disease*, Berlyne GM (ed), Baltimore, The Williams & Wilkins Co., pp 23-37, 1968.
19. Gotch FA, Krueger KK (eds): Proceedings of a conference on adequacy of dialysis, Kidney Int 7 (Suppl 2), pp S1-S266, 1975.
20. Scribner BH, Farrell PC, Milutinovic J, Babb AL: Evolution of the middle molecule hypothesis, in *Proceedings of the Fifth International Congress of Nephrology*, Villarreal H (ed), Basel, Karger, pp 190-199, 1974.
21. Bergstrom J, Furst P: Uremic middle molecules. Clin Nephrol 5:143, 1976.
22. Lowrie EG, Laird NM, Parker TF, Sargent JA: Effect of the hemodialysis prescription on patient morbidity: Report from the National Cooperative Dialysis Study. N Engl J Med 305:1176-1181, 1981.
23. Merrill JP, Legrain M, Hoigne R: Observations on the role of urea in uremia. Am J Med 14:519, 1953.
24. Johnson WJ, Hagge WW, Wagoner RD, Dinapoli RP, Rosevear JW: Effects of urea loading in patients with far-advanced renal failure. Mayo Clin Proc 47:21-29, 1972.
25. Hakim RM: Nephrology forum: Assessing the adequacy of dialysis. Kidney Int 37:822-832, 1990.
26. Depner TA, Rizwan S, Stasi T: Prepump vacuum effect of hemodialysis blood flow: in vitro and in vivo studies. Trans Am Soc Artif Intern Organs, in press, 1990.
27. Craddock PR, Fehr J, Brigham KL, Kronenberg RS, Jacob HS: Complement and leukocyte-mediated pulmonary dysfunction in hemodialysis. New Engl J Med 296:770-774, 1977.
28. Hakim RM, Lowrie EG: Hemodialysis-associated neutropenia and hypoxemia: the effect of dialyzer membrane materials. Nephron 32:32, 1982.
29. Van Stone JC, Bauer J, Carey J: The effect of dialysate sodium concentration on body fluid compartment volume, plasma renin activity and plasma aldosterone concentration in chronic hemodialysis patients. Am J Kidney Dis 11:58, 1982.
30. Lowrie EG, Teehan BP: Principles of prescribing dialysis therapy: Implementing recommendations from the National Cooperative Dialysis Study. Kidney Int 23 (Suppl 13):S113-122, 1983.
31. Ilstrup K, Hanson G, Shapiro W, Keshaviah P: Examining the foundations of urea kinetics. Trans Am Soc Artif Intern Organs 31:164-168, 1985.

32. Hume R, Weyers E: Relationship between total body water and surface area in normal and obese subjects. J Clin Pathol 24:234-238, 1971.
33. Watson PE, Watson ID, Batt RD: Total body water volumes for adult males and females estimated from simple anthropometric measurements. Am J Clin Nutr 33:27-39, 1980.
34. Heymsfield SB, Olafson RP, Kutner MH, Dixon DW: A radiographic method of quantifying protein calorie undernutrition. Am J Clin Nutr 32:693, 1979.
35. Lewis DS, Rollwitz WL, Bertrant HA, Masoro EJ: Use of NMR for measurement of total body water and estimation of body fat. J Appl Physiol 60:836, 1986.
36. Heymsfield SB, Rolandelli R, Casper K, Settle RG, Koruda M: Application of electromagnetic and sound waves in nutritional assessment. J Parenter Enter Nutr 11:64S, 1987.
37. Presta E, Segal KR, Rutin B, Harrison GG, Van Itallie TB: Comparison in man of total body electrical conductivity and lean body mass derived from body density: validation of a new body composition method. Metabolism 22:524, 1983.
38. Mitch WE, Klahr S (eds): *Nutrition and the Kidney*, Boston, Little, Brown, 1988.
39. Borah MF, Schoenfeld PY, Gotch FA, Sargent JA, Wolfson M, Humphreys MH: Nitrogen balance during intermittent dialysis therapy of uremia. Kidney Int 14:491-500, 1978.
40. Cottini EP, Gallina DK, Dominguez JM: Urea excretion in adult humans with varying degrees of kidney malfunction fed milk, egg, or an amino acid mixture: Assessment of nitrogen balance. J Nutr 103:11-21, 1973.
41. Sargent JA: Control of dialysis by single-pool urea model: the National Cooperative Dialysis Study. Kidney Int 23 (Suppl 13):S19-25, 1983.
42. Laird NM, Berkey CS, Lowrie EG: Modeling success or failure of dialysis therapy: the National Cooperative Dialysis Study. Kidney Int 23 (Suppl 13):S101-106, 1983.

Chapter 4

SINGLE-COMPARTMENT MODEL

MODELS FOR HEMODIALYSIS UREA KINETICS

During the past 40 years, several mathematical models have appeared that describe urea kinetics during and between hemodialyses (1,2,3,4,5,6,7,8,9,10). These models compute changes in urea concentration with time and are based on logical expressions of urea mass balance. The basic equations are straightforward, but the mathematics become more complex as additional variables are added to improve accuracy (10). Any one of these models can generate a predictable profile of urea concentrations on the vertical axis and time on the horizontal axis for each patient. To create the profile, only two or three urea or urea nitrogen concentrations are measured, one or two before dialysis and one at the end of dialysis. Figure 4.1 shows the pattern of changes in the blood urea nitrogen (BUN) that urea modeling techniques can generate for an entire week. The weekly pattern repeats itself, just as the patient's dialysis schedule repeats itself at one-week intervals. The asymmetry of the weekly pattern is a consequence of the patient's requirement for more than one and less than seven dialyses per week and our ancestors' adoption of the biblical seven-day week.

How can we make such predictions based on only two or three measurements of BUN? This question is best answered by looking at the factors that determine urea concentration. For the patient undergoing periodic dialysis, the major determinants of urea concentration are the dialysis schedule and duration, the patient's size, urea generation rate and residual clearance, the rate of fluid gain/loss, and the dialyzer's urea clearance. Most of these parameters are known or measurable, leaving only two in the above list, the patient's size and urea nitrogen generation rate, as unknowns. Modeling techniques require fewer real data points to resolve a limited number of unknown parameters. This will become more evident in the discussions that follow.

Table 4.1 lists the models used to measure hemodialysis urea kinetics up to now.

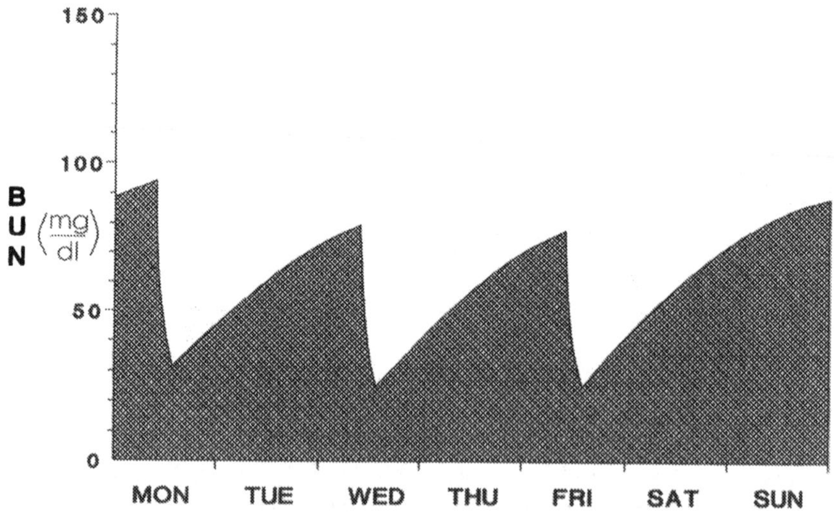

Figure 4.1. A week of urea oscillations. Urea nitrogen concentrations fall rapidly during dialysis and rise gradually between dialyses. This repeating pattern is constant from week to week.

Table 4.1 Hemodialysis urea models

I. One compartment models
  - Two-BUN method (10,11)
    - a. constant volume
    - b. variable volume*
  - Three-BUN method
    - a. constant volume (12)
    - b. variable volume (4)

II. Two-compartment models
  - Two-BUN method
    - a. constant volume
    - b. variable ECF volume (isosmotic)*
    - c. variable ECF and ICF volume (osmotic)
  - Three-BUN method

III. Three- (or more) compartment models

IV. Dialysate collection model (6)

* Reviewed in this book.

This table highlights four standard approaches to urea modeling. The first three use measurements taken from the blood side of the dialyzer, assuming one or more compartments for urea distribution. The fourth depends less on blood-side measurements but instead relies on measurements via collected dialysate of what is removed. In its other aspects, this dialysate collection model is similar to the first three models. All models have variants requiring one or two BUN measurements, and more recent models make allowance for expansion and contraction of compartment volumes. This book places major emphasis on those models marked with an asterisk. Single-compartment models will be reviewed in this chapter. Double-compartment models will be reviewed in chapter 5.

Overview of Kinetic Analysis

The term *kinetics* denotes movement. When applied to the physiology of body solutes, it usually refers to movement of solutes to or from a body compartment. This usually causes a change in solute concentration. Kinetic analysis is traditionally the realm of the pharmacologist, who is intimately familiar with drug movement within the body. The pharmacologist's principal tool for study of drug kinetics (pharmacokinetics) is a detection device that quantifies drug concentrations in extracellular fluid. The pharmacologist measures the drug concentration while controlling factors that affect changes in concentration, such as dosage, absorption from sites of administration, scheduling of doses, patient size, and age.

The study of urea kinetics can be considered a special application or extension of pharmacokinetics. Although urea is not an exogenously administered drug, the marked variations in concentration and large fluxes that occur during and after hemodialysis can be described using classic pharmacokinetic expressions. These expressions describe the mass balance of drugs within the body using rates of elimination or rates of movement between compartments. The kinetic terms or rates are classified according to a logical scheme that describes the interaction between the forces that drive the movement of a solute and the principal barrier to its movement. The latter is usually a membrane composed of cells or the cell membrane itself. This classification scheme is based on the order of magnitude of the driving force. The concept of order is a mathematical ranking determined by the highest exponent in the basic equation that describes the kinetics. Following are descriptions of zero- and first-order kinetic expressions.

Zero-order kinetics refers to movement of solute where the *rate of solute movement is constant*. In other words (and the reason for the designation *zero*), the variables that affect the rate of movement have been eliminated by raising them to the zero power. Any number raised to the zero power is one.

$$ds/dt = k \qquad 4.1$$

s is the amount of solute present
t is time
k is a constant

The production of urea, an event that occurs primarily in the liver, is uninfluenced by the forces that determine movement of urea between body compartments and can be considered constant during dialysis. The kinetic expression of urea generation, then, is a zero-order expression adequately described by equation 4.1.

For diffusion across semipermeable membranes, the driving force is the concentration itself (see discussion of Fick's law below), a force that is rarely constant. Diffusion must therefore be described in terms of first-order kinetics. For first-order diffusion kinetics, the *rate of solute movement is proportional to the concentration.* This is expressed mathematically as

$$ds/dt = k{\cdot}C \qquad 4.2$$

C is solute concentration

For systems with a fixed volume of distribution (V), C may be expressed as s/V. For first-order diffusion kinetics under these circumstances, a fixed fraction of the solute is removed per unit of time:

$$\frac{ds/s}{dt} = k/V \qquad 4.3$$

k/V is constant

Similarly, if V is fixed, the rate of change in concentration is a first-order process:

$$dC/dt = k \cdot C/V = k'C \quad 4.3a$$

A second-order kinetic expression would appear as

$$ds/dt = k \cdot C^2 \quad 4.4$$

For second-order kinetics, the *rate of solute movement is proportional to the square of the concentration.*

First-order kinetics are observed more commonly than zero-order or second-order kinetics in biological systems. One reason for this is that movement of solutes and drugs across cell membranes is often a passive process governed by simple diffusion. Simple diffusion is always a first-order process, i.e., the driving force for movement of solute is the concentration gradient itself. The same rule applies to filtration, also known as convective transport as opposed to diffusive transport. During filtration, if the driving force is a constant hydrostatic or osmotic pressure, the rate of movement of solute across the filter is proportional to the solute concentration. Under these circumstances, filtration is also a first-order process. Even active transport, classically simulated with Michaelis-Menten kinetic models, is often proportional to the concentration of solute substrate when substrate is present in limited amounts (13).

Laws of diffusion

More detailed examination of the constant (k) in equation 4.2 discloses other variables that govern simple diffusion. Incorporation of these variables leads to a more precise expression of diffusion rates known as Fick's first law of diffusion. This law reiterates the first-order kinetics of simple diffusion: flux (ds/dt) is directly proportional to the concentration gradient in the direction of diffusion.

$$\frac{ds}{dt} = -(D_s \cdot A)\frac{dC}{dx} \quad 4.5$$

For diffusion of solutes across semipermeable membranes

D_s is the diffusion coefficient for solute (s) in cm^2/sec

A is the area membrane area in cm^2

x is the membrane thickness in cm

The term ($k \cdot C$) in equation 4.2 has been broken into its individual components, D_s, A, and dC/dx.

Equation 4.5 indicates that the rate of solute movement across semipermeable membranes is directly proportional to the membrane area (A), to the membrane-solute diffusion coefficient (D_s) and to the concentration gradient (dC), and inversely proportional to the membrane thickness (dx). The diffusion coefficient for solute s, D_s defines the permeability of the membrane to a specific solute at a given temperature and is independent of membrane area and thickness. For hemodialyzers, membrane area and thickness are constant. These constant terms may be combined with the diffusion coefficient to form the mass transfer area coefficient (*KA*). Equation 4.5 then simplifies to

$$\frac{ds}{dt} = \mathrm{KA} \cdot dC \qquad 4.6$$

KA, the mass transfer area coefficient, has units of cm^3/min or ml/min; it is discussed more thoroughly in chapter 7. Equation 4.6 indicates that the movement of solute is dependent on only two factors, the membrane permeability reflected by *KA*, and the concentration gradient across the membrane (*dC*). If the dialysate concentration is zero, as it is for single-pass systems, *dC* is equal to *C* and equation 4.6 has a form identical to equation 4.2, where $KA = k$.

Equation 4.6 defines movement of solute in a static system. As we introduce flow of blood on one side of the membrane and flow of dialysate on the other, a more complex relationship emerges. These variables, however, can be incorporated into the mathematical expressions allowing precise definition of solute removal. The equations that describe solute movement in terms of blood and dialysate flow are discussed further in chapter 7.

First-order kinetics: clearance, rate constant, half-life, and exponential decline

Nephrologists are intimately familiar with the term *clearance*, shown graphically in figure 4.2 (14). Clearance is a useful expression that usually describes the

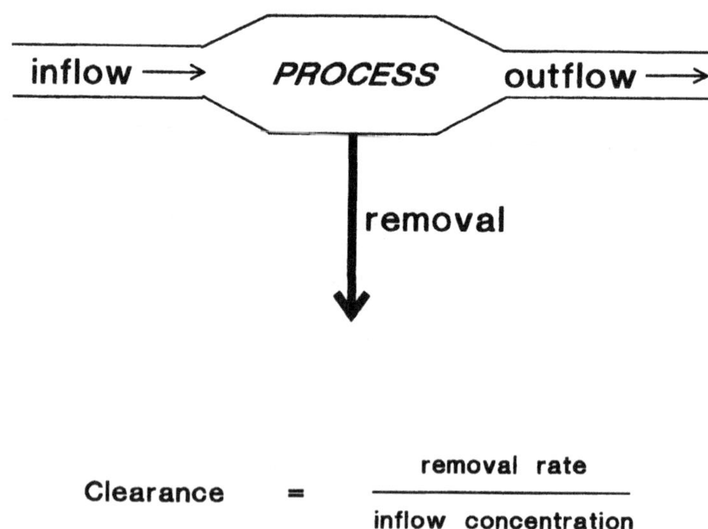

Figure 4.2. Clearance is constant for simple first-order processes.

movement of solute out of a compartment, factoring for the solute concentration within the compartment:

$$\text{Clearance} = \frac{\text{removal rate}}{\text{inflow concentration}} \qquad 4.7$$

The word expression of clearance in equation 4.7 is similar to equation 4.3, which describes first-order kinetics. Clearance is a practical way of describing any first-order process, e.g., glomerular filtration. If the process is first-order, the term *clearance* becomes a measure of the process itself, independent of the solute concentration, and has units of flow, e.g., ml/min. The clearance of heat by a nuclear reactor in thermal units per hour or the clearance of carbon dioxide by the lungs in liters/hour can be expressed by equation 4.7; both are equally well depicted by figure 4.2. In both cases there is no need to specify the temperature (for the nuclear reactor) or the concentration of CO_2 (for the lungs) when the clearance is measured, because for true first-order processes, clearance is constant throughout a wide range of these variables.

Dialysis follows the laws of diffusion, a first-order process. For a hemodialyzer such as the hollow fiber kidney we could simply measure the amount of urea removed per minute or per hour to compare dialyzers. However, in each case we are required to specify the concentration of urea in the arterial blood, since the rate of urea removal is directly proportional (first-order) to the compartment concentration. By dividing the removal rate by the arterial concentration, we eliminate that requirement, and the result is a clearance. We can then compare dialyzers by examining their clearances of urea or other solutes without need to specify the urea concentration. Two other variables, blood flow and dialysate flow, have an equally significant influence on dialyzer clearance. The discussion of dialyzer clearance in chapter 7 includes a somewhat more complicated mathematical expression to factor these variables as well. The result, described in detail in chapter 7, is an expression of dialyzer efficiency that uses the dialyzer's mass transfer area coefficient for urea (*KA*) described above. *KA* defines the dialyzer's capacity to remove urea independent of all three of the above variables.

The solution to equation 4.3a is based on mathematical integration of the expression dx/x to yield $\ln(x) + c$. $\text{Ln}(x)$ refers to the natural logarithm of x, and c is the constant of integration. Taking antilogs of both sides of the resulting equation, we get

$$C = C_0 \cdot e^{-k \cdot t} \qquad 4.8$$

This is a familiar expression for the first-order logarithmic decline in concentration of, for example, a drug after a loading dose. C_0 is the starting concentration at time zero; k (k' in equation 4.3a) is the rate constant for elimination of the drug (V is constant); C is the new concentration at time t. The elimination rate constant, k, determines the rapidity of change in concentration. It is measured in units of time^{-1}, a unit of measurement that is difficult to conceptualize. The numerator in this fractional unit is not missing; it is itself a fraction, a fraction of the loading dose at time zero or of total body drug at time t. The rate constant (k) can be defined as the fractional removal of solute per unit of time. This concept is expressed mathematically in equation 4.3. The inverse of k is the turnover rate, the time required to remove all the solute at the initial rate. When the fraction removed is 1/2, $C = C_0/2$ and equation 4.8 reduces to

$$t_{1/2} = \ln 2/k \qquad 4.9$$

The variable $t_{1/2}$ is the half life of solute s. It is important to emphasize that k is not a function of the compartment alone or of the solute alone but is rather a function of both, i.e., it describes the movement of a specific solute to or from a specific compartment.

A comparison of rate constants and clearance is helpful. Because the rate constant describes fractional removal, calculation of either the instantaneous or cumulative amount removed requires knowledge of the amount present at the start. This requires knowledge of the distribution volume of the solute and its initial concentration. Clearance, as explained above, is a term that gives us the ability to describe the rate of removal without knowing the volume of distribution or concentration. This is made possible because clearance is expressed in *volumes removed* / unit of time. To convert clearance to actual units of solute removed, one need only know the concentration. The mathematical relationship between these two terms is

$$k = \text{clearance}/V \qquad 4.10$$

The single-compartment model

An expression similar to equation 4.2 can be derived for urea removal from the body during and between dialyses (4). Counteracting and complicating this simple removal process is a constant generation of urea during and between dialyses. The one-compartment model assumes a single volume for urea distribution, as shown in figure 4.3.

The word equation depicted by figure 4.3 is
change in urea content = urea generation - urea removal

The mathematical equation for the same expression is

$$d(\text{urea content})/dt = G - K \cdot C \qquad 4.11$$

G is the urea generation rate (mg/min)
K is the total clearance of urea (ml/min) and must be distinguished from the rate constant, k, as discussed above.

Equation 4.11 is the fundamental expression of single-compartment urea kinetics; it describes urea movement both during and between dialyses with only a minimum error (see chapter 5 for more discussion of the single-compartment error). The term *urea content* can be expressed as concentration of urea (C), multiplied by urea volume (V):

$$d(V \cdot C)/dt = G - K \cdot C \qquad 4.12$$

Equation 4.12 can be integrated to obtain an expression that gives a specific concentration C at time t. The solution is simplified if we consider V to be constant. This, of course, is seldom the case, because patients tend to lose weight during dialysis and to gain weight between dialyses. However, the error introduced by assuming V is constant, and simply adding and subtracting the volume lost and gained for the cycle under study, is usually less than 15%. The error is greater for patients with large weight changes during and between dialyses. When V is constant, equation 4.12 rearranges to

$$dC/(G - K \cdot C) = dt/V \qquad 4.13$$

Integration yields

$$C = C_0 \cdot e^{-K \cdot t/V} + \frac{G}{K}(1 - e^{-K \cdot t/V}) \qquad 4.14$$

If instead of clearance (K) we use the rate constant (k), equation 4.14 becomes:

$$C = C_0 \cdot e^{-k \cdot t} + \frac{G}{k \cdot V}(1 - e^{-k \cdot t}) \qquad 4.15$$

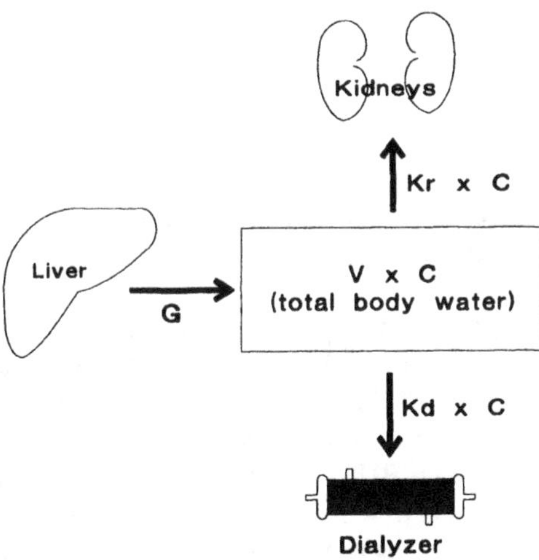

Figure 4.3. Single-pool model of urea mass balance. Hepatic urea generation is represented by G, native kidney removal by K_r x C, dialyzer removal by K_d x C.

This expression is familiar to pharmacologists as the constant infusion model for administration of a drug such as heparin. The first term in equation 4.15 describes the simple logarithmic decline in concentration due to first-order clearance. The second more complex term describes the contribution of ongoing generation (G) or infusion in the case of drug pharmacokinetics.

During hemodialysis the clearance term (K) is actually composed of two elimination pathways, the dialyzer clearance (K_d) and the patient's own residual renal clearance (K_r), if any remains.

$$C = C_0 \cdot e^{-(K_r + K_d)t/V} + \frac{G}{K_r + K_d}\left[1 - e^{-(K_r + K_d)t/V}\right] \qquad 4.16$$

Constant-volume model, three BUN measurements

Determining urea volume (V) and generation rate (G)

Equation 4.16 describes urea concentration (BUN) at any time during or between dialyses, provided that no change in weight (V) occurs. For the interval between dialyses, K_d is zero. Equation 4.16 contains two unknown and unmeasurable variables, V and G. Resolution of these variables is obtained by rearranging equation 4.16 and substituting C_1 and C_2 for C_0 and C, respectively:

$$V = \frac{(K_r + K_d)T_d}{\ln\left[\dfrac{G - C_1(K_d + K_r)}{G - C_2(K_d + K_r)}\right]} \qquad 4.17$$

C_1 is the BUN before dialysis (mg/ml)
C_2 is the BUN after dialysis (mg/ml)
T_d is the duration of dialysis (min)

$$G = K_r\left[\frac{C_3 - C_2\, e^{-K_r \cdot T_i/V}}{1 - e^{-K_r \cdot T_i/V}}\right] \qquad 4.18$$

C_3 is BUN before the second dialysis (mg/ml)
T_i is time between dialyses (min)

Sensitivity analysis of equation 4.16 shows that the decrease in BUN from C_1 to C_2 is primarily determined by the dialyzer clearance, K_d. Residual clearance (K_r) and generation rate (G) play relatively minor roles. Therefore resolution of V is accomplished by examining the decrease in BUN during dialysis according to equation 4.17. In contrast, the rise in BUN from C_2 to C_3 between dialyses is primarily determined by the ratio of G to V, K_r playing a significant role. Resolution of G is accomplished by examining the change in BUN from C_2 to C_3.

Each of the above equations contains the other variable. To obtain real values for V and G, two sets of data must be obtained during or between dialyses. This is most conveniently done by measuring the BUN before, after, and again before the next dialysis. The second BUN is used twice to provide the two sets of data. Solution of equations 4.17 and 4.18 for the unknowns V and G is best done by iteration using a calculator or computer. Since G is a minor factor in equation 4.17, the intradialysis data are used to solve for V. This number is then used to solve for G from equation 4.18 and the second set of data between dialyses. The resulting value for G is then substituted in equation 4.17, etc. After three iterations, G and V become relatively constant, within 5% of their final convergence values. Further iterations provide higher degrees of precision.

Determining dialyzer clearance (K_d)

Another rearrangement of equation 4.17 gives an expression for K_d, dialyzer clearance:

$$K_d = \frac{V}{T_d} ln \left[\frac{G - C_1(K_d + K_r)}{G - C_2(K_d + K_r)}\right] \qquad 4.19$$

This equation is used to calculate the effective urea clearance if V is previously known. G is determined from equation 4.18. Values for V to insert here are an average of those previously calculated from urea kinetic analyses or an estimate of total body water using established formulas (15,16). Formulas for estimating total body water are derived from measurements in normal populations. These can lead to large errors when applied to uremic populations, especially in malnourished or edematous patients starting a maintenance dialysis program. Because equation 4.19 does not provide an explicit solution for K_d (K_d is present on both sides of the equation) an iterative approach must be taken. Values for K_d are repeatedly inserted into equation 4.19 until its value stabilizes. This usually requires three or four iterations.

Determining ideal dialysis time

Further rearrangement of equation 4.17 gives an equation for T_d, time on dialysis:

$$T_d = \frac{V}{K_d + K_r} ln \left[\frac{G - C_1(K_d + K_r)}{G - C_2(K_d + K_r)}\right] \qquad 4.20$$

This is the dialysis time required to lower the BUN from C_1 to C_2 in a patient with urea volume V, generation rate G, and residual clearance K_r, using a dialyzer with urea clearance K_d.

If we assume regular dialysis intervals by dividing the week into three equal parts (if the dialysis frequency is three times/week), it becomes possible to calculate the average predialysis or postdialysis BUN. The average predialysis and postdialysis BUN should closely match corresponding BUN values measured at midweek for patients dialyzed three times/week. By substituting appropriate intradialysis and interdialysis values into equation 4.16, the following expression for average predialysis BUN evolves:

$$C_{1av} = \frac{G\left[\frac{e^{-K_r \cdot T_a/V}}{K_r + K_d}(1 - e^{-(K_r + K_d)T_d/V}) + \frac{1 - e^{-K_r \cdot T_a/V}}{K_r}\right]}{1 - e^{-K_r \cdot T_a/V}e^{-(K_r + K_d)T_d/V}} \qquad 4.21$$

C_{1av} is average predialysis BUN (mg/ml)
T_a is the average time between dialyses (min) = 10,080/N - T_d
N = number of dialyses/week

Average postdialysis BUN can be determined from the average predialysis concentration and the interdialysis interval:

$$C_{2av} = \frac{C_{1av} - G(1 - e^{-K_r \cdot T_a/V})/K_r}{e^{-K_r \cdot T_a/V}} \qquad 4.22$$

C_{2av} is the average postdialysis BUN (mg/ml)

Time-averaged BUN was the criterion used by the National Cooperative Dialysis Study (NCDS) to measure the effect of dialysis, as discussed in chapter 3 (17). Since BUN falls steeply during dialysis and then more than doubles between dialyses, the average value would seem to be a reasonable gauge of uremic exposure. A true measure of average BUN over a period of time would require measurement of the area under the BUN/time curve. For practical purposes, the mean of average predialysis BUN and average postdialysis BUN is adequate. This simplification involves two assumptions:

1. The dialysis schedule is nearly uniform (the week is evenly divided)
2. The interdialysis curve of BUN versus time is linear

The latter assumption is true for nearly all patients, since most have little or no residual clearance and weight gain is small relative to V. The former assumption causes a slight error, as we will see in the next section.

If we make the above assumptions, time-averaged BUN can be expressed

$$TAC = (C_{1av} + C_{2av})/2 \qquad 4.23$$

Rearranging,

$$C_{1av} = 2 \cdot TAC - C_{2av} \qquad 4.24$$

TAC = time-averaged BUN
C_{1av} = average predialysis BUN
C_{2av} = average postdialysis BUN

In chapter 8, determinants of ideal TAC (ITAC) are examined in detail. If ITAC, V, and G are known, it is possible to compute ideal average postdialysis BUN (IC_{2av}):

$$IC_{2av} = \frac{2 \cdot \text{ITAC} - G\,(1 - e^{-K_r \cdot T_a/V})/K_r}{1 + e^{-K_r \cdot T_a/V}} \qquad 4.25$$

With IC_{2av} in hand, we can derive an expression that provides an estimate of ideal time on dialysis. This is a slightly modified version of equation 4.20:

$$T_b = \frac{V}{K_d + K_r} \ln \left[\frac{G - IC_{1av}(K_d + K_r)}{G - IC_{2av}(K_d + K_r)} \right] \qquad 4.26$$

T_b = ideal time on dialysis to achieve ITAC (min)
IC_{1av} = ideal average predialysis BUN (mg/ml)
IC_{2av} = ideal average postdialysis BUN (mg/ml)

If an estimate of ideal time-averaged BUN (ITAC) is available for a patient with established volume and generation rate, it is possible to predict the duration of dialysis that will achieve, at steady state, a time-averaged BUN equivalent to the ideal value. The variables required to accomplish this estimate are V, G, K_d, K_r, and number of dialyses/week. V and G have already been calculated (equations 4.17 and 4.18), and the remaining variables can be arbitrarily assigned. Equation 4.26 gives T_b, but notice that IC_{2av} is required. The equation that provides IC_{2av} (equation 4.25) requires input of time between dialyses (T_a), which depends on T_b. Since algebraic resolution is not possible, an iterative technique can be applied similar to that used for calculation of V, G, and K_d. The value for T_b obtained from equation 4.26 is used to calculate T_a in equation 4.25. IC_{2av} from equation 4.25 is used for calculation of T_b, etc. Three or four iterations are usually sufficient to provide a stable value for T_b within 1% of steady state.

Evaluation of the constant-volume model

Error in V due to weight loss during dialysis

When fluid is removed during dialysis, the constant-volume model described by equation 4.16 overestimates V. V is falsely inflated because its value is assumed to be the same at the end of dialysis as it was at the beginning. A more accurate estimate of V is obtained by simply subtracting the volume lost during dialysis from equation 4.17's estimate of V; but even after subtraction, the volume is too large. Figure 4.4 shows the errors in V, G, PCRn, and ideal time on dialysis introduced by the constant-volume model as weight loss during dialysis varies from 0 to 10 kg. With each increment in time during dialysis, the amount of urea removed is overestimated because the change in concentration is multiplied by an inflated volume. To fit the overestimated amount removed during the entire dialysis to the measured concentrations, a larger urea volume must be assumed. This accounts for the additional error in V, shown in figure 4.4, that accrues beyond the actual volume lost during dialysis.

Error in G due to weight gain between dialyses

The potential inflation of *G* consequent to the overestimation of *V* by the constant-volume model is more than offset by the gain in weight between dialyses. Underestimation of *G* occurs because the change in concentration from C_2 to C_3 is lessened by the dilution that occurs consequent to fluid gain. Because of the smaller change in BUN between dialyses, the equations predict that *G* must be less than it actually is. The final result is a significant underestimation of *G* using the constant-volume model (figure 4.4).

The constant-volume model, then, is applicable only when little or no weight change occurs during and between dialyses or when rough estimates of *V* and *G* are desired. For the single-compartment model, the constant-volume method has been largely replaced by the variable-volume method described below.

VARIABLE-VOLUME MODEL, THREE BUN MEASUREMENTS

If volume is assumed to vary but at constant rates during and between dialyses, equation 4.12 becomes

$$C \cdot dV/dt + V \cdot dC/dt = G - K \cdot C \tag{4.27}$$

dV/dt = constant rate of change in volume

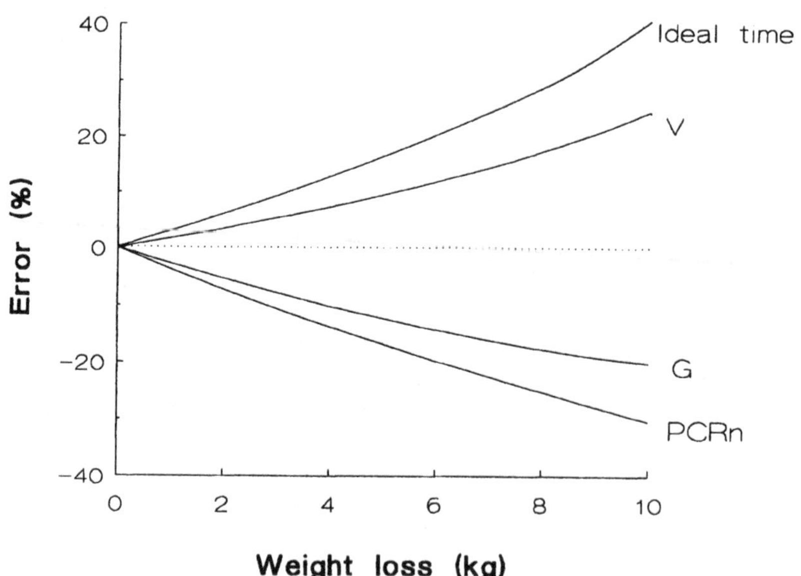

Figure 4.4. Fixed-volume errors. Errors appear in *V*, ideal dialysis time, *G*, and PCRn due to failure to consider weight (volume) changes during and between dialyses. K_d = 180 ml/min; T_d = 4 hours; Schedule = three dialyses/week; predialysis V = 40 liters.

Integration of equation 4.27 gives the following expression for C:

$$C = C_0\left[\frac{V + B\cdot t}{V}\right]^{-(K_r+K_d+B)/B} + \left[\frac{G}{K_r + K_d + B}\right]\left[1 - \left[\frac{V + B\cdot t}{V}\right]^{-(K_r+K_d+B)/B}\right] \qquad 4.28$$

$B = dV/dt$, the rate of change in volume, i.e., weight change/minute, which is different for the intradialysis versus interdialysis interval.

C_0 is the BUN when $t = 0$

A set of equations for V, G, K_d, and T_b analogous to equations 4.17, 4.18, 4.19, and 4.26 can be derived from equation 4.28:

$$V = \frac{B\cdot T_d}{\left[\frac{C_2(K_r + K_d + B) - G}{C_1(K_r + K_d + B) - G}\right]^{\{B/(K_r + K_d + B)\}} - 1} \qquad 4.29$$

B is rate of weight gain during dialysis (usually negative) (g/min)

$$G = (K_r + B)\left[\frac{C_3 - C_2\left[\frac{V + B\cdot T_i}{V}\right]^{-(K_r+B)/B}}{1 - \left[\frac{V + B\cdot T_i}{V}\right]^{-(K_r+B)/B}}\right] \qquad 4.30$$

B is rate of weight gain between dialyses (g/min)

$$K_d = \left[\frac{B}{\ln\,(V + B\cdot T_d)/V}\right]\ln\left[\frac{G - C_1(K_r + K_d + B)}{G - C_2(K_r + K_d + B)}\right] - K_r - B \qquad 4.31$$

B is rate of weight gain during dialysis (usually negative) (g/min).

$$T_b = \frac{V}{B}\left[\left[\frac{G - C_{2av}(K_r + K_d + B)}{G - C_{1av}(K_r + K_d + B)}\right]^{-(K_r+K_d+B)/B} - 1\right] \qquad 4.32$$

B is rate of weight gain between dialyses (g/min)

Source code for the variable-volume model, three BUN values

Appendix A shows an algorithm for calculating urea volume (V) and generation (G) using three blood samples and three patient weight measurements. The series of ten iterations takes into consideration volume changes during and between dialyses. An accuracy of ± 5% is usually achieved after three iterations, so the number of loops can be reduced if timing is important and the computer operates slowly. The mathematical equations used to generate these program lines are taken directly from the above equations with only minor symbol modifications.

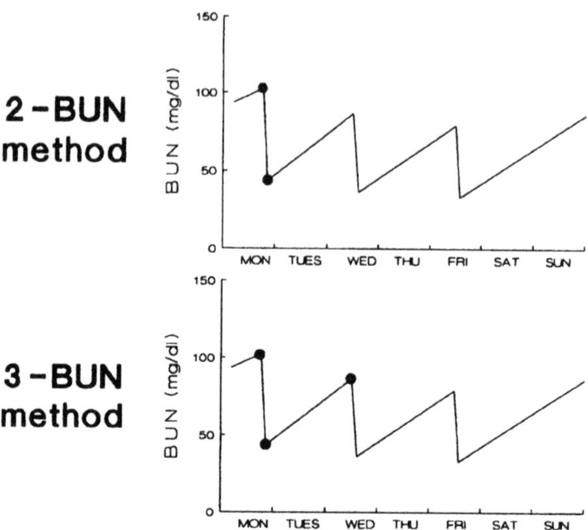

Figure 4.5. The weekly profile of BUN versus time can be generated from two or three data points. Urea generation, calculated using the three-BUN method, is heavily dependent upon the difference between the second and third BUN. Urea generation, calculated using the two-BUN method, is primarily determined by the actual value of the predialysis BUN.

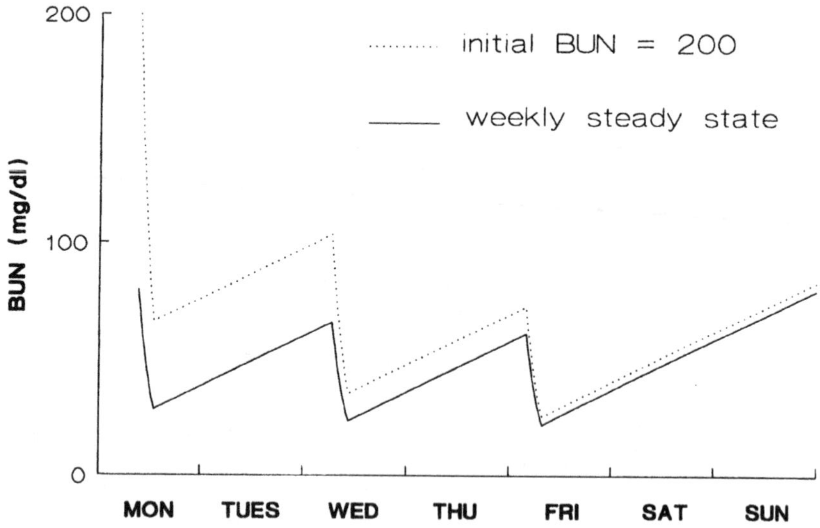

Figure 4.6. Demonstration of time required to reach a steady-state equilibrium after starting hemodialysis in a patient with initial BUN = 200 mg/dl. Solid line represents the steady-state, dotted line actual BUN values. After three dialyses, the two nearly overlap.

Two-BUN method, variable-volume

The algorithm shown in appendix A requires three measurements of BUN and weight. It is possible to model only two measurements: a single predialysis BUN and a single postdialysis BUN (10,11). This simplifies urea modeling and allows the study to be completed during a single dialysis. The measured data points are illustrated in figure 4.5 for both methods.

The weekly steady state

To apply the two-BUN method for measuring G, the patient's weekly urea concentration profile (figure 4.1) must be constant, i.e., the patient must be in a weekly steady state. Estimation of V and projected clearance will be unaffected by non-steady-state conditions because V is primarily determined by intradialysis modeling of predialysis and postdialysis BUN. Urea generation, however, is highly influenced by the predialysis BUN value. A steady state means that the dialysis prescription is unchanged for at least three dialyses. Protein intake should also be controlled, but as shown below, modest variance from dialysis to dialysis will not affect estimations of G when the two-BUN method is used. Establishment of weekly steady-state conditions will ensure stability of the predialysis BUN on any particular day from week to week and an accurate estimate of average urea generation (G). If urea concentrations are not in a weekly steady state, as often occurs for example in patients with acute renal failure, the three-BUN method must be used to avoid a potentially significant error in G. It is worth noting, however, that the value of G calculated in nutritionally unstable patients will be accurate only for the single interdialysis interval studied. Figure 4.6 shows that BUN values stabilize quickly even when the starting BUN is 200 mg/dl and the predialysis BUN is less than 100 mg/dl at equilibrium. It is not necessary to wait several weeks after starting or changing a maintenance hemodialysis regimen to model urea kinetics using the two-BUN technique. With a constant dialysis prescription, urea concentrations will generally approach steady state after two or three dialyses.

Technique for modeling with two BUN values

The change in BUN induced by dialysis determines V, which is calculated using equation 4.29 in exactly the same manner as the three-BUN method. Therefore V determined by the two-BUN method will differ little from V obtained using three BUN values. From this point forward, the two methods diverge. If the patient is in a true weekly steady state, the urea nitrogen generation rate can be derived by imposing a periodic solution to the equations describing C_1 and C_2 (equation 4.28 during and between dialyses)(11). For the two-BUN technique, we must know the dialysis weekly schedule and the weekday of the study, e.g., Monday. After calculation of V, predialysis and postdialysis BUN values for each dialysis day are calculated sequentially using equation 4.28 until a new predialysis BUN is obtained for the day of BUN measurement one week later. G is adjusted until the calculated predialysis BUN at the end of the period one week later matches the measured predialysis BUN within 1%. Figure 4.7 shows this technique for periodic solution of G in graphic form. Note that the calculated predialysis BUN overestimates the measured value in figure 4.7A and underestimates it in figure 4.7B. Figure 4.81 shows a flow diagram of the loops that eventually lead to solution of V and G.

If the variable-volume method is used, either the patient's weight must be measured before and after dialysis or another assessment of fluid gained and lost must be adopted. Treatment of fluid gained between dialyses and lost during dialyses is discussed under *Variable-volume considerations* below.

Estimation of K_d from two BUN measurements

A similar technique is used to estimate K_d when the user provides an assessment of V. Instead of calculating V at the top of the loop in figure 4.8, we substitute the calculation for K_d from equation 4.31. Otherwise the algorithm is identical.

Estimation of ideal time on dialysis using two BUN measurements

The estimation of ideal time on dialysis (T_b) using the two-BUN method differs from the three-BUN technique described above. Instead of breaking up the week into equally spaced intervals to estimate average predialysis and postdialysis BUN, the true weekly schedule is used. The area under the curve (AUC) of BUN versus time is measured, obviating the need for equations to calculate average predialysis

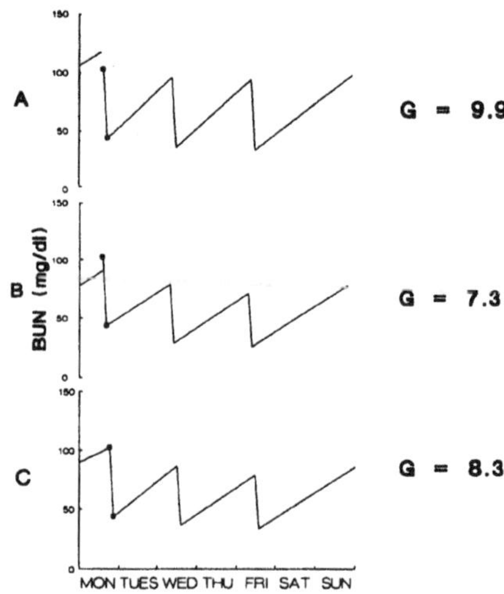

Figure 4.7. Two-BUN method for resolution of G. This diagram shows the iterative technique for determining urea nitrogen generation rate (G) using the two-BUN method. An arbitrary value for G is selected (9.9 mg/min) in (A) and predialysis BUN is determined for the seventh day (one week later). This value is found to be too high, so G is adjusted downward to 7.3 mg/min and a new profile is generated in (B). After multiple iterations, predialysis BUN matches the calculated value one week later when G = 8.3 mg/min as shown in (C). Reproduced with permission from Trans Am Soc Artif Intern Organs 35:500, 1989.

and postdialysis BUN. This technique is illustrated in figure 4.9. Time-averaged BUN is defined as

$$\text{TAC} = \text{AUC}/10080 \tag{4.33}$$

The denominator is the number of minutes in a week.

The integration technique illustrated in figure 4.9 for measuring AUC is combined with equation 4.33 to calculate TAC. Each time interval during or between dialyses is divided into trapezoidal blocks. For the single-compartment model, five blocks are sufficient to impart an accuracy that is within 0.1% of the same technique using 1000 blocks per interval. TAC computed this way is the true time-averaged BUN. After TAC is calculated by this method, T_b is adjusted, using the algorithm shown in figure 4.10, until TAC matches ideal TAC (ITAC) within 1%. Ideal TAC is calculated from the patient's protein catabolic rate using the target therapy modeling line, a process developed at length in chapter 8.

Variable-volume considerations

With each measurement of BUN, patient weight and time is recorded. For the three-BUN method, the rate of weight gain between dialyses is determined by subtracting the postdialysis weight from the second predialysis weight and dividing by the interdialysis time interval. This rate, expressed in liters/day, is assumed to be constant from day to day and from week to week in each patient as a reflection of fluid intake. For the two-BUN method, the weight lost during dialysis is used to

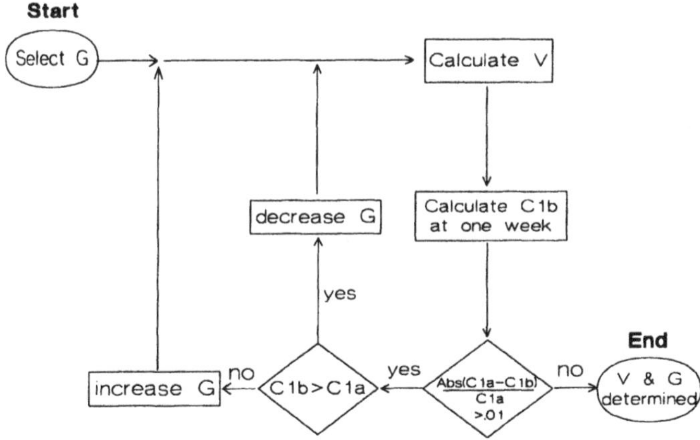

Figure 4.8. Algorithm for calculating urea volume (V) and generation rate (G) using the two-BUN method. An arbitrary value is selected for G. Equation 4.29 is used to calculate V. Predialysis BUN ($C1$b) is then calculated at seven days and compared to the measured predialysis value ($C1$a). An appropriate adjustment is made in G, and the process is repeated until the predialysis BUN values match within 1%. Reproduced with permission from Trans Am Soc Artif Intern Organs 35:501, 1989.

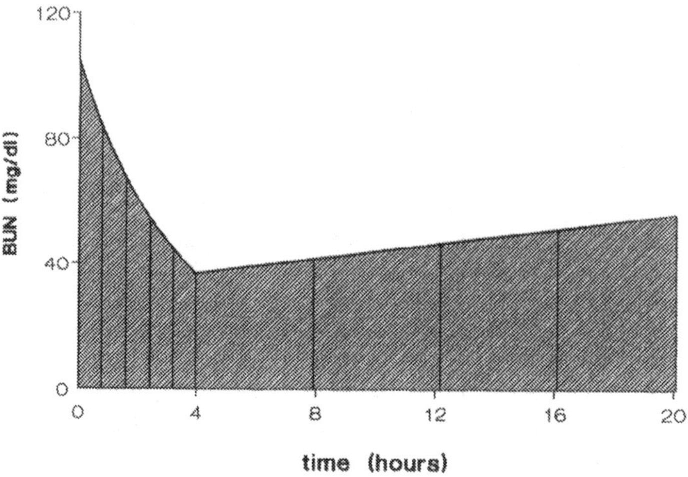

Figure 4.9. Manual integration technique used to estimate time-averaged BUN (TAC). TAC is the area under the curve divided by the time interval. The area for each interval is calculated using the trapezoidal area formula. This minimizes the number of divisions required for an accurate estimate of area.

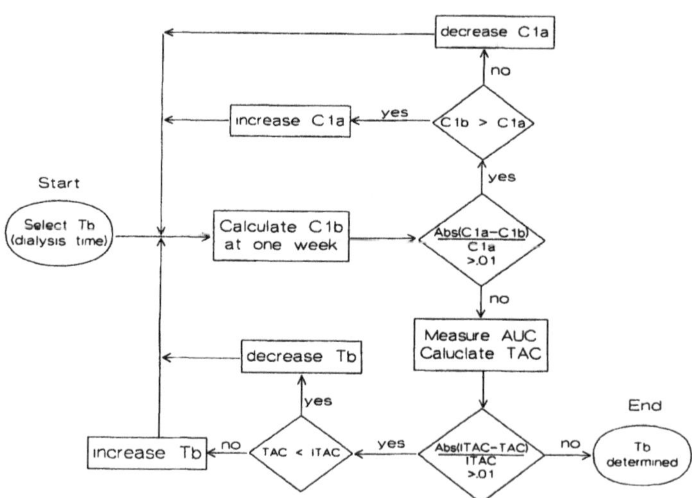

Figure 4.10. Two-BUN method: double-loop algorithm for calculating ideal time on dialysis (T_b). The duration of dialysis is adjusted until TAC matches ITAC (ideal time-averaged BUN), as shown in the lower loop. After each adjustment of dialysis time, predialysis and postdialysis BUN are recalculated and then adjusted until weekly equilibrium is reached. The latter is determined when predialysis BUN at one week (C_{1b}) matches the starting value (C_{1a}), as depicted in the upper loop. This algorithm may be applied to both one- and two-compartment models.

determine the rate of weight gain between dialyses. The latter is assumed to be constant and is determined by dividing the weight lost during dialysis by the preceding time interval between dialyses. The ultrafiltration rate is adjusted for each dialysis to bring the patient's weight postdialysis to the measured postdialysis value. The latter is assumed to be the patient's *dry* weight. These adjustments for volume usually have minor effects, but can be significant in patients with substantial weight gain between dialyses.

COMPARISON OF THE TWO-BUN METHOD WITH THE THREE-BUN METHOD

In many respects, the two-BUN method is simpler and more accurate than the three-BUN method. The basic mathematical expression (equation 4.28) is relied upon heavily, and the true weekly schedule is used to calculate T_b instead of the averaging technique described previously for the three-BUN method.

An argument in favor of the two-BUN method is that the periodic solution to the expressions of C_1 and C_2 is less dependent on a single interdialysis generation rate than on the average generation rate during the preceding week (see chapter 3, *importance of average values*). In other words, G determined by the two-BUN method is less influenced by a high or low protein intake during a single short interval and is therefore less likely to deviate far from the true mean for G. Because the two-BUN model must fit a value of G for the entire weekly profile, it is less likely to be influenced by a single interdialysis interval. The three-BUN method ignores the actual values of C_1 and C_2 and uses instead another data point (C_3, the third BUN) to calculate G using the difference between C_3 and C_2. This places complete reliance on a single interval for a variable with a well known large variance. The three-BUN method also gives the patient an opportunity to interfere with the results. If the patient observes the drawing of the first two blood samples, signaling that a kinetic study is under way, he or she may tighten dietary controls. This dietary adjustment lowers the value of the third BUN, measured when the patient returns for dialysis, and falsely decreases both G and PCRn. The ultimate result is an inappropriate reduction in dialysis intensity. Patient-controlled skewing of urea kinetics is avoided with the two-BUN method because there is no forewarning that a study is about to be done.

The proposed improvement in accuracy must be questioned because whenever the number of measured data points is reduced for modeling in general, accuracy is usually compromised. If three data points are better than two data points for modeling in general, why is this not true for urea modeling? This question is partially answered above, i.e., not all of the information available from three data point measurements is used. It is theoretically possible to improve the three-BUN method to provide better estimates of V and G than the two-BUN method. This would require an algorithm similar to that used for the two-BUN method (figure 4.8) and comparison of calculated C_{1b} to both C_1 and C_3. This technique would provide a slightly more accurate determination of G but would incur more work.

Theoretical comparison of methods

Provided that the patient is in a weekly steady state, the technique that uses two instead of three BUN values has less chance for error as discussed above. Urea nitrogen generation (G) is the most inherently unstable parameter of the four major determinants of urea kinetics (K_d, K_r, V, and G). Instead of relying on one postdialysis-to-predialysis interval (C_2 to C_3) for estimation of G, the two-BUN technique uses the real value of C_1, a concentration that is determined by an average of preceding generation rates. To illustrate this point, consider an average 70 kg patient dialyzed Monday, Wednesday, Friday for four hours with a dialyzer that clears urea at 180 ml/min (whole-blood clearance). This patient's steady-state BUN profile at midweek is shown as the solid line in figure 4.11. For simplicity we will consider that no weight is gained or lost during dialysis and that no residual urea clearance remains. This average patient has a 36.4 liter urea volume and a generation rate of 6.6 mg/min (PCRn = 1.15 g/kg/day). Between Wednesday and Friday he strays from his usual diet, boosting his urea generation by 30% from 6.6 to 8.6 mg/min. Consequently, his predialysis BUN on Friday increased from its steady-state value of 74 mg/dl to 88 mg/dl ([2mg/min x 2640 min]/364 dl = 14 mg/dl increment). The new pattern for increase in BUN from Wednesday to Friday is shown by the dotted line from C_2 to C_3' in figure 4.11.

If an analysis of urea kinetics were done from Wednesday to Friday of this week using the three-BUN technique, an error would occur because of the temporary nonrepresentative increase in PCRn and G. Table 4.2 shows the consequences of this error when modeling is done by the two-BUN versus the three-BUN method. The increased protein intake has little effect on V because V is primarily determined by intradialysis modeling of the first predialysis and postdialysis BUN (C_1 and C_2).

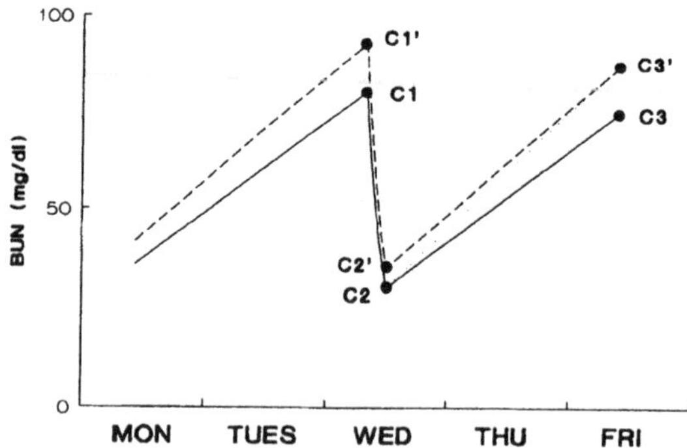

Figure 4.11. Two-BUN versus three-BUN model: effect of a transient real increase in G on model estimate of G. Solid line C_1—C_2—C_3 represents real steady-state conditions. Dashed line C_1'---C_2'---C_3' is a new modeled steady-state profile when C_1 or C_3 is temporarily inflated by excessive protein intake. The steeper slope of dotted line C_2---C_3' reflects the value for (G) determined by the three-BUN method. The slope of C_2'---C_3', used by the two-BUN method, is closer to the slope of C_2–C_3. Reproduced with permission from Trans Am Soc Artif Intern Organs 35:501, 1989.

Calculated ideal time on dialysis increases from 4.0 to 5.0 hours because of the apparent increase in PCRn.

Table 4.2 Errors due to transient increase in PCRn. Comparison of three-BUN and two-BUN methods

Variable	Steady state	After increase in PCRn 3-BUN method	2-BUN method
K_d (ml/min)	180	180	180
K_r (ml/min)	0	0	0
T_d (hours)	4	4	4
Wt (kg)	70	70	70
BUN-pre (mg/dl)	80	80	94
BUN-post (mg/dl)	30	30	35
BUN-pre (mg/dl)	74	88	88
G (mg/min)	6.6	8.6	7.7
V (liters)	36.4	35.7	36.4
PCRn (g/kg/day)	1.13	1.47	1.31
T_b (hours)	4.0	5.0	4.6

If the two-BUN method is used, no error occurs because the third data point is not used in the calculations. This is not an equivalent comparison, however, so we will consider that the same increase in protein intake occurs one cycle earlier, from Monday to Wednesday, causing the predialysis BUN on Wednesday to rise from its steady-state value of 80 mg/dl to 94 mg/dl. The new steady state profile established by the two-BUN method is shown as the dashed line in figure 4.11. The patient's post-dialysis measured BUN would be 35 instead of 30 after this change in predialysis BUN if the urea volume (V) remains constant. The new profile with inflated values of C_1 and C_2 produces the errors shown in table 4.2. The errors in G and PCRn are significantly less than those generated by the three-data-point technique (16% versus 30%). This is best demonstrated graphically as the difference in slopes of the dotted versus dashed lines from Wednesday to Friday in figure 4.11. Estimated time on dialysis is increased to 4.6 hours instead of 5.0 hours. Similar results are obtained when the increase in PCRn occurs between Wednesday and Friday. The improvement in accuracy is a consequence of the technique used to calculate G that depends on fitting an entire week of calculated predialysis and postdialysis BUN values to the measured values, rather than relying on a single postdialysis-to-predialysis interval, as explained above.

Clinical studies comparing methods

The previous discussion argues that the two-BUN technique may paradoxically present more accurate results than the three-BUN technique. To examine this question more closely, we compared the actual predialysis BUN with that predicted by the two-BUN model three days, one week, and two months after the initial modeling in 29 patients (177 paired measurements). The mean error determined from these studies is shown in figure 4.12. At two days the error was 7.4 ± 5.6% (SD), at one week 8.6 ± 8.8%, and at two months 16.6 ± 12.7%. These data support the two-BUN method as a predictor of future urea concentrations and justify its use in the estimation of urea generation rate. Since urea volume is nearly identical when measured by either of these two techniques, there is usually little to be gained by routinely adding the third set of data, a task that significantly increases the complexity of kinetic modeling. If the patient is not in a weekly steady state, the third data point is required for an accurate estimation of *G*.

Source code for the variable-volume model, two BUN values

Appendix B shows an algorithm for calculating *V* and *G*, using the variable-volume two-BUN method. Although the code is more extensive, the iterations require less than one second to complete by contemporary microcomputers. The results are more exact than those obtained from the algorithm in appendix A, because the actual interdialysis time intervals and intradialysis weight losses are used to calculate *G*, rather than the averages used by the three-BUN method.

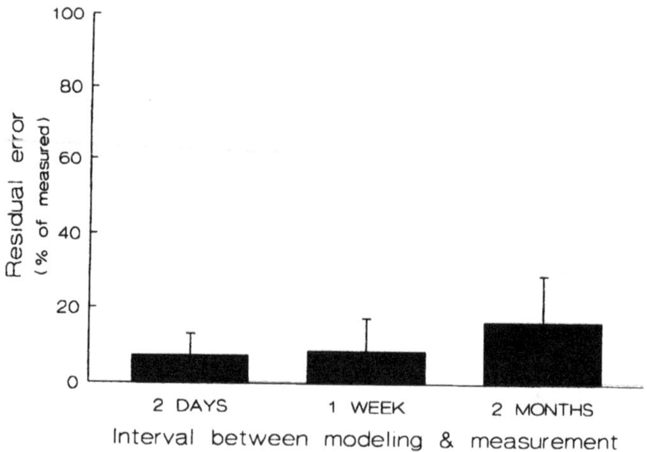

Figure 4.12. Urea modeling with two BUN measurements. Mean error ± SD is shown for measured versus predicted predialysis BUN at three different time intervals following the kinetic study. The absolute value of the error is expressed as a percentage of the measured value. At two days the mean error is 7.4% ± 5.6%. At one week the mean error is 8.6% ± 8.8%. At two months the mean error is 16.6% ± 12.7%.

References

1. Wolf AV, Remp DG, Kiley JE, Currie GD: Artificial kidney function: kinetics of hemodialysis. J Clin Invest 30:1062-1070, 1951.
2. Dombeck DH, Klein E, Wendt RP: Evaluation of two pool model for predicting serum creatinine levels during intra- and inter-dialytic periods. Trans Am Soc Artif Intern Organs 21:117-124, 1975.
3. Frost TH, Kerr DNS: Kinetics of hemodialysis: A theoretical study of the removal of solutes in chronic renal failure compared to normal health. Kidney Int 12:41-50, 1977.
4. Sargent JA, Gotch FA: Mathematic modeling of dialysis therapy. Kidney Int 18:S2-10, 1980.
5. Sargent JA, Lowrie EG: Which mathematical model to study uremic toxicity—National Cooperative Dialysis Study. Clin Nephrol 17:303-314, 1982.
6. Malchesky PS, Ellis P, Nosse C, Magnusson M, Lankhorst B, Nakamoto S: Direct quantification of dialysis. Dial Transplant 11:42-44, 1982.
7. Sprenger KBG, Kratz W, Lewis AE, Stadtmuller U: Kinetic modeling of hemodialysis, hemofiltration, and hemodiafiltration. Kidney Int 24:143-151, 1983.
8. Guthke R, Gunther K, Stein G, Knorre WA: Two-pool model analysis of data in hemodialysis by means of programmable pocket calculator TI 59. Comput Prog Biomed 19:189-195, 1985.
9. Heineken FG, Evans MC, Keen ML, Gotch FA: Intercompartmental fluid shifts in hemodialysis patients. Biotechnol Progr 3:69-73, 1987.
10. Gotch FA: Kinetic modeling in hemodialysis, in *Clinical Dialysis* (2ed), Nissensen AR, Gentile DE, Fine RN (eds), Norwalk CT, Appleton and Lange, pp 118-146, 1989.
11. Depner TA, Cheer AY: Modeling urea kinetics with two vs three BUN measurements: A critical comparison. Trans Am Soc Artif Intern Organs 35:499-502, 1989.
12. Sargent JA, Gotch FA: The study of uremia by manipulation of blood concentrations using combinations of hollow fiber devices. Trans Am Soc Artif Intern Organs 20:395-401, 1974.
13. Neame KD, Richards TG: *Elementary Kinetics of Membrane Carrier Transport*, Oxford, Blackwell Scientific Publishers, 1972.
14. Pitts RF: The nature of glomerular filtration, in *Physiology of the Kidney and Body Fluids* (3ed), Chicago, Year Book Medical Publishers, pp 49-70, 1974.
15. Hume R, Weyers E: Relationship between total body water and surface area in normal and obese subjects. J Clin Pathol 24:234-238, 1971.
16. Watson PE, Watson ID, Batt RD: Total body water volumes for adult males and females estimated from simple anthropometric measurements. Am J Clin Nutr 33:27-39, 1980.
17. Lowrie EG, Laird NM, Parker TF, Sargent JA: Effect of the hemodialysis prescription on patient morbidity: Report from the National Cooperative Dialysis Study. N Engl J Med 305:1176-1181, 1981.

Chapter 5

MULTICOMPARTMENT MODELS

Urea compartments in normal humans

Urea is a small, uncharged, yet highly soluble organic solute. Because of its high diffusibility, it finds its way easily across nearly all biological membranes by simple or facilitated diffusion. The result is rapid and complete equilibration among all aqueous body compartments (1). Transport of urea from its site of production in the liver is relatively unimpeded and, at equilibrium, tissue water concentrations are equal throughout the organism with one exception.

The kidney normally maintains a marked urea concentration gradient from cortex to medulla to help with water conservation. Restrictive transport of urea in the medullary loop of Henle together with active sodium transport in the thick ascending portion of the loop ultimately produces a high solute concentration at the papillary tip. The corticomedullary urea gradient results from countercurrent flow in the loop that multiplies the effect of active sodium transport in the ascending limb. In humans, the highest concentration of urea at the tip of the papilla reaches levels of 600 mosm (36 g/l) in the water-conserving antidiuretic state (2). This compares with 4-6 mosm (0.24 to 0.36 g/L) in the blood or a greater than 100:1 papilla-to-blood ratio. The urea distribution space in normal people must therefore include this small but highly concentrated pool in the renal medulla that has a slight inflating effect on the apparent volume of distribution.

This medullary pool of urea is diminished or absent in patients with end-stage renal failure for several reasons. Various insults to the kidney disrupt the intricate tubular architecture and its geometry of varying urea permeability; loss of nephrons diminishes the power of the kidney to drive the gradient multiplier; high single nephron filtration rates effectively wash out potential concentration gradients; and shrinkage of renal medullary mass decreases the size of the pool. Consequently, the urea space of distribution in patients with end-stage renal disease is very close to total body water. This simplifies the approach to urea kinetic modeling, permitting a single-compartment model to satisfy most needs. Other characteristics of urea that justify the single-pool model are excellent water solubility and high diffusivity that permit rapid distribution of urea throughout all body water compartments (1,3).

LIMITATIONS OF THE SINGLE-COMPARTMENT MODEL

Modern hemodialyzers clear urea 4-7 times faster than two normal adult kidneys. For this reason, and because body urea concentration prior to dialysis averages 4-8 times higher than in normals, hemodialysis causes a tremendous change in urea concentration (see figure 4.1). This change occurs within the short duration of a single dialysis (2-4 hours) and through a relatively narrow funnel, the intravascular space, the only space the hemodialyzer can access. This small space equilibrates rapidly with the larger interstitial space. These two extracellular pools equilibrate with the intracellular space, a larger compartment where most of the body pool of urea resides. Because of its rapid diffusibility, urea equilibrates promptly among these compartments. In chapter 4, when the single-compartment model was derived, the assumption was made that urea equilibration in all body compartments is instantaneous. This means that cell membrane and vascular permeability to urea is infinite, permitting intravascular, interstitial, and intracellular pools to be considered as a single pool, with no significant urea gradients forming between them. In reality, urea permeability is finite. During dialysis, urea movement across cell walls and tissue compartments cannot keep up with the flux through the dialyzer, and a small but discernible concentration gradient develops (4,5).

Most authorities envision the location of this concentration gradient not at the level of capillary endothelium but at the cell wall, between intracellular and extracellular pools (4,6,7,8). Capillary endothelium has large pores and possibly other transport pathways that should pose no barrier to urea movement, i.e., capillary permeability to urea might be considered infinite. But the rapid fall in urea concentration that occurs within minutes after starting high-flux dialysis suggests that a gradient develops here as well. Cell walls have a finite permeability to urea that is limiting for urea diffusion, especially during rapid hemodialysis-induced solute flux. The location of urea gradients within the body has received little attention in the past but, there is evidence that the concentration differences that form during hemodialysis are not uniform (9,10,11). Since we know little about relative permeabilities of various cell membranes to urea, we cannot predict the gradient patterns; some organs and tissues likely participate more than others. We know that the red cell is highly permeable to urea and probably does not participate in the gradient formation (12,13). One study suggests that urea concentration differences are greater in certain organs due to slower removal of urea from poorly perfused vascular beds (11).

The urea concentration gradient that develops between blood and cerebrospinal fluid (CSF) during dialysis is thought to cause symptoms that sometimes appear toward the end of dialysis or within a few hours after stopping dialysis (14,15). The urea gradient and symptoms associated with it have been termed *urea disequilibrium* (16). There is little question that urea movement from the CSF and other body compartments is delayed; the important question is how significant the delay is. If we assume that the body consists of two compartments, and if we assign a finite mass transfer coefficient for the barrier between them, we can predict the size of the urea concentration gradient that will develop during dialysis. We can also predict the changes in extracellular urea concentration that occur during and between dialyses. These predictions compare favorably with actual measurements.

Description of the Two-Compartment Model

The above considerations suggest that a two-compartment or multicompartment model might better describe hemodialysis-associated urea kinetics. To better understand the advantages and complexity of adding another compartment, we will examine the model diagram more closely. For purposes of illustration and consistency, we will refer to the two compartments as *extracellular* or *first* and *intracellular* or *second*, acknowledging that this represents a simplification of multiple gradients that may develop in a nonuniform pattern among several body compartments. The extracellular compartment equilibrates immediately with the dialyzer; the intracellular compartment is remote from the dialyzer. Figure 5.1 shows a simplified diagram of a two-compartment model that assumes free diffusion of urea within the extracellular and intracellular compartments but a finite diffusion rate between them. This model differs from the single-compartment model shown in figure 4.3 only by addition of *KC*, the mass transfer area coefficient for diffusion of urea between compartments. This coefficient is the same for diffusion of urea in either direction, into or out of the compartment, since asymmetric transport has not been shown for urea in humans. The equations that describe mass balance kinetics for urea in both compartments are straightforward. From equation 4.12, if we assume that extracellular compartment volume (V_e) is constant,

$$V_e \frac{dC_e}{dt} = G - KC(C_e - C_i) - C_e(K_d + K_r) \qquad 5.1$$

$$V_i \frac{dC_i}{dt} = KC(C_e - C_i) \qquad 5.2$$

V_e = extracellular compartment volume (ml)
V_i = intracellular compartment volume (ml)
C_e = extracellular urea nitrogen concentration (mg/ml)
C_i = intracellular urea nitrogen concentration (mg/ml)
G = urea nitrogen generation rate (mg/min)
KC = intercompartment mass transfer coefficient (ml/min)
K_d = dialyzer urea clearance (ml/min)
K_r = patient's residual kidney clearance (ml/min)
t = time (min)

There are four components to the movement of urea into or out of the extracellular compartment, dC_e/dt, described by equation 5.1 and illustrated in figure 5.1:

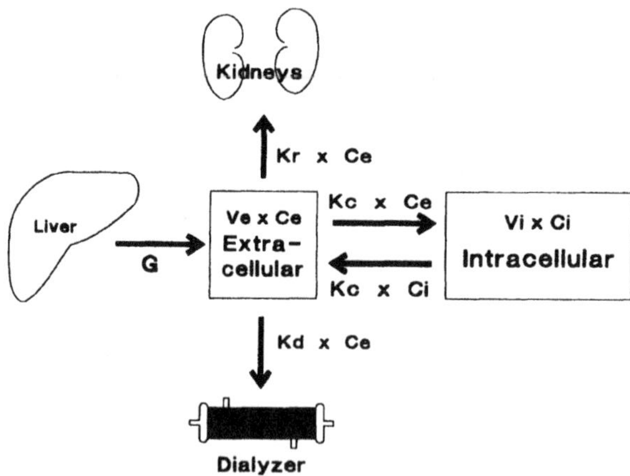

Figure 5.1. Urea mass balance is shown for the two-compartment, fixed-volume model. Hepatic urea generation is represented by G, native kidney removal by K_r x C_e, dialyzer removal by K_d x C_e. KC is the mass transfer area coefficient for urea diffusion across cell membranes.

1. Generation of urea
2. Dialyzer urea clearance
3. Patient residual urea clearance
4. Diffusion of urea between the two compartments

In contrast, only one component, number 4 in the above list, is required to describe the diffusion of urea into or out of the intracellular compartment. Movement of urea to and from the intracellular compartment is expressed as dC_i/dt in equation 5.2.

Equation 5.1 is similar to equation 4.12 with one additional parameter, $-KC(C_e - C_i)$. This expression describes the transmembrane movement of urea as the product of the intercompartment mass transfer factor (KC), and the diffusion gradient for urea ($C_e - C_i$). The direction of movement is negative (away from the extracellular space) when the gradient is positive (extracellular concentration is greater than intracellular). The two-compartment model depicted in figure 5.1 assumes that total urea volume and the ratio of intracellular to extracellular volume are fixed and do not vary during dialysis (fixed-volume model).

If urea is generated only in the extracellular space as explained below, the movement of urea into or out of the intracellular space can be simply expressed as diffusion across cell membranes. Hence, equation 5.2, which describes changes in intracellular urea, contains only the cell diffusional component, $KC(C_e - C_i)$ on the right side. This equation shows that the intracellular concentration will increase when extracellular concentration is greater than intracellular concentration between dialyses; conversely, during dialysis when extracellular concentration plummets, the intracellular concentration also falls. Figure 5.2 shows the urea concentrations in both compartments during and between dialyses as predicted by the model. The lines are obtained from integration of equations 5.1 and 5.2. During dialysis, extracellular concentration (solid line) is significantly lower than intracellular concentration (dotted line). For about 20-30 minutes after dialyzer blood flow ceases, the intracellular urea concentration continues to fall while the extracellular urea concentration is steeply rising. At about 20-30 minutes, the two concentrations cross and then both rise more slowly.

Site of urea generation

Between dialyses, figure 5.2 shows that intracellular concentration falls slightly below extracellular concentration because urea first enters the extracellular space and then diffuses intracellularly. The generation of urea is best considered an extracellular event, even though it occurs intracellularly in the liver. After it is synthesized, urea diffuses out through the hepatic vein to the remainder of body water space. Because the liver makes up a small fraction (about 5%) of total body mass, nearly all the generated urea is delivered to nonhepatic peripheral tissues from the blood, an extracellular compartment. Therefore, the equations that describe urea generation account for urea appearance only in the extracellular space; from there it diffuses into the intracellular space.

POSTDIALYSIS REBOUND IN UREA CONCENTRATION

The best objective evidence for urea disequilibrium is found in measurements of postdialysis rebound. Rebound refers to the sharp increase in extracellular urea nitrogen concentration (BUN) observed immediately after stopping dialysis, as described above. This rapid rise in BUN can be readily demonstrated with frequent blood samplings postdialysis (5). Figure 5.3 shows an example of postdialysis rebound following four hours of standard dialysis in an adult. The data points shown are BUN measurements at 5-10 minute intervals immediately after dialysis. The solid line is the curve of best fit to the data using the two-compartment model. The rapid phase is complete in 20-30 minutes, after which the BUN rises slowly in accordance with the patient's urea generation rate, fluid gain, and intrinsic renal clearance (17). The magnitude of rebound is rarely more than 5 mg/dl even in patients treated with high-flux dialyzers with high urea permeability operating at high blood and dialysate flow rates (5). The two-compartment model predicts this small brief change in BUN, a consequence of passive urea diffusion from the intracellular pool into the smaller extracellular space. A 5 mg/dl change in concentration reflects the movement of approximately 0.5-1.0 grams of urea nitrogen in an adult whose total body urea content averages 10-20 grams.

Rebound is complete after 30-40 minutes, but the gradient that caused the rebound is established early in dialysis. Figure 5.4 shows that the urea gradient is significant immediately after the start of dialysis. In fact, the largest concentration difference between compartments is apparent early in the dialysis treatment, because urea nitrogen concentrations are highest then, and flux through the dialyzer is at a maximum when blood urea nitrogen concentrations are highest. The patient is subject to disequilibrium during the entire duration of dialysis, but the concentra-

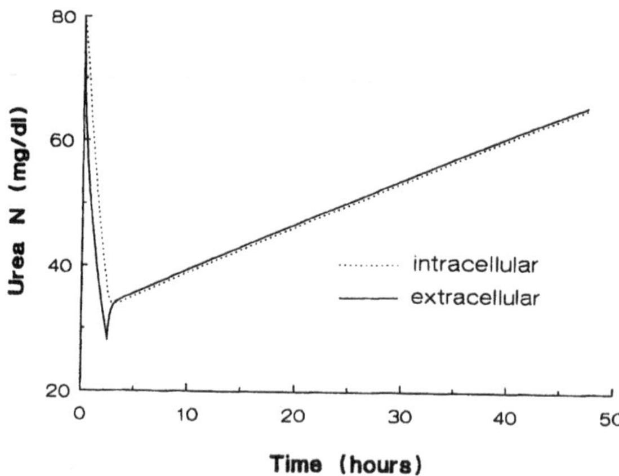

Figure 5.2. A profile of urea nitrogen concentrations generated by the two-compartment model is shown for a single dialysis and interdialysis interval. The dotted line is intracellular urea concentration (mg/dl); the solid line is extracellular urea nitrogen concentration (mg/dl).

tion gradient and disequilibrium diminish with time on dialysis when urea clearance is constant. The clinical consequences of urea disequilibrium are uncertain, but it has been suggested as a cause of dialysis-induced tissue swelling. Swelling can occur as water moves down the osmotic gradient caused by dialysis-induced urea flux. This effect is illustrated in figure 5.5. Water movement into the cell dissipates the osmotic gradient. The patient is at risk for swelling during the entire dialysis treatment or as long as the urea gradient exists. Intracellular tissue swelling is thought to be a problem only for the central nervous system where the restraining confinement of the bony cranium may cause pressures to rise and symptoms of brain swelling to appear. Much doubt, however, exists about the significance of dialysis-induced tissue swelling. The magnitude of tissue swelling is further discussed below.

TWO-COMPARTMENT MODELING TECHNIQUES

Addition of a second compartment adds at least two more variables, *KC* and the ratio V_i/V_e, that the two-compartment model must resolve in addition to *V* and *G*, which the one-compartment model resolves (figure 5.1). More variables appear if the compartment volumes are allowed to expand and to contract. To resolve just two more variables, additional data points (BUN measurements) must be obtained and at least four independent sets of equation must be solved to obtain discrete values for four variables in each patient. This becomes an impractical logistic task as well as a difficult mathematical feat to be applied routinely. Several BUN measurements would be required during and immediately following dialysis, with precise timing of each; the many-hundredfold iterations would require too much time to complete

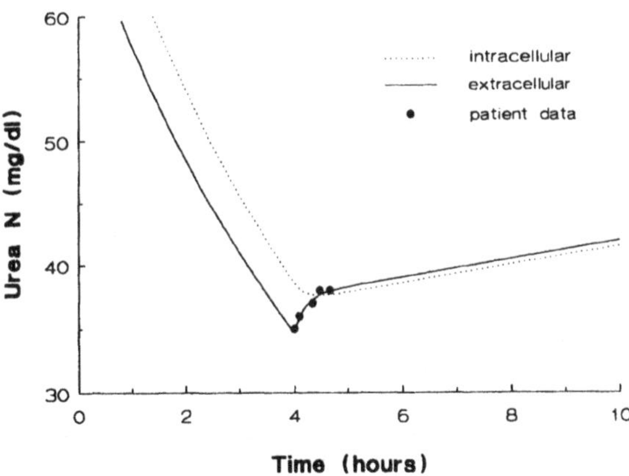

Figure 5.3. BUN rebound postdialysis is shown for a single patient treated with conventional dialysis. The solid line is the line of best fit using the two-compartment model. The dotted line is the corresponding intracellular urea nitrogen concentration. K_d = 148 ml/min, K_r = 0.

without a super computer. To render this task more practical, we have simplified the technique of two-compartment modeling by assuming a fixed value for *KC* and a fixed value for V_i/V_e. Sensitivity analysis shows that V_i/V_e can vary over a wide range without significantly affecting *V* and *G*. The rationale for selection of *KC* is explained below. Accepting an average value for *KC* represents the weakest assumption in the simplified technique. Studies are needed to examine *KC* in a large number of patients to establish its variability and the factors that determine *KC*.

Solutions to Equations for the Two-Compartment Model

Equations 5.1 and 5.2 can be solved analytically for C_i and C_e by several methods. These include solutions for systems of ordinary differential equations, Laplace transformations, or matrix techniques. Following is an analytical solution to the two-compartment model offered by Sargent and Gotch (18). This series of equations was originally developed for middle molecules but can also be applied to urea. The equations assume that dV/dt is zero (fixed-volume model):

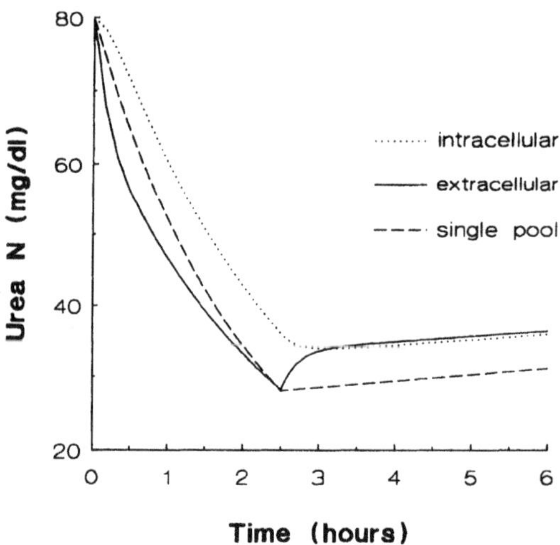

Figure 5.4. Closeup view of the intradialysis interval shows urea nitrogen concentrations during high-flux dialysis. Concentrations are predicted by the two-compartment model (solid and dotted lines) and by the one-compartment model (dashed line). The maximum gradient between intracellular and extracellular urea nitrogen concentration is 14 mg/dl and occurs at 30 minutes. The mean gradient is 12 mg/dl.

The higher serum concentrations predicted by the single-compartment model tend to exaggerate *V*. This is offset by the rebound in urea concentration that occurs after dialysis, a phenomenon that causes the single-compartment model to underestimate *V*. The result is that *V* is predicted equally well by either model.

$$C_e = \left[\frac{C_{i0} \cdot KC}{V_e} + \frac{C_{e0} \cdot KC}{V_i} + \frac{G}{V_e} - \frac{G \cdot KC}{V_e \cdot V_i \cdot B} - C_{e0} \cdot B\right]\frac{e^{-B \cdot t}}{A - B} - \left[\frac{C_{i0} \cdot KC}{V_e} + \frac{C_{e0} \cdot KC}{V_i} + \frac{G}{V_e} - \frac{G \cdot KC}{V_e \cdot V_i \cdot A} - C_{e0} \cdot A\right]\frac{e^{-A \cdot t}}{A - B} + \frac{G \cdot KC}{V_e \cdot V_i \cdot A \cdot B} \quad 5.3$$

$$C_i = \left[\frac{C_{e0} \cdot KC}{V_i} + \frac{C_{i0}(KC + K_d + K_r)}{V_e} - \frac{G \cdot KC}{V_e \cdot V_i \cdot B} - C_{i0} \cdot B\right]\frac{e^{-B \cdot t}}{A - B} - \left[\frac{C_{e0} \cdot KC}{V_i} + \frac{C_{i0}(KC + K_d + K_r)}{V_e} - \frac{G \cdot KC}{V_e \cdot V_i \cdot A} - C_{i0} \cdot A\right]\frac{e^{-A \cdot t}}{A - B} + \frac{G \cdot KC}{V_e \cdot V_i \cdot A \cdot B} \quad 5.4$$

$$S1 = \frac{1}{2}\left[\frac{KC}{V_i} + \frac{KC + K_d + K_r}{V_e}\right] \quad 5.5$$

$$S2 = \frac{1}{2}\sqrt{\left[\frac{KC}{V_i} + \frac{KC + K_d + K_r}{V_e}\right]^2 - 4\frac{KC\,(K_r + K_d)}{V_e \cdot V_i}} \quad 5.6$$

$$A = S1 - S2 \quad 5.7$$

$$B = S1 + S2 \quad 5.8$$

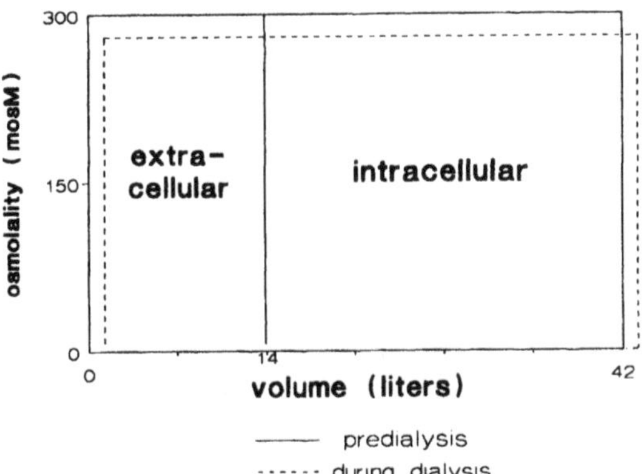

Figure 5.5. Water shifts during dialysis. This area plot shows the theoretical shift in compartment volumes that occurs during hemodialysis when a urea concentration gradient forms. Dashed lines show the new conditions that exist after the gradient-driven water shift. Osmolality falls first in the extracellular space because of dialyzer urea removal. Water then moves from the extracellular compartment to the intracellular compartment. The ECF volume contracts in the absence of dialyzer ultrafiltration.

C_{i0} = initial intracellular BUN
C_{e0} = initial extracellular BUN

No explicit solutions are given for V_i, V_e, G, or K_d, but these parameters are easily resolved by numerical iteration. Figure 5.6 shows the algorithm for iterative solution of V and G using the above two-compartment model and two data points. V_i and V_e are set to fixed percentages of V, usually 2/3 and 1/3, respectively. If we assume initial values for C_{i0} and C_{e0} equal to the predialysis BUN, we can calculate the postdialysis BUN from equations 5.3 and 5.4. The calculated postdialysis BUN (C_{2b}) is compared to the measured postdialysis BUN (C_{2a}). Adjustments are then made in either V or K_d and the process repeats, as shown in the upper loop of figure 5.6, until the calculated postdialysis BUN agrees with the measured value within predefined limits (usually 1%).

A similar iterative approach is applied to the estimate of G, as illustrated in the bottom loop of figure 5.6. If only two data points are available, calculation of predialysis and postdialysis BUN can be repeated sequentially using equations 5.3 and 5.4 for a one-week cycle. The calculated value of predialysis BUN (C1b in figure 5.6) is then compared with the measured predialysis BUN (C1a). If the calculated BUN does not match measured predialysis BUN within 1%, G is adjusted and the entire cycle repeats. This technique assumes that the patient is in a weekly steady

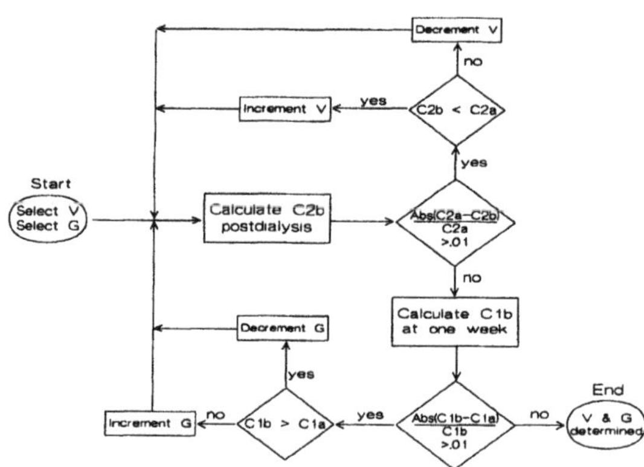

Figure 5.6. This double-loop algorithm determines urea volume (V) and generation rate (G) using the two-compartment model and two data points. V is adjusted, as shown in the upper loop, until the calculated postdialysis BUN (C_{2b} from equation 5.3) matches the measured postdialysis value (C_{2a}) within 1%. G is then adjusted, as shown in the lower loop, until the calculated value for predialysis BUN (C_{1b} from equation 5.3) at one week matches the measured value (C_{1a}) within 1%. The process then repeats, starting with the upper loop, until both values for V and G satisfy the above conditions.

state for urea balance. If a second predialysis BUN is available (three-BUN method), only one calculation of predialysis BUN is done before adjusting G. Refer to the discussion of two-BUN versus three-BUN analysis techniques in chapter 4.

After G is obtained by either of the above techniques, its value must be substituted into equations 5.3 and 5.4, and V calculated again by iteration (top loop in figure 5.6). This process continues until values for both V (or K_d) and G are found that, when entered into equations 5.3 and 5.4, produce values for calculated predialysis and postdialysis BUN that match the measured values. A match is determined according to predefined specifications, usually 1% of the measured values. Although the above processes appear complex and time-consuming, most desktop computers can complete multiple iterations and deliver accurate values for V and G within one second.

The computer algorithm for resolving C_e and C_i is shown in appendix C. The language is BASIC but the code is easily adaptable to Pascal, C, or other programming languages. Equations 5.3 to 5.8 are embedded in the code.

Graphic description of the two-compartment model

Figure 5.7 shows the profile of urea nitrogen concentrations generated by the above two-compartment model for a standard dialysis. For comparison, the dashed line shows the profile generated by the single-compartment model under the same conditions. Figure 5.8 shows a closer view of the intradialysis interval. The major impact of delayed intercompartment diffusion begins shortly after the onset of

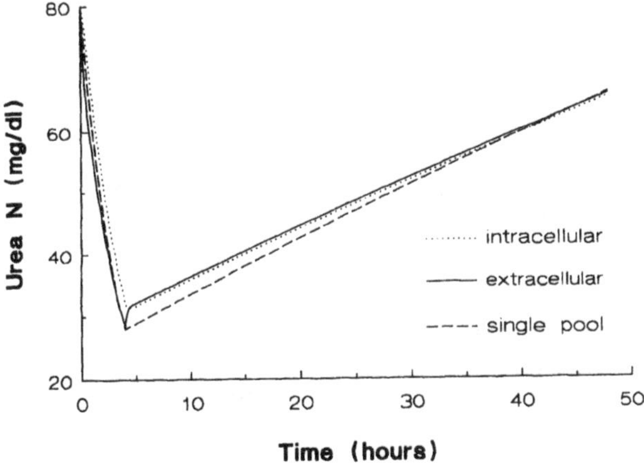

Figure 5.7. Two-compartment versus one-compartment model of a standard four-hour dialysis. Intracellular and extracellular urea nitrogen concentrations predicted by the two-compartment model are shown as dotted and solid lines, respectively. The dashed line is the profile of urea nitrogen concentrations predicted by the single-compartment model for the same predialysis and postdialysis concentrations. $K_d = 200$, $K_r = 2.8$ ml/min, patient weight = 70 kg.

dialysis and extends to the early interdialysis phase. Extracellular urea concentration, the only measurable component, is lower throughout dialysis because of restricted diffusion from the intracellular compartment. The difference between extracellular concentrations predicted by the two models is most marked approximately 30 minutes after the start of dialysis, becoming less pronounced and finally disappearing shortly after the end of dialysis. Throughout dialysis, the unmeasurable intracellular concentration (dotted line in figures 5.7 and 5.8) is significantly higher than extracellular concentration predicted by either model. The gradient from intracellular to extracellular concentration is significant throughout dialysis but is somewhat more pronounced at the start, when concentrations in both compartments are highest. This gradient then disappears, as discussed above, approximately 30-40 minutes after stopping the blood pump. The rebound in extracellular urea nitrogen concentration results from continued diffusion of urea from the larger intracellular compartment after dialysis ceases. During this short period, intracellular concentrations continue to fall. Thereafter, during the entire interdialysis cycle, intracellular concentrations are slightly lower than extracellular concentrations because of delayed diffusion of extracellularly generated urea into the intracellular space.

HIGH-FLUX DIALYSIS AND TWO COMPARTMENTS

Figures 5.9 and 5.5 show the contour of BUN concentrations generated by the one-compartment and two-compartment models during high-flux dialysis. Here K_d, the dialyzer urea clearance, is set at 315 ml/min, a clearance that is achievable by modern high-porosity dialyzers with a large surface area. High-flux dialysis

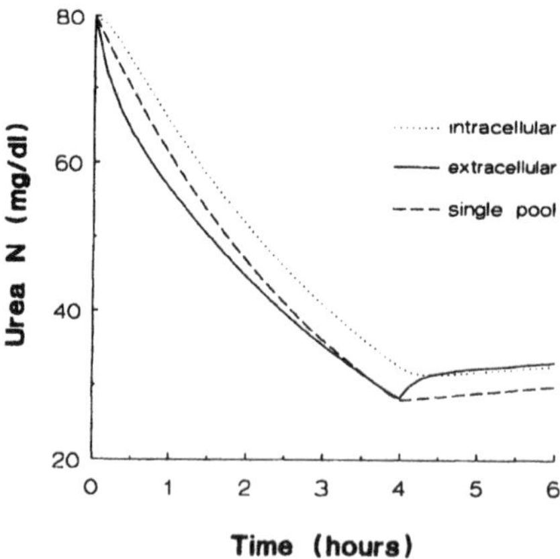

Figure 5.8. Closer view of the intradialysis interval during a standard four-hour dialysis. Legends are the same as figure 5.7. Compare this graph with the shortened high-flux profile depicted in figure 5.4.

magnifies the differences between the one-compartment and two-compartment models. The most pronounced differences again appear during the early interdialysis interval. Both the shorter dialysis duration (2.5 versus 4.0 hours) and the higher solute removal rate enhance the rebound in extracellular urea nitrogen concentration following high-flux dialysis. The two-compartment effect can have a significant impact on predicted values for urea volume and generation rate, as discussed below. During dialysis there is only a small difference in extracellular urea nitrogen concentration predicted by the one-compartment versus the two-compartment model.

Solutes with Low Mass Transfer Coefficients

Figure 5.10 shows two-compartment modeling of another solute with tissue permeability (intercompartment mass transfer coefficient) one tenth that of urea. The theoretical solute depicted here has a dialyzer clearance of 180 ml/min, similar to standard dialysis of urea, but it diffuses much more slowly from the intracellular compartment. The profile of concentrations shown in figure 5.10 would be expected for a solute like creatinine or phosphate when a very permeable dialyzer is used. The rebound following dialysis is more than half the decline in concentration observed during dialysis. The marked delay in diffusion from the intracellular compartment decreases the effectiveness of dialyzer solute removal. If single-compartment kinetics are assumed for this true two-compartment model, the apparent volume of urea distribution is underestimated. The two models predict markedly different extracellular concentrations between dialyses, as shown in

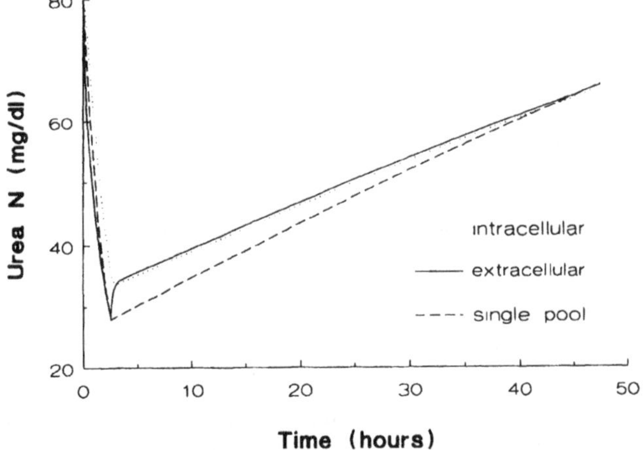

Figure 5.9. Similar to figure 5.7, the profiles of intracellular and extracellular urea nitrogen are shown during and between high-flux dialyses as predicted by the two-compartment model. The profile of urea nitrogen concentrations predicted by the single-pool model deviates further, especially following dialysis. $K_d = 350$ ml/min, $K_r = 0$, $t_d = 2.5$ hours, patient weight = 70 kg.

figure 5.10. In contrast, both models predict similar extracellular concentrations during dialysis. Clearly, the single-compartment predictions, shown as the dashed line, for the interdialysis period are inadequate. A multicompartment model is mandatory for predicting concentrations and kinetic parameters of solutes with low intercompartment diffusion rates. Because solutes like this are much less concentrated in body fluid compartments than urea, little or no osmotic effect causing intercompartmental shifting of water would be expected from this disequilibrium state (see discussion of volume shifts below).

Measuring the Intercompartment Mass Transfer Area Coefficient

To use the two-compartment model described above, an estimate of the average intercompartment permeability to urea must be available. This can be expressed as an average urea/membrane permeability coefficient similar to the dialyzer membrane permeability coefficient (equation 4.5). If the membrane surface area is included, permeability can be expressed as a *mass transfer area coefficient* for urea (*KC*) similar to the coefficient *KA*, developed for dialyzers (equation 7.2). The two-compartment model described above uses the latter coefficient. Hemodialyzer mass transfer coefficients can be directly measured by sampling arterial and venous blood during dialysis (see chapter 3). Direct measurement of cellular mass transfer coefficients would require sampling the intracellular compartment. Diffusion rates can be measured in vitro, for instance with red blood cell preparations, but extrapolating to in vivo conditions then involves several assumptions about surface

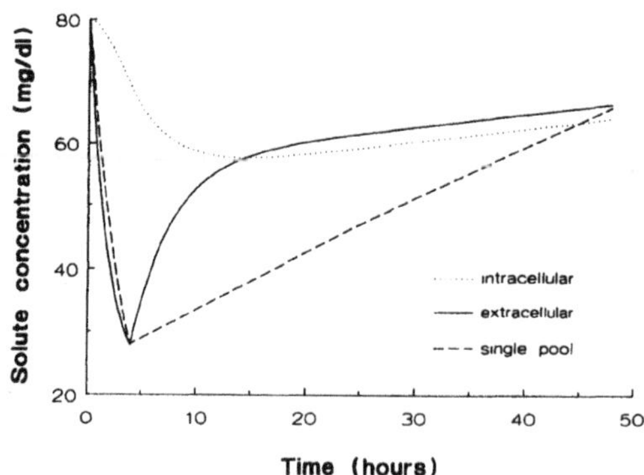

Figure 5.10. Slowly-diffusible solute: two-pool versus one-pool model. The two-compartment model predicts that intracellular and extracellular concentrations will differ markedly during and after dialysis for a solute with cellular mass transfer area coefficient (*KC*) of 80 ml/min. This compares with 800 ml/min, the *KC* for urea. The dashed line shows solute concentrations predicted by the single-compartment model for comparison. K_d = 200 ml/min, K_r = 2.8 ml/min, t_d = 4 hours, patient weight = 70 kg. See text for further discussion of this graph.

area and perfusion rates of various tissues. A better estimate can be obtained during actual dialysis of the patient. Since intracellular concentrations cannot be easily measured during dialysis, an indirect technique using the measured rebound in extracellular urea nitrogen concentration postdialysis has been used to estimate *KC* in vivo (5,10,19). Figure 5.11 shows an example of this technique using a magnified view of the immediate postdialysis interval following standard hemodialysis. Superimposed on three curves of urea nitrogen concentrations predicted by the two-compartment model are mean values for patient BUN values collected at 5-10 minute intervals postdialysis. The data used to generate the curves are an average composite of the patient's dialysis parameters. The theoretical curve fits the patient data best when *KC* is 800 ml/min. This agrees with values obtained by other investigators using this and other techniques (5).

The value for *KC* obtained from curve-fitting techniques illustrated in figure 5.11 is an effective *KC*, an average value for all tissues in the body. The true mass transfer area coefficient for urea movement between compartments during dialysis is likely to be lower than the value obtained using these techniques. But because *KC* is measured during the actual dialysis procedure, it is more useful to the clinician than true *KC* or *KC* extrapolated from in vitro measurements. Other factors may influence or modulate the effective urea transfer coefficient such as body size, tissue perfusion, and compartment volume changes during dialysis. The only factor that two-compartment programs usually acknowledge is body size. Although there are no data to confirm this, *KC* is assumed to vary directly with body surface area. Surface area is calculated from body height and weight using standard formulas. A

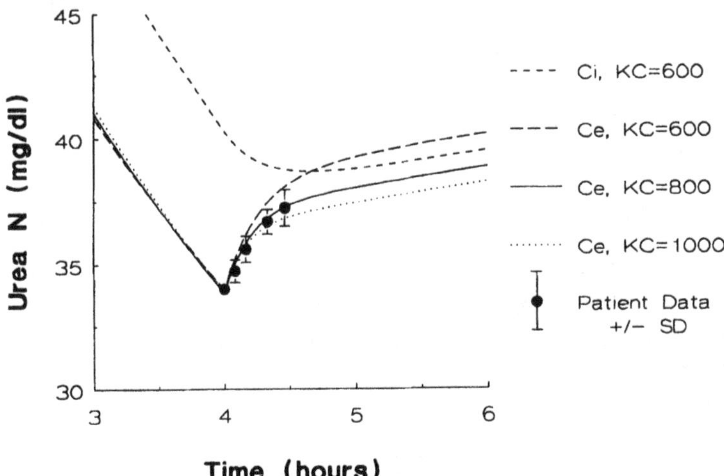

Figure 5.11. Calculation of *KC* from postdialysis rebound. Mean data collected from eight patients following standard four-hour dialyses shows that the rebound in BUN closely approximates that predicted by the two-compartment model. The solid line is the line of best fit. K_d = 175 ml/min, K_r = 0. See text for discussion.

KC value of 800 ml/min is commonly assumed for a standard 70 kg person with body surface area of 1.76 square meters.

A COMPARISON OF ONE-COMPARTMENT WITH TWO-COMPARTMENT MODELS

The apparent urea volume can be measured by analyzing dialysate urea concentrations using the direct quantification method (20,21). Equation 5.9 gives total mass balance during dialysis, assuming no generation and no volume change:

$$V = \frac{\text{amount removed}}{C_1 - C_2} \qquad 5.9$$

Analysis of dialysate gives the amount of urea removed during dialysis. This quantity divided by the change in blood urea concentration during dialysis ($C_1 - C_2$) gives the urea volume (V). The value for V calculated in this way is often significantly lower than total body water, i.e., less urea is removed than expected, given the change in BUN (20,21,22). Since we know that urea freely distributes in all body water compartments, and since radioisotope dilution techniques show that the space of urea distribution is total body water (23,24), another cause of the diminished apparent V must be invoked. One possible cause of this discrepancy is a delay in urea diffusion from a second body compartment.

Extracellular urea nitrogen concentrations calculated by the two-compartment model during dialysis are similar to those calculated by the one-compartment model using a reduced urea volume (V). This effect is the result of delayed urea diffusion from the intracellular space to the blood space where the dialyzer does its job. Clearance of urea from the blood space continues at a high rate, but the amount of urea available to the dialyzer for removal is less. A significant amount remains in the second (intracellular) space and is not available to the dialyzer. The resistance to urea diffusion is reflected in the concentration gradient that appears between the two compartments during dialysis, as illustrated in figures 5.3 and 5.5. Because of the resistance to diffusion between compartments, it appears to the dialyzer that there is less urea available for removal than would be available if no barrier to diffusion existed between compartments. The BUN falls more rapidly and postdialysis BUN is lower, just as it would be in a smaller patient with smaller urea volume dialyzed with the same dialyzer under the same conditions. This appears to the single-compartment model and to the direct quantification method as a diminution in the space of urea distribution (V).

THE DIRECT QUANTIFICATION METHOD

The error in V, incurred by assuming single-compartment kinetics when double-compartment kinetics are more appropriate, also occurs with the direct quantification method. Use of total dialysate collection to measure urea removal is subject to error because of the nonequilibrated BUN postdialysis. This inflates the change in

BUN induced by dialysis ($C_1 - C_2$ in equation 5.9) and causes a false lowering of the calculated urea volume (25). Much of the literature that critically evaluates the single-compartment model makes the inappropriate assumption that the direct quantification method is a gold standard (26,27,28). Both the direct quantification method and the single-compartment method can underestimate *V*. The error is greater for the direct quantification method because there is no offsetting error (see discussion below). This means that when the two methods are compared, the single-compartment method gives a higher and more accurate estimate of *V*. Only when the true equilibrium value of postdialysis BUN is measured, or a two-compartment model is used, can the direct quantification method provide an accurate estimate of *V*.

The direct quantification method can introduce other errors. This method will overestimate *V* if urea generated during dialysis and urea removed by convection (ultrafiltration) are not subtracted from the numerator in equation 5.9. Generated urea especially during short dialyses has a small impact on *V* but failure to include the ultrafiltration component will cause a significant overestimation of *V*. Omission of urea removed by residual kidney function from the numerator in equation 5.9 usually has a small effect during dialysis because of the high ratio of dialyzer clearance to residual kidney clearance.

Impact of Pool Number on Calculated Variables

More difficult to perceive is the effect of applying the two-compartment model to a set of measurements taken from a patient. Modeling uses a form of inverse reasoning. Instead of predicting concentrations during or between dialyses, we are given the concentrations and are asked to calculate a volume and generation rate that fit these values. To model urea kinetics, we measure two or three urea nitrogen concentrations and then try to fit a urea volume (*V*) and generation rate (*G*) that will predict the second (and third) from the first concentration using a mathematical model. Figure 5.9 illustrates these conditions and the effect of both models. The first data point is 80 mg/dl and the second is 28 mg/dl. Both models must provide extracellular urea nitrogen concentrations that pass through these points.

Figure 5.9 illustrates different BUN values predicted by the two models at various times but does not give insight into the differences in *V* and *G* predicted by the two models. Table 5.1 summarizes the errors introduced by one-compartment modeling. Table 5.2 shows the two model predictions for *V* and *G* in a single patient whose BUN falls from 80 to 28 during dialysis under a variety of conditions. It shows that *V* is accurately determined by either the one-compartment model or the two-compartment model in all cases. But the generation rate (*G*) is overestimated by the one-compartment model unless significant residual function remains or when marked weight gain occurs between dialyses. These two conditions tend to diminish the two-pool effect on *G*. Greater errors in *G* occur during high-flux dialysis. Because most patients have modest weight gains between dialyses, the single-compartment variable-volume model usually gives an accurate prediction of both V and G. It is important to note that these data are valid only if the cellular mass-transport coefficient (KC) is relatively high (800 ml/min in these simulations). There is evidence to suggest that KC may be much lower in some patients who have poor peripheral circulation (29). This will cause a larger gradient between the two slowly equilibrating compartments, further decreasing the efficiency of dialysis and magnifying the discrepancies between the two model predictions for V and G. When KC is low, the one-compartment model gives a value for V that is too low and a value for G that is too high.

Table 5.1 Errors caused by one-compartment modeling

1. Slight underestimation or overestimation of V
2. Consistent moderate overestimation of G

Table 5.2 Impact of one- versus two-pool modeling on patient variables

Conditions	Input				Pools	Output	
	K_d ml/min	K_r ml/min	dW kg	T_d hrs		V liters	G mg/min
Standard	180	0	0	4.0	1	37.8	7.13
					2	37.8	6.77
High flux	315	0	0	2.2	1	37.8	6.85
					2	38.7	6.31
High residual clearance	180	6	0	4.0	1	37.3	10.9
					2	37.3	10.6
Large weight change	188	0	7	5.1	1	37.2	9.22
					2	37.8	9.30

K_d = dialyzer clearance
K_r = residual clearance
dW = weight loss during dialysis
T_d = duration of dialysis (adjusted to give approximately equal V)
V = volume of urea distribution (urea volume)
G = urea generation rate

Schedule = Mon Wed Fri
Study date = Wed
BUN predialysis = 80 mg/dl
BUN postdialysis = 28 mg/dl
Patient weight = 70 kg

The solid line in figure 5.9 shows the pattern of decline in BUN during dialysis as is predicted by the two-compartment model. Since a measurable rebound in urea concentration occurs postdialysis, we can presume that this model is more accurate in describing the real curve of BUN versus time. The single-pool model predicts the fall in BUN shown by the dashed line in figure 5.9. Despite the discrepancies in predicted urea nitrogen concentrations, the one-compartment model closely approximates V for urea predicted by the two-compartment model. The reason for this accuracy is the fortuitous occurrence of two offsetting errors in the one-compartment model assumptions:

1. BUN is falsely elevated during dialysis
2. Postdialysis BUN, assuming complete equilibration, is falsely low

Error 1 leads to a falsely elevated prediction of urea removal rate throughout dialysis but especially early in dialysis when the BUN is high and urea removal rates are also high. If we ignore urea generation and residual clearance during dialysis, the equation for V simplifies to

$$V = \frac{K_d \int_{C_1}^{C_2} dC}{C_1 - C_2} \qquad 5.9a$$

Constant dialyzer clearance at inflated urea concentrations increases the numerator in equation 5.9a, causing a falsely high prediction of V. The effect of this error is similar to that of a poorly functioning dialyzer (one that clears less than expected). Both cause false elevations in apparent V because to account for the larger-than-real amount of urea removed, a larger urea volume must be invoked to arrive at the same urea concentration postdialysis.

Error 2, similar to error 1, results from disequilibrium between intracellular and extracellular compartments. The true equilibrium BUN postdialysis is better reflected in the post-rebound value (after dialysis ceases and sufficient time passes for mixing). This value can be approximated more closely by extrapolating the flat linear portion of the interdialysis curve (figure 5.9) back to the ending time of dialysis. Error 2 inflates the denominator in equation 5.9a, falsely lowering predicted V. The one-compartment model accepts the postdialysis BUN as the true equilibrium value. This assumes that BUN has fallen more than it actually has at equilibrium. To account for this fall, the model must assume removal from a smaller volume. The two-compartment model avoids this pitfall by acknowledging that urea is not in equilibrium at the end of dialysis.

Error 2 predominates but usually only slightly. The error becomes more significant when KC is low, when urea flux is high and when dialysis is shortened. Thus the one-compartment model slightly underestimates V during high flux dialyses.

There is no offsetting compensation for the error in G and protein catabolic rate (PCR); both tend to be overestimated. G is determined primarily by the change in BUN between dialyses. If there is no residual clearance and no weight gain, the expression for G from total mass balance between dialyses is

$$G = V \cdot (C_3 - C_2)/t \tag{5.10}$$

t is time between dialyses
C_2 is postdialysis BUN
C_3 is predialysis BUN

The low nonequilibrated BUN postdialysis produces a steeper slope in the line depicting BUN versus time between dialyses (figure 5.9) for the one-compartment model. Equation 5.10 shows that this slope, $(C_3 - C_2)/t$, is directly correlated with urea generation rate (G). The result is a significant overestimation of G (5% to 15%) in nearly all patients by the single-compartment model. Normalized PCR (PCRn) is factored for V, but there is little or no error in V, so an inflation of PCRn also occurs commensurate with the elevation in G.

Several studies of urea generation during dialysis using the one-compartment model have concluded that dialysis is a catabolic process because urea generation rates are higher on dialysis days than on nondialysis days (30,31,32,33,34). This conclusion results in part from failure to consider urea disequilibrium during dialysis and the rebound in urea nitrogen concentration that subsequently occurs (5). A multicompartment approach is necessary to model this aspect of hemodialysis accurately.

Determinants of postdialysis urea rebound

When used to analyze patient data, the single-compartment model can cause an error in V because it does not account for urea disequilibrium and postdialysis rebound. If we could estimate the magnitude of rebound, we could adjust the single-compartment profile to force a better fit with the actual data points. This would allow a more accurate estimate of patient parameters with the simpler model.

Examining the determinants of postdialysis urea rebound helps to show why this is not possible. The gradient in extracellular/intracellular urea concentration at the end of dialysis determines the absolute magnitude of rebound. The magnitude of this gradient depends solely on the absolute rate of decline in extracellular urea concentration at the time. However, because the absolute rate of urea removal at a constant clearance depends on blood urea concentration, the rate of decline in urea concentration depends heavily on the end-dialysis BUN. The BUN at the end of dialysis depends on the starting concentration, the dialyzer clearance, the urea volume, and the duration of dialysis. Table 5.3 lists the direction of influence by each of these factors. If the other factors are constant, the BUN should rebound higher in patients with higher postdialysis BUN values, especially if treatment time is reduced. For a high-flux dialysis in a large patient, the postdialysis BUN will be relatively high and the single-compartment model will significantly underestimate V. Conversely, if BUN falls below 30 mg/dl postdialysis, there is less rebound and the single-compartment calculations of V and G correlate more closely with results from two-compartment modeling (table 5.3). Figure 5.12 illustrates this point in a high-flux dialysis prolonged to eight hours. If, in this hypothetical patient, dialysis stops prematurely, the gradient in intracellular/extracellular urea concentration will determine the magnitude of rebound. Diminishing height of the vertical arrows with time reflects a reduction in the rebound potential as the BUN falls. Carried to an extreme, equilibrium will eventually occur when the BUN no longer falls and intracellular levels catch up with extracellular levels. At this point, no intercompartment gradient will exist, no rebound will occur, and the single-compartment model will accurately predict the interdialysis BUN profile. However, V will be overestimated due to error 1 discussed above, which is not compensated by error 2. G will be overestimated because of the error in V (equation 5.10). Since all four factors in table 5.3 play major roles in determining the magnitude of rebound, it is difficult to derive a simple correction factor for postdialysis *equilibrium BUN* that would allow more accurate application of the single-pool model. In short, there is no substitute for the two-compartment model.

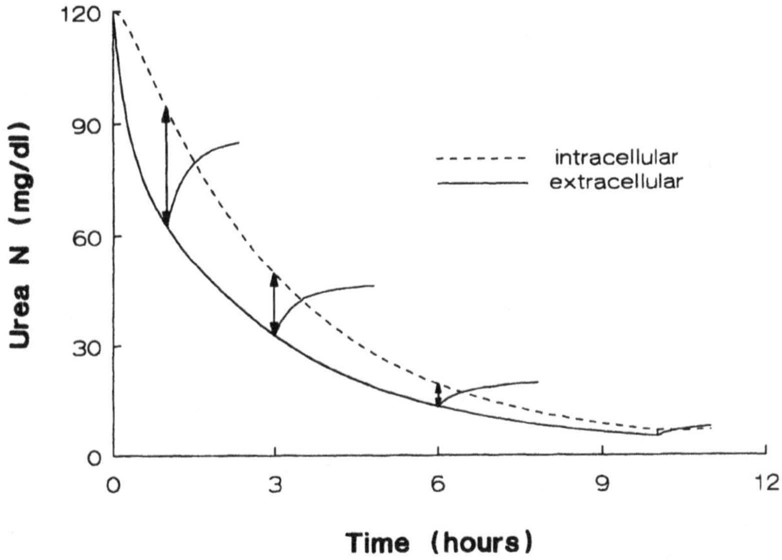

Figure 5.12. Postdialysis BUN determines rebound. Intracellular (dashed line) and extracellular (solid line) urea nitrogen concentrations diverge shortly after starting dialysis. As dialysis progresses they begin to converge. The rebound in BUN after stopping dialysis at one hour, three hours, six hours, and ten hours is related to the intracellular/extracellular urea gradient (vertical arrows).

Table 5.3 Factors that determine the magnitude of rebound in urea concentration after dialysis

Factor	Direction of influence
Low postdialysis BUN	Decrease
High dialyzer urea clearance	Increase for short dialysis
Duration of dialysis	
Short	Increase
Long	Decrease
Large volume of urea distribution	Decrease

TWO-COMPARTMENT MODEL WITH VARIABLE ECF VOLUME

Nearly all patients gain weight between dialyses, so they must lose weight during dialysis. The weight gain represents fluid intake not compensated by fluid losses and is usually matched, perhaps driven, by sodium intake (35). If the fluid gained between dialyses and lost during dialysis (*dV* in equation 4.6) is isosmotic with respect to sodium, it will be confined to the extracellular space. Similar to the single-compartment model discussed in chapter 4, accurate prediction of urea concentrations and mass balance during two-compartment modeling should include these ECF volume shifts that tend to lower the predialysis BUN and enhance the removal of urea during dialysis. Failure to include ECF gain and loss, especially in patients with large increments in weight between dialyses, will cause a significant underestimation of G and an overestimation of V.

Figure 5.13 depicts the two-compartment, variable ECF volume model. Analytical solutions to the equations that describe this model are not possible. Highly accurate approximate solutions can be obtained with series expansions or numerical methods (36). A simple numerical method is derived in appendix D. These techniques are satisfactory in the experimental setting, especially when high-speed mainframe computers are available, but are often too slow when adapted to desktop computers. Fortunately, an approximate solution can be obtained by other means, as described below.

The gain in weight that invariably occurs between dialyses dilutes the urea space, tempering the rise in urea concentration. When weight is gained, predialysis BUN is lower than it would be in the absence of fluid gain. This effect of fluid gain is similar but not exactly the same as the effect of residual clearance. For modest fluid gain within the usual range seen clinically (0-5 kg/dialysis), we can simulate the variable volume model by augmenting K_r using a fraction of the rate of fluid gain between dialyses.

During dialysis, fluid lost is nearly isosmotic to plasma. Failure to incorporate fluid lost during dialysis into the model causes an inflation in V. An adjustment in K_d, similar to the above adjustment in K_r, minimizes this error. The adjustment is made by adding a fraction of Qf, the ultrafiltration rate to K_d. The more exact solution illustrated in appendix D was used to generate data for this chapter.

The variation in extracellular volume (dW) that occurs during and between dialyses has more impact when two-compartment kinetics are in effect. In patients who gain large volumes between dialyses, extracellular volume may vary as much as 30% when total body water varies only 10%. So, not only does extracellular concentration change more rapidly during dialysis, but extracellular volume also changes more, both during and between dialyses. The blood (and extracellular volume) is the site of action by the dialyzer; the intracellular response is more sluggish. Table 5.4 shows the influence of dW on calculations of V and G. The gold standards for V and G, derived from true two-compartment, variable-volume modeling, appear in bold typeface. The error in V incurred by failure to include fluid removed during dialysis is greater than the error committed by single-compartment

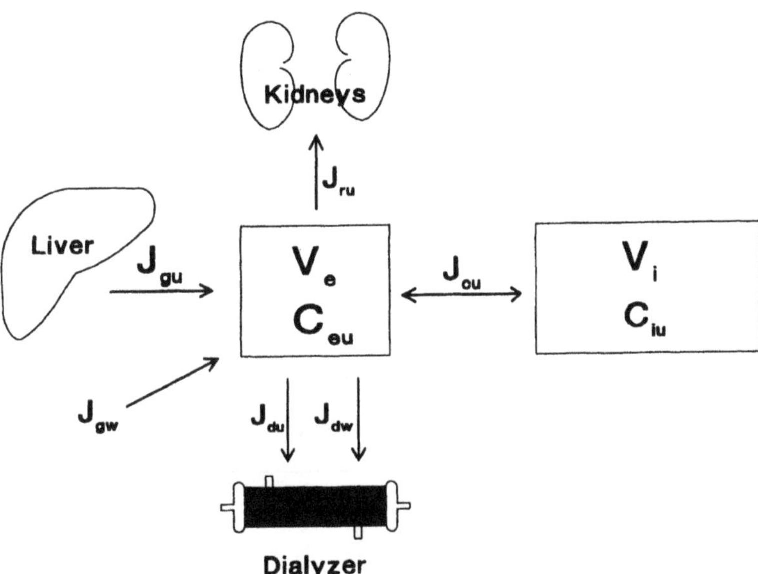

Figure 5.13. Two-compartment, variable-ECF-volume model. The mass balance diagram shows urea kinetics and extracellular volume shifts during and between dialyses. Solute and water flux are represented by the symbol J. J_{gw} is isosmotic fluid intake that is presumed to stay within the extracellular space. Similarly, J_{dw} is isotonic loss of water from the extracellular space that exits through the dialyzer. J_{gw}, J_{dw} and V_i (intracellular volume) are constant. J_{ru}, J_{gu}, J_{du} and J_{cu} represent flux of urea through the native kidney, from intake, across the dialyzer and between compartments, respectively.

modeling even if the volume removed is subtracted from V predialysis, as it is in table 5.4. Though the same weight change occurs during as between dialyses, the effect is more pronounced during dialysis when the extracellular/intracellular gradient is exaggerated. The consequences of these weight changes is to diminish the gradient between ECF and ICF, an effect that brings the results of two-compartment modeling closer to that of one-compartment modeling (see table 5.4). If ECF volume changes are not considered during two-compartment modeling, the error introduced is often greater than the single-compartment error. Therefore, if two-compartment modeling is applied to urea kinetic analysis, some accounting for ECF volume shifts must be included; otherwise, one-compartment modeling is preferable. One-compartment modeling is often sufficient to predict V and G accurately in these cases and may be preferred when seeking kinetic parameters alone. For accurate prediction of urea concentrations during and immediately after dialysis, the two-compartment model is required, and it must include changes in ECF volume.

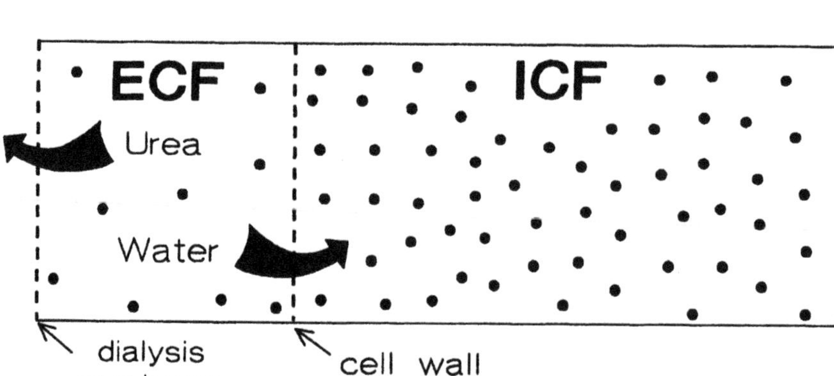

Figure 5.14. Osmotic forces during hemodialysis. Delayed movement of urea from the intracellular space during dialysis causes water to move in the opposite direction into the cell. The urea gradient, shown as relative density of dots, is exaggerated in this diagram.

Table 5.4 Effect of extracellular fluid retention (*dW*) on *V* and *G* for fixed-volume (fv) versus variable-volume (vv) models

	V#			*G*#		
	vv		fv	vv		fv
dW	1-Pool	2-Pool		1-Pool	2-Pool	
Low postdialysis BUN: C_1 = 80 mg/dl, C_2 = 28 mg/dl						
Standard dialysis: K_d = 180 ml/min, t = 240 min, Q_b = 300 ml/min						
0*	37.8	**37.9**	37.8	7.12	**6.78**	6.77
1	36.6	**36.7**	37.2	7.19	**6.90**	6.63
2	35.4	**35.4**	36.6	7.17	**7.03**	6.47
4	33.1	**32.9**	35.4	7.42	**7.29**	6.21
7	29.7	**29.4**	33.8	7.66	**7.70**	5.83
10	26.3	**26.1**	31.8	7.91	**8.11**	5.44
High-flux dialysis: K_d = 315 ml/min, t = 150 min, Q_b = 450 ml/min						
0	42.7	**43.7**	43.5	7.81	**7.15**	7.14
1	41.4	**42.2**	42.7	7.89	**7.28**	7.00
2	40.2	**40.9**	42.0	7.90	**7.40**	6.85
4	37.7	**38.2**	40.6	8.01	**7.63**	6.55
7	34.0	**34.2**	38.3	8.22	**7.98**	6.12
10	30.4	**30.4**	36.1	8.40	**8.34**	5.71
High postdialysis BUN: C_1 = 100 mg/dl, C_2 = 40 mg/dl						
Standard dialysis: K_d = 180 ml/min, t = 210 min, Q_b = 300 ml/min						
0	38.4	**39.1**	39.1	8.18	**7.79**	7.82
1	37.1	**37.7**	38.5	8.24	**7.95**	7.67
2	35.9	**36.4**	38.0	8.37	**8.11**	7.52
4	33.4	**34.1**	36.9	8.56	**8.45**	7.24
7	29.7	**30.1**	35.3	8.86	**8.92**	6.84
10	26.0	**26.1**	33.6	9.18	**9.40**	6.43
High-flux dialysis: K_d = 315 ml/min, t = 130 min, Q_b = 450 ml/min						
0	42.8	**45.1**	45.0	8.87	**8.18**	8.18
1	41.4	**43.6**	44.3	8.93	**8.32**	8.03
2	40.0	**42.1**	43.6	9.00	**8.46**	7.85
4	37.4	**39.0**	42.3	9.13	**8.74**	7.57
7	33.3	**34.6**	40.0	9.34	**9.16**	7.08
10	29.4	**30.2**	38.1	9.61	**9.59**	6.65

* *dW* = intradialysis weight loss (liters/dialysis)
\# units: *V* = liters, *G* = mg/min

TWO-COMPARTMENT MODEL WITH VARIABLE ECF AND ICF VOLUME

The ultimate model would include all variables that could possibly impinge on urea transfer during and between dialyses. Another obvious variable not yet considered is the osmotically driven movement of water between compartments. Until now we assumed that urea moves down its concentration gradient during dialysis from intracellular to extracellular locations and that intracellular water stays put during and between dialyses. We know that water moves freely across cell membranes in response to osmotic forces like that induced by the decrease in extracellular urea concentration during dialysis. So the above model that includes isosmotic fluid shifts during dialysis fails to consider the additional shifting of fluid intracellularly during dialysis that is depicted in figure 5.14.

More complex models have been derived for simulating urea removal during and between dialyses that consider osmotic and isosmotic fluid shifts (36). Figure 5.15 shows a mass balance diagram of the movements of urea, sodium, and water that

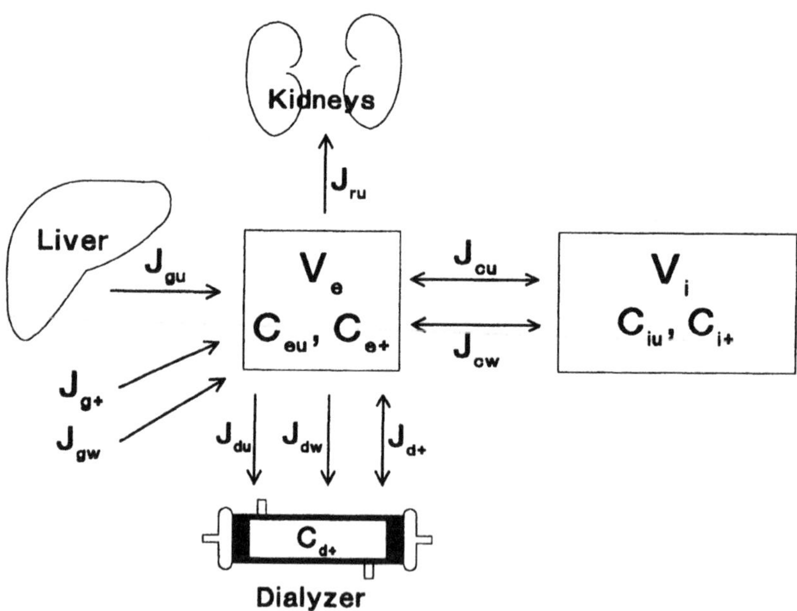

Figure 5.15. This mass balance diagram shows a two-compartment model that allows both extracellular and intracellular volumes to vary. Solute and water flux are represented by the symbol *J*; *g* = generation, u = urea, w = water, + = cation, r = residual kidney function, d = dialyzer function, c = cell wall, e = extracellular, i = intracellular. See appendix E for a more detailed description.

must be computed simultaneously using this model. Figure 5.15 differs from figure 5.1 by the addition of water and cation flux. Solute and water flux are expressed as the symbol *J* instead of the product of clearance and concentration used in figure 5.1. The differential equations describing shifts of urea, sodium, and water are straightforward but require numerical analysis for solution. Stepwise solution of a system of simultaneous differential equations takes much more computer time, so these models have been applied only in the experimental setting. Appendix E lists the differential equations required to solve the two-compartment model with variable ECF and ICF volume. Following is a discussion of the magnitude of this shift in a model that incorporates osmotic fluid balance and the variables included in all the models considered so far.

THE MAGNITUDE OF INTRACELLULAR SWELLING

If tissue swelling occurs because of osmotically driven water movement as shown in figure 5.14, the ratio of intracellular to extracellular volume will increase and concentrations in both compartments will change independent of the modeled parameters discussed so far. The two-compartment variable-ECF volume model does not consider such water shifts and assumes essentially that water is impermeable to cell membranes. This assumption is, of course, incorrect, and it is a potential source of error in the data generated by the model. The magnitude of this error is apparently not great because the two-compartment variable-ECF model accurately predicts changes in urea concentration postdialysis (37). Compensation for intracellular swelling is partially accomplished by overestimating *KC*, the intercompartment mass transfer coefficient for urea. Movement of water into the intracellular compartment during dialysis tends to dissipate the concentration gradient, a phenomenon that can be mimicked in the model by increasing the diffusion coefficient for urea. Other diffusible solutes tend to follow these water shifts into the intracellular space, further diminishing the osmotic gradient and tissue swelling. *KC*, then, is a composite parameter representing the relative permeabilities of water, urea, and other solutes at the cell surface. The major determinant of *KC* is the cellular urea permeability. Other factors such as intravascular and interstitial hydrostatic pressure play a role in determining intracellular volume during dialysis.

The magnitude of intracellular water accumulation can be estimated by considering some simplified examples. Figure 5.5 shows a diagram of the extracellular to intracellular water movement and osmolarity changes that can occur during dialysis. The dotted line represents new conditions that exist postdialysis in a patient whose weight does not change during dialysis (predialysis weight = postdialysis weight). Despite lack of ultrafiltration, extracellular volume can be expected to decrease in response to osmolar forces, as this diagram shows.

Consider, for example, a hypothetical case where cell membranes are permeable to water but not to urea. Table 5.5 lists the assumptions about cell membrane permeability to urea and water, for one-compartment and two-compartment models, for the example given here, and for our best estimate of reality.

Table 5.5 Cell membrane permeability to urea and water: assumptions by various models

	Urea	Water
One-compartment model (variable ECF volume)	∞	0 or +
Two-compartment model (variable ECF volume, constant ICF volume)	+	0
Two-compartment model (variable ECF and ICF volume)	+	or + ∞
This example	0	∞
Reality	+	+

0 impermeable
+ finite permeability
∞ infinite permeability
ECF extracellular fluid
ICF intracellular fluid

For the one-compartment model, cell membrane water permeability is not a factor, since we assume urea permeability is infinite and no osmotic gradient can develop across cell membranes. For our hypothetical model, the converse set of conditions exists, where no urea exits from the intracellular space but water diffuses freely, i.e., cell walls are impermeable to urea but completely permeable to water. The example strays far from reality but serves to maximize the volume shift, and illustrate a point. During a typical dialysis, BUN may fall as much as 70%. Assuming near complete equilibration, this means that the amount of urea removed from the body during a single dialysis exceeds that contained in the entire extracellular space. To exaggerate the effect of dialysis on transcellular water movement, let us consider that all urea is removed from the extracellular space in the above example. If the predialysis BUN is 80, the decline in urea nitrogen concentration during dialysis is 80 mg/dl, representing a 28 milliosmolar decline or a 9% fall in total extracellular osmolarity. A relatively small percentage change in osmolarity occurs because sodium, not urea, is the major solute contributing to extracellular osmolarity. If cell membranes are permeable to water but not to urea, osmotic equilibrium will be maintained during dialysis by movement of water from the smaller extracellular space into the larger intracellular space to eliminate the 28 milliosmolar gradient. If free water distributes approximately two thirds intracellularly and one third extracellularly, a 3% increase in intracellular volume would be matched by a 6% decrease in extracellular volume to maintain isosmotic equilibrium. At the end of dialysis, extracellular urea concentrations would be zero and intracellular urea concentration would be 3% lower than its value at the start of dialysis.

We know that urea diffuses readily from the intracellular space and that the maximum urea gradient during dialysis even during high-flux dialysis at high blood and dialysate flow rates is much lower than 80 mg/dl. For the example of high-flux dialysis illustrated in figure 5.5, the maximum urea nitrogen gradient is 15.1 mg/dl or 5.4 mosm. This represents a 1.8% osmolar gradient. Reasoning as we did above, this gradient has the potential to cause a maximum decrease of 1.2% in extracellular volume and a maximum increase of 0.6% in intracellular volume. Also, water that has moved into the cells during dialysis will move back out with urea during the 20-30 minute reequilibration that occurs postdialysis. Thus, changes in compartment volumes induced by urea disequilibrium are small and transient even in patients undergoing high-flux dialysis. These volume changes are unlikely to contribute either to symptoms during and after dialysis or to significant errors in our two-compartment model predictions.

It must be emphasized that the above examples estimate the maximum change in osmolarity that can be expected from removal of urea. A high dialysate sodium concentration and/or movement of other solutes into the intracellular space in response to water movement will diminish the effect. In fact, the concentration gradient for sodium, the major extracellular solute contributing to serum osmolality, has a much greater potential for causing water shifts than does the urea gradient (38,39). Combined sodium and urea modeling during dialysis has been used to predict these changes (36,40). Varying dialysate sodium concentration during dialysis can be used to support blood and extracellular volume to benefit the patient (41,42).

Potassium is a small, easily dialyzable solute that accumulates in anuric states. Potassium removal during dialysis has the potential for causing intracellular swelling. Because the extracellular potassium concentration is low (4-6 mmol/l predialysis) and removal rates compared to urea are low, the magnitude of this osmotic effect is negligible. In addition, the volume-regulatory mechanisms that appear to be an inherent property of most tissues cause potassium to leak out of the cell when swelling occurs (43,44). This cell volume-maintaining response would reduce dialysis-induced cell swelling and facilitate removal of potassium by maintaining serum potassium levels during dialysis.

Because the dilutional effects of intracellular water shifts are minimal, *KC* correlates closely to the average intercompartment permeability of urea, and the two-compartment, variable-ECF model accurately predicts extracellular and intracellular urea concentrations during and between dialyses. When significant sodium gradients develop during dialysis, a more complex model may be required.

Additional Compartments

We have simplified the two-compartment model by considering it purely as an extracellular/intracellular phenomenon. In reality, multiple gradients probably develop among intravascular, interstitial, and intracellular compartments in a nonuniform pattern throughout the body. It might be expected that *KC*, the intercompartment mass transfer area coefficient, would vary among different body organs, for example the gut versus muscle. Three to five compartments have been used to model dialysis urea kinetics by investigators in the past (10,11). Bowsher et al. applied a three-compartment model to data from hemodialysis in the dog and found a reduction in the average intercompartmental urea clearance as time progressed during dialysis (11). The authors suggested that the cause of this decrease in transport between compartments was a decrease in tissue perfusion that correlated with the known decrease in cardiac output and tissue perfusion induced by dialysis with cellulose membranes. A similar change in clearance during dialysis in humans was noted for the drug metabolite, N-acetylprocainamide (45). There are no studies of this effect on human urea kinetics, and there are no comparative studies with modern synthetic membranes. Most studies show that urea removal rates are not significantly affected by these additional compartment parameters. For today's dialysis, these more complex models offer little additional benefit for the clinician who is interested in predicting urea nitrogen concentrations and mass balance in his/her patients. For models of other solutes, or for investigators looking for a more precise description of urea kinetics, these models may be required.

Recommendations Regarding Modeling with One Versus Two Compartments

Table 5.6 lists the relative indications for application of one-compartment or multicompartment models. Although the one-compartment model fails to provide an accurate description of the BUN profile during and shortly after dialysis, it usually does an excellent job of predicting V and a reasonable job of predicting G. The accurate prediction of V is due to the offsetting errors discussed above. Current applications of urea kinetic modeling place emphasis on predicting values for ideal or target duration of dialysis. These predictions are usually based on maintaining a constant linear relationship between normalized PCR (PCRn) and a measure of dialysis effectiveness such as time-averaged BUN or Kt/V (see chapter 8). The absolute value of V or K has little effect on their ratio, the prime determinant of dialysis outcome. If errors in both the numerator and denominator of these proportions are equivalent, they tend to cancel, leaving an accurate estimate of ideal dialysis time. The single-compartment model is therefore reasonably accurate for most applications, especially where determining ideal time on dialysis is the principal goal.

Table 5.6 Which model to use?

One-compartment model

1. For most applications where:
   KC (urea) is greater than 600 ml/min
   The major goal is to predict ideal dialysis time
   Slight overestimation of *G* does not pose a problem
2. To simplify modeling and to improve speed of calculations
3. To allow strict comparison of modeling results with those of the National Cooperative Dialysis Study
4. When no consideration of compartment volume change is available for the multicompartment model and the patient gains fluid between dialyses

Multicompartment model

1. To obtain an accurate profile of BUN during and immediately after dialysis
2. To obtain a very accurate estimate of *G*
3. When *KC* (urea) is less than 600 ml/min
4. To model kinetics for solutes other than urea

The two-compartment, constant-volume model introduces an additional error because it fails to consider both osmotic and isosmotic volume shifts during dialysis. Failure to include these shifts can lead to significant overestimation of *V* and underestimation of *G* in patients who lose weight during dialysis. Most of the two-compartment models that are currently in use, including one model described here (equations 5.3-5.8), do not include dV/dt in the calculations. We have shown that simply subtracting the weight loss during dialysis will resolve a major part of the error, but a significant error remains due to the real decrease in *V* that is ongoing during dialysis. Failure to consider this ongoing loss causes urea removal to be overestimated so, calculated *V* is larger than real *V*. The variable-volume models described in chapter 4 and in this chapter avoid this error by modeling volume changes during and between dialyses.

Those who require an accurate BUN profile during and immediately after dialysis should use the two-compartment variable-volume model shown in figure 5.13 or, even better, the more complex model that includes osmotically driven water shifts depicted in figure 5.15. Table 5.3 shows that the two-pool model offers little improvement over the one-compartment model for estimation of *V* (urea) in most patients. The simplicity and speed (less computer computation time) of the one-compartment model are attractive. The two-compartment model employed by programs such as the one described in chapter 6 takes 2 to 3 times as long to complete its mathematical iterations. This is less a problem for the faster, more recent computer models or those central processing units (CPUs) that make use of a math computer models or those central processing units (CPUs) that make use of a math coprocessor. For standard single CPU computers, the time constraints can be prohibitive. Another argument favoring the one-compartment model is that it was the model used in the National Dialysis Cooperative Study (46). Conclusions reached by this study group that outlined domains of "ideal" dialysis and "inadequate" dialysis may not apply if another model is used. Even though V and G may be measured more accurately by the two-compartment model, comparison of outcome measurements with the NCDS data may not be valid. The inaccuracies in measurement of V and G by the one-compartment model are greater for high-flux dialysis. Since the dialysis centers that participated in the Cooperative Study did not use high-flux dialysis, their results can only be extrapolated in a general way to this dialysis setting, regardless of the model employed.

For solutes other than urea or when KC is very low, there is usually little question. Lower transcellular transport rates exaggerate the rebound phenomenon and cause error 2 to become predominant (e.g., creatinine, middle molecules, vitamin B12). Multicompartment models are more appropriate for these solutes.

Blood sampling techniques and precautions

Timing and methods for drawing blood samples are crucial to the success of urea modeling. Failure to observe simple precautions will cause wide variances in apparent outcome that will lead to inappropriate actions by the nephrologist and the dietitian. For either model, and especially if the two-compartment model is used, care must be taken to draw the second blood sample (postdialysis) just before or simultaneous with stopping the blood pump. There should be no reason to slow the blood pump gradually. If equilibration is allowed to occur (90% complete in 20 minutes) the one-compartment error (error 1 above) will be nullified, and V will be falsely increased due to the single uncompensated effect of error 2, described above. If the two compartment model is in effect, an even greater error in the same direction will occur because of spurious elevation in the measured postdialysis BUN.

The nurse or technician must be carefully instructed to draw blood from the arterial port. Large errors will result from low BUN values if blood is mistakenly drawn from the venous port. Our new employees are often confused by the difference between urea clearance measurements and urea kinetic measurements. For clearance measurements, blood is drawn simultaneously predialyzer and postdialyzer. This procedure must be distinguished from blood drawn predialysis and postdialysis for urea modeling. If the laboratory reports BUN values below 10 mg/dl, dialyzer venous blood sampling must be suspected.

The arterial port must be located upstream from the heparin infusion port to avoid dilution of blood by heparin. Similarly, other infusions must be given downstream from the blood sampling port to avoid dilution artefacts.

To avoid recirculation artefacts, the predialysis blood sample must be drawn before the blood pump is started, preferably soon after the access device is cannulated. For patients with fistulas, we prefer to draw the predialysis sample as soon as the first needle is placed, before any dilution with heparin or saline can occur.

Similarly, care must be taken to avoid recirculation when the postdialysis blood sample is drawn. Postdialysis blood sampling is even more important because the BUN has often fallen to a very low level where small changes (e.g., as little as 1 or 2 mg/dl) have a relatively large impact on the modeled results. To avoid false depression of the postdialysis BUN from recirculation, we slow the blood pump to its lowest setting, usually 25-50 ml/min, and clamp the venous line. We then set the venous pressure alarm to its maximum setting and wait until it alarms. At that point, we unclamp the venous line, allow the pump to turn two more revolutions, stop the pump and sample from the arterial port. This technique may require modification depending on the distance between the arterial needle tip and sampling port. Ports placed closer to the needle leave less "dead" space that must be washed out before sampling.

A standard instruction sheet placed at the bedside containing blanks for data entry helps to avoid confusion at the time of blood sampling.

References

1. Colton CK, Lowrie EG: Hemodialysis: Physical principles and technical considerations, in *The Kidney*, Brenner BM, Rector FC Jr (eds), Philadelphia, Saunders, pp 2425-2489, 1981.
2. Pitts RF, Windhager EE: Mechanisms of reabsorption and excretion of ions and water, in *Physiology of the Kidney and Body Fluids* (3ed), Chicago, Year Book Medical Publishers, pp 99-139, 1974.
3. Steffenson KA: Some determinations of the total body water in man by means of intravenous injections of urea. Acta Physiol Scand 13:282, 1947.
4. Shackman R, Chisholm GD, Holden AJ, Pigott RW: Urea distribution in the body after haemodialysis. Br Med J 2:355-58, 1962.
5. Pedrini LA, Zereik S, Rasmy S: Causes, kinetics and clinical implications of post-hemodialysis urea rebound. Kidney Int 34:817-824, 1988.
6. Lee CJ, Chang YL: On the solution of equations for feel-better hemodialysis. Comput Biol Med 17:161-172, 1987.
7. Maeda K, Kawaguchi S, Kobayashi S, Niwa T, Kobayashi K, Saito A, Iyoda S, Ohta K: Cell-wash dialysis. Trans Am Soc Artif Intern Organs 26:213-218, 1980.
8. Schindhelm K, Farrell PC: Patient-hemodialyzer interactions. Trans Am Soc Artif Intern Organs 24:357-365, 1978.
9. Grossmann DF, Kopp KF, Frey J: Transport of urea by erythrocytes during haemodialysis. Proc Europ Dial Transplant Assoc 4:250-253, 1968.
10. Popovich RP, Hlavinka DJ, Bomar JB, Moncrief JW, Decherd JF: The consequences of physiological resistance on metabolite removal from the patient-artificial kidney system. Trans Am Soc Artif Intern Organs 21:108-115, 1975.

11. Bowsher DJ, Krejcie TC, Avram MJ, Chow MJ, del Greco F, Atkinson AJ: Reduction in slow intercompartmental clearance of urea during dialysis. J Lab Clin Med 105:489-497, 1985.
12. Kaplan MA, Hays L, Hays RM: Evolution of a facilitated diffusion pathway for amides in the erythrocyte. Am J Physiol 226:1327-1332, 1974.
13. Brahm J: Urea permeability of human red cells. J Gen Physiol 82:1-23, 1983.
14. Harris CP, Townsend JJ: Dialysis disequilibrium syndrome. West J Med 151:52-55, 1989.
15. Krane NK: Intracranial pressure measurement in a patient undergoing hemodialysis and peritoneal dialysis. Am J Kidney Dis 13:336-339, 1989.
16. Arieff AI, Massry SG, Barrientos A, Kleeman CR: Brain water and electrolyte metabolism in uremia: effects of slow and rapid hemodialysis. Kidney Int 4:177, 1973.
17. Keller F, Offermann, Scholle J: Kinetics of the redistribution phenomenon after extracorporeal elimination. Int J Artif Organs 7:181-188, 1984.
18. Sargent JA, Gotch FA: Principles and biophysics of dialysis, in *Replacement of Renal Function by Dialysis* (3ed), Maher JF (ed), Dordrecht, Kluwer Academic Publishers, pp 87-143, 1989.
19. Rastogi SP, Frost T, Anderson J, Ashcroft R, Kerr DNA: The significance of disequilibrium between body compartments in the treatment of chronic renal failure. Proc Europ Dial Transplant Assoc 5:102-115, 1968.
20. Malchesky PS, Ellis P, Nosse C, Magnusson M, Lankhorst B, Nakamoto S: Direct quantification of dialysis. Dial Transplant 11:42-44, 1982.
21. Lankhorst BJ, Ellis P, Nosse C, Malchesky P, Magnusson MO: A practical guide to kinetic modeling using the technique of direct dialysis quantification. Dial Transplant 12:694, 1983.
22. Ellis PW, Malchesky PS, Magnusson MO, Goormastic M, Nakamoto S: Comparison of two methods of kinetic modeling. Trans Am Soc Artif Intern Organs 30:60-64, 1984.
23. Hume R, Weyers E: Relationship between total body water and surface area in normal and obese subjects. J Clin Pathol 24:234-238, 1971.
24. Moore FD, Olesen KH, McMurrey JD, Parker HV, Ball MR, Boyden CM: *The Body Cell Mass and its Supporting Environment*, Philadelphia and London, Saunders, 1963.
25. Gotch FA: Kinetic modeling in hemodialysis, in *Clinical Dialysis* (2ed), Nissensen AR, Gentile DE, Fine RN (eds), Norwalk CT, Appleton and Lange, pp 118-146, 1989.
26. Aebischer P, Schorderet D, Juillerat A, Wauters JP, Fellay G: Comparison of urea kinetics and direct dialysis quantification in hemodialysis patients. Trans Am Soc Artif Intern Organs 31:338-341, 1985.
27. Keshaviah P, Ilstrup K, Shapiro W, Hanson G: Hemodialysis urea kinetics is not single pool (abstract). Kidney Int 27:165, 1985.

28. Ilstrup K, Hanson G, Shapiro W, Keshaviah P: Examining the foundations of urea kinetics. Trans Am Soc Artif Intern Organs 31:164-168, 1985.
29. Tsang HK, Leonard EF, LeFavour GS, Cortell S: Urea dynamics during and immediately after dialysis. ASAIO Journal 8:251-260, 1985.
30. Ward RA, Shirlow MJ, Hayes JM, Chapman GV, Farrell PC: Protein catabolism during hemodialysis. Am J Clin Nutr 32: 2443-2449, 1979.
31. Farrell PC, Hone PW: Dialysis-induced catabolism. Am J Clin Nutr 33:1417-1422, 1980.
32. Borah MF, Schoenfeld PY, Gotch FA, Sargent JA, Wolfson M, Humphreys MH: Nitrogen balance during intermittent dialysis therapy of uremia. Kidney Int 14:491-500, 1978.
33. Gutierrez A, Alvestrand A, Bergstrom J, Wahren: Blood-membrane interaction without dialysis induces increased protein catabolism in normal man. Artif Organs 9:9-13, 1985.
34. Farrell PC: Adequacy of dialysis; marker molecules and kinetic modeling. Artif Organs 10:195-200, 1986.
35. Kimura G, Gotch FA: Serum sodium concentration and body flud distribution during interdialysis: importance of sodium to fluid intake ratio in hemodialysis patients. Int J Artif Organs 7:331-336, 1984.
36. Heineken FG, Evans MC, Keen ML, Gotch FA: Intercompartmental fluid shifts in hemodialysis patients. Biotechnol Progr 3:69-73, 1987.
37. Pastan S, Colton C: Transcellular urea gradients cause minimal depletion of extracellular volume during hemodialysis. Trans Am Soc Artif Intern Organs 35:247-250, 1989.
38. Fleming SJ, Wilkinson JS, Greenwood RN, Aldridge C, Baker LRI, Cattell WR: Effect of dialysate composition on intercompart-mental fluid shift. Kidney Int 32:267-273, 1987.
39. Fleming SJ, Wilkinson JS, Aldridge C, Greenwood RN, Muggleston SD, Baker LRI, Cattell WR: Dialysis-induced change in erythrocyte volume calculated from packed cell volume. Clin Nephrol 29:63-68, 1988.
40. Wehle B, Asaba H, Castenfors J, Gunnarsson B, Bergstrom J: Influence of dialysate composition on cardiovascular function in isovolaemic haemodialysis. Proc Europ Dial Transplant Assoc 18:153-159, 1981.
41. Swartz RD, Somermeyer MG, Hsu CH: Preservation of plasma volume during hemodialysis depends on dialysate osmolality. Am J Nephrol 2:189-194, 1982.
42. Van Stone JC, Bauer J, Carey J: The effect of dialysate sodium concentration on body fluid compartment volume, plasma renin activity and plasma aldosterone concentration in chronic hemodialysis patients. Am J Kidney Dis 11:58, 1982.
43. Cala PM, Mandel LJ, Murphy E: Volume regulation by Amphiuma red blood cells: cytosolic free Ca and alkali metal-H exchange. Am J Physiol 250:C423-429, 1986.

44. Grinstein S, Rothstein A, Sarkadi B, Gelfand EW: Responses of lymphocytes to anisotonic media: volume-regulating behavior. Am J Physiol 246:C204-215, 1984.
45. Stec GP, Atkinson AJ Jr., Nevin MJ, Thenot J-P, Ruo TI, Gibson TP, Ivanovich P, del Greco F: N-Acetylprocainamide pharmacokinetics in functionally anephric patients before and after perturbation by hemodialysis. Clin Pharmacol Ther 26:618-628, 1979.
46. Sargent JA: Control of dialysis by single-pool urea model: the National Cooperative Dialysis Study. Kidney Int 23 (Suppl 13):S19-25, 1983.

Chapter 6

A PRACTICAL SOLUTION: UREAKIN

"It is unworthy of excellent men to lose hours like slaves in the labor of calculations...."

Gottfried Wilhelm Von Leibnitz, 1671

Von Leibnitz was a German philosopher-mathematician who gave us our concept of kinetic energy. With the help of Issac Newton, he developed the first theories of differential calculus over 300 years ago. Early in the seventeenth century Von Leibnitz invented a calculator that probably inspired the above quote.

Since its introduction in the early 1970s, the programmable digital calculator has been extensively applied to clinical analysis of urea flux in patients who depend on this life-sustaining therapy (1,2,3). It extended the opportunity to quantify the outcome of hemodialysis from major research institutions to all dialysis centers, but it was used in very few. Only after the appearance of high-flux dialysis and its demand for individualized treatment did the popularity of the pocket calculator advance to significant levels. Use of the calculator in dialysis centers is somewhat hampered by its slow speed and limited storage capacity, which require repetitive entries and limit the number and variety of reports that can be generated.

VALUE OF THE COMPUTER PROGRAM

As an extension of the calculator, the computer can considerably ease the pain of long and repetitive mathematical calculations. It also allows the clinician and other health care personnel who may be untrained in mathematics to use the powerful modeling tools of the mathematician. If used on a regular basis, the computer program can help clinicians to better visualize and understand the dialysis effect. Unlike the calculator, user-friendly features can be incorporated into computer programs that allow editing of data, graphic portrayal of urea concentrations and model "fiddling" to predict the effect of changes in the prescription before applying it to the patient. The model serves as a testing ground for new concepts and equipment for the individual patient in particular and for the dialysis industry in general. Changes in the dialysis prescription, including filtration rate, dialyzer surface area and porosity, and dialysis duration and frequency, will have different effects in different patients depending upon the patient's size, urea nitrogen generation rate and residual clearance. The model incorporates each patient's individual variables to accurately predict effects in one patient that may be very different from effects in another. As an administrative tool, the computer program can be applied to examine and compare patients within a program or to compare the average outcome of the entire program with the outcome of other centers. As such it becomes a vital part of an overall dialysis quality assurance program. Measurement of the effectiveness of dialysis should be the central focus of any program that proposes to measure quality of care in hemodialysis units. We know of no better way to do this than to examine the kinetic flux of urea, the most abundant solute removed from each patient by the dialyzer. Using properly designed computer software, the clinician can quickly and easily accomplish this otherwise difficult task.

Description of the program

UREAKIN is a series of programs that were developed over the past six years at the University of California at Davis for evaluation of urea kinetics in hemodialyzed patients. The source code is written in BASIC (Beginner's All-purpose Semi-Instructional Code) for microcomputers operating under MSDOS (TM Microsoft Corporation, Redmond, WA), the operating system for IBM personal computers (TM International Business Machine Corporation). It was originally written for computers operating under CP/M (TM Digital Research Corporation, Scotts Valley, CA), an older operating system for microcomputers. In 1986 the programs were revised to operate using MSDOS, since it had largely begun to replace CP/M for business applications. At the same time, the source code was compiled using a more powerful version of BASIC to improve speed and handling. These revisions considerably enhanced program execution, increased speed, facilitated input, and eliminated tedious installation procedures for screen control and printers. In 1988 another major revision reduced the number of required data points to two instead of three and added an optional two-pool model. Since then, numerous additions and improvements have been added, and we expect that future changes will be added as hardware and software advances appear. Although the mathematical model is not limited to urea and could be used to model other solutes, the programs focuses intently on urea kinetics; no efforts were extended to incorporate other solutes.

Theoretical basis for the program

The equations outlined in chapter 4 (single-compartment model) and in chapter 5 (two-compartment model) are used to calculate urea volume and generation when a predialysis BUN and a postdialysis BUN are provided by the user. Additional parameters are requested and others are generated, but these are the major input and output variables. The program goes on to calculate projected dialyzer clearance (K_{dp}), protein catabolic rate (PCRn), time-averaged BUN (TAC), and Kt/V using the formulas detailed in chapters 4 and 5. It then calculates a value for ideal time-averaged BUN (ITAC) from which ideal dialysis duration is derived. The basis for ideal outcome parameters is the data provided by the National Cooperative Dialysis Study (NCDS), the rationale for which is discussed in chapter 8.

Conventions and assumptions

The following conditions and conventions are presumed to exist or are set by the program:

1. Dialysis occurs at the same time each day. The graphs of individual patient BUN profiles show dialysis starting at about 10:00 AM, but the actual time is not important
2. Except for the index dialysis (day of study), dialysis duration is constant from dialysis to dialysis.
3. Dialyzer clearance (independent of ultrafiltration) and patient residual clear-

ance ($K_r + K_d$) are constant from dialysis to dialysis.

4. All weight loss is via ultrafiltration, i.e., no fluid is given during dialysis.
5. If weight is gained instead of lost during dialysis, Q_f, the ultrafiltration rate, is set to zero (it is never a negative number).
6. The interdialysis rate of weight gain (due to fluid accumulation) is constant each day.
7. Postdialysis weight (dry weight) is constant.
8. Because the week is asymmetric, predialysis weight will differ depending on the interdialysis time interval.
9. Ultrafiltration rates are not constant, but vary in proportion to the length of the interdialysis time interval (higher after the weekend).
10. Clearances are always reported as blood water clearance plus the ultrafiltration component.
11. The ultrafiltration component of reported clearance is the average of ultrafiltration components for the week.
12. The dialyzer mass transfer coefficient is calculated from blood water clearance and is therefore a *blood water* coefficient.
13. If the index dialysis (study dialysis) has a longer or shorter duration than the average dialysis, the weight postdialysis is still considered to be dry weight and equal to the postdialysis weight after an average dialysis. The ultrafiltration rate is adjusted appropriately by UREAKIN during the index dialysis.
14. A ceiling for ideal time-averaged BUN is set at 70 mg/dl.

FILES AND FILE EXTENSIONS USED BY UREAKIN

Table 6.1 lists the disk files that constitute the program, followed by a brief description of each file.

UREAKIN.DTA contains information about dialyzers, the default disk drive and directory, a bell toggle, the printer port, and the selected printer's control codes for compressed print, enlarged print, left margin offset, etc. The utility file, UK-UTIL.EXE includes procedures for altering the above data file. Another printer may be selected with the utility function from a list of popular printers, or the printer's codes may be entered individually. Multiple copies of report forms and instruction sheets can be printed from UK-UTIL.EXE. Each of these forms may be edited to suit the needs of the user.

Additional support functions offered by UK-UTIL.EXE include simple techniques for calculating dialyzer mass transfer coefficients, calculation of urea nitrogen removal and volume of distribution during a single dialysis, estimation of initial dialysis duration, and calculation of average blood flow.

The main program is designed to be used on a regular basis by dialysis personnel who are not required to have extensive knowledge of computers or programming. Some understanding of the operation of microcomputers will be helpful.

Table 6.1 Description of files

Program files	
UREAKIN.EXE	Main program
UK-UTIL.EXE	Support utilities for UREAKIN
Data files	
EXAMPLE.KIN	Example of a patient data file
BLANK.SUM	Blank summary file to avoid an error message when listing these files
Help files	
These can be called from within the program:	
HELP1.TXT	For main menu
HELP5.TXT	For function #5
HELP6.TXT	For function #6
HELP8.TXT	For reports menu
HELPU.TXT	For utilities menu
Report forms	
These text files can be printed from UK-UTIL.EXE:	
INSTRUCT.TXT	Nurse/tech instruction sheet with blanks for kinetic data
CLEARANC.TXT	Instruction sheet for measuring dialyzer clearance and recirculation
HEADINGS.TXT	Listing of headings for complete report
KRFORM.TXT	Instruction sheet with blanks for residual clearance measurements
Miscellaneous	
UREA123.WK1	Lotus 123 (TM) template for dumping UREAKIN data to a spreadsheet format
UREAKIN.DTA	Configuration file, modified by UK-UTIL; loaded by main program at start-up
UPRINT.DRV	Printer control codes for popular printers used by UK-UTIL

The main program file is titled UREAKIN and is the only command that the user need remember to start the program. All other features are accessed via menus. A data file name may follow the UREAKIN command if desired, e.g., UREAKIN EXAMPLE. This will cause the program to load the file EXAMPLE.DAT before presenting the main menu and allow the user to quickly examine the program's features. The sign-on message is skipped and a list of surrogate patient names, TEST1, 1ST through TEST12, 12TH appears. These may be selected individually by pressing the space bar or ENTER key. The F10 key selects all patients; F9 deselects all; the default is all. Pressing ESC exits this routine. The main menu then appears as shown in table 6.2.

Table 6.2 MAIN MENU

	F1 = HELP
1) ENTER DATA FROM KEYBOARD	drive:\directory
2) RETRIEVE FILE FROM DISK	C:\UFILES\
3) SAVE SELECTED PATIENTS TO DISK	
	file
4) LIST AND SELECT PATIENTS	UK90-03
5) VIEW/EDIT PATIENT DATA	
6) ADJUST DIALYSIS VARIABLES	
7) SORT SELECTED PATIENTS	printer
	EPSON
8) PATIENT REPORTS	
9) UTILITIES	
D) DELETE FILE	
L) LOG (set) DISK\directory FOR DATA FILES	
Q) QUIT	

Options available from the main menu

Each of the options depicted in table 6.2 is selected by entering the number or letter at the left corresponding to the desired function. Keys F2 through F9 also serve as numbers; key F1 prints a help file on the screen. Following are detailed descriptions of each of the menu options.

Enter Date From Keyboard (1)

This function is used to enter new patient data, creating a new data file on the disk. It can also be used to change data in old files.

Routine new patient entry

Follow the instructions displayed on the screen. If a file has already been selected (opened) you will be asked to "A"dd to it, open a "N"ew file, or "E"dit the current file. Enter the first letter of your selection. If *E* for Edit is selected, you will be asked to enter the starting record number (e.g., if your file contains 20 patients and you would like to start editing patient number 10, enter *10*. This routine also makes it possible to enter a new kinetic analysis simply by editing a previous study (see below).

Default values, values automatically entered when you press ENTER, are shown above the cursor. Occasionally they will appear in brackets. Brackets [ ] always indicate defaults; optional choices are shown in parentheses ( ).

If you enter the wrong choice and have moved on to the next selection, the UP arrow key can be used to return to previous entries. This key should be used sparingly because it does not always return to the previous function and subsequent default values are changed.

Following are brief descriptions of each entry field:

PATIENT NAME

Suggested entry format: JONES, ALICE. Space for seventeen characters is available. Each patient must have a name, and the first eight letters must be unique. This requirement is imposed by the summary file routine (see below) that produces a separate file for each patient. It is suggested that for patients with identical last names, a number be inserted in place of the eighth letter (e.g., Johnson1, Johnson2, Johnson3, etc.). All entry is in upper case regardless of the keyboard switch setting.

ID NUMBER

An identifying number or combination of letters and numbers. The maximum number of characters is eight. This number is used to verify patient identification during the summary file generation routine (main menu #8).

DATE OF STUDY

This must be the date of the blood sampling for BUN. You can reenter the previous patient's study date by pressing the END key. The END key facilitates entry of dates when modeling is done for a number of patients on the same day. The program checks for proper format, MM/DD/YY or MM-DD-YY and numerical range, e.g., 1 to 12 for months and 1 to 30 for days. You may edit the displayed date using the cursor keys rather than type over all of the characters. The day of the week, determined from this date entry, is used to predict the dialysis schedule. If the schedule entered does not include this day, you will be asked to reenter the date and schedule.

SCHEDULE

Days of the week are numbered starting with Sunday. A chart is displayed to help select the series of numbers corresponding to the days of dialysis. These numbers must be entered sequentially, in ascending order. Errors such as nonnumerical entries, numbers above 7 or below 1, double numbers, or a nonascending sequence are detected by the program, which then prompts for another entry. Do not include commas or spaces between these numbers.

DIALYZER MODEL

In Edit mode, the previous dialyzer, blood flow and dialysate flow will be displayed followed by a question regarding whether or not to change it. When entering a new patient, or if the answer to the above question is yes, a table of dialyzers is shown (see table 6.3). Select the number corresponding to the dialyzer in use. If your dialyzer is not listed, press ENTER and you will be asked to enter its name. Space for eight characters is available for the name. Selecting a dialyzer from the list provides the added benefit of automatic clearance calculation. Each dialyzer in the list has its urea mass transfer coefficient stored with it. The program uses this coefficient, the blood flow, and the dialysate flow to calculate the dialyzer clearance. To add more dialyzers or to alter the mass transfer coefficient, select Utilities function #4.

Table 6.3 Sample listing of dialyzers and mass transfer coefficients (*KA*)

Dialyzer	*KA*	Manufacturer
CF-1511	360	Baxter (Travenol)
CF-1211	300	Baxter (Travenol)
CF-2308	460	Baxter (Travenol)
TORAY10	390	Toray
TEST	***	Used for testing the system
CD3500	220	CD Medical
CD4000	290	CD Medical
CD-90	420	CD Medical
CD-135	475	CD Medical
CD-DUO	435	CD Medical
FR-F40	370	Fresenius
FR-F60	600	Fresenius
FR-F80	900	Fresenius
FR-F6	540	Fresenius
FR-F8	660	Fresenius
HOSP-8	300	Hospal
HOSP-10	340	Hospal
HOSP-12	400	Hospal
HOSP-16	540	Hospal

BLOOD FLOW

Enter dialyzer whole-blood flow in ml/min. The default is 300 ml/min.

DIALYSATE FLOW

Enter dialysate flow in ml/min. This defaults to 500 ml/min, a common dialysate flow for single-pass systems.

RESIDUAL CLEARANCE

This is the patient's native kidney or endogenous urea clearance. After approximately six months of hemodialysis, the majority of patients have no residual clearance, but if urine output is in excess of one liter/day it is reasonable to measure urea clearance. This can be done directly by measuring urinary urea nitrogen or by estimating the urea clearance from the creatinine clearance, a test more familiar to the clinical laboratory. See Edit function (#5) in the main menu for calculation of residual clearance (K_r) avoiding additional blood sampling for BUN measurement.

STARTING AND ENDING TIMES

Time entries must be in 24-hour format. Either three or four digits are allowed. Defaults are taken from the patient's previous study or, if the patient is new, a three-hour dialysis is assigned. After the second time is entered, a check is made for total dialysis time. If this value is less than 30 minutes or greater than ten hours, a bell sounds and you are asked to reenter both time values.

AVERAGE DURATION OF DIALYSIS

If the dialysis under study is not typical (longer or shorter than usual), another time entry is provided for the usual duration (in hours). This value is used to estimate average predialysis BUN and time-averaged BUN. The default value for average duration of dialysis is the period computed from your entries of starting and ending times above (rounded integer). If the study duration differs from the average duration, the graph of BUN versus time will reflect this. In such a case the line representing BUN will not stop at the second BUN value but will pass through it or stop short of it.

WEIGHT

Enter the patient's weight in kilograms. If you elect not to consider weight changes in these calculations only the first weight need be entered. The weight following dialysis will then default to the first (no weight change is assumed). Weight is a required entry because it becomes a divisor in subsequent calculations of V as percent body weight and cannot be zero.

ESTIMATED VOLUME OF DISTRIBUTION (V)

You should enter your best guess of the patient's urea volume (V) or total body water. If previous kinetic studies have been done for this patient, the average calculated V will be the default if you have run the summary calculation (Reports Menu, selection #4). This estimate or average is used for reports and is explained more fully in the Reports Menu help file.

PATIENT HEIGHT

Enter the patient's height in inches. The program converts height in inches to centimeters and uses it and the patient's weight to calculate surface area. Surface area is used to normalize the cellular mass transfer coefficient for urea (KC) to patient size.

2-POOL

Enter yes or no to this question to select either a single-pool model (default) or a two-pool model for the kinetic analysis. In most cases the single-pool model is preferred.

Periodic entry simplified

The ENTER DATA FROM KEYBOARD function (#1) is the preferred way to repeat kinetic studies done on a regular basis for the entire dialysis center (or shift of patients). This avoids reentering data and saves time if your dialysis population is stable. To edit an old file, select the following main menu functions in the order given:

Menu item	Function
2)	Load file from disk
4)	Select patients
3)	Save selected patients to disk under a new name
2)	Load new file from disk
1)	Edit file

See instructions for menu items #3 and #4 below. For menu item #4, you should select patients who continue to be dialyzed in your center; you can press ENTER for *all* if no patients have left since your last urea kinetic analysis. When asked for a file name under menu item #3, we enter year and month, e.g., UK88-07. This format gives an orderly sorted listing in chronological order by the program's file lister (main menu function #2). You must follow the operating system (MSDOS) rules for naming disk files.

Retrieve File From Disk (2)

A single file of patients may be selected from the alphabetical listing simply by moving the cursor to the file name and pressing ENTER. You can move about in the list by using the cursor control keys (UP, DOWN, RIGHT, LEFT, PageUp, PageDown, HOME, END) or by pressing a letter key. Pressing a letter key will move the cursor to the next file with a first letter that matches the letter pressed. If your file is on another disk or subdirectory, press ESC to return to the main menu and reset the disk drive/directory by pressing L (see instructions below).

If a file has been opened previously, the question ""A"dd to the file or "N"ew file"?" will appear. If "N" is selected for New file, the previous file will be closed and the program will prompt for a new file to be opened. If "A" is selected to join two files, you will be asked for the name of another file that will be appended to the current file. Several files may be loaded sequentially in this manner. Each set of patients will be appended to the end of the previous set. This allows a mix-and-match approach to generate any combination of patients in a given file; e.g., a single patient may be selected from multiple files to produce a file containing all previous kinetic studies of that patient. An easier way to do this, however, is to select the summary file technique, Reports Menu function #8.

The maximum number of patients allowed in a file is 500. This is an unwieldy number; smaller files should be used for large units by dividing patients according to nursing shifts or weekday schedules.

A patient data file may be designated on the command line when first loading UREAKIN, e.g., UREAKIN FILENAME.

It is not necessary to include the file extension. The file will be loaded, bypassing the sign-on message, and then the main menu will appear.

Save Selected Patients To Disk File (3)

There is no need to save your working file to the disk because the program does this automatically. The purpose of this option is to save selected patient records in another file. If the file name you enter is identical to one already on the selected disk you will be notified by the program and asked whether to overwrite or add to the existing file. It is important to understand that if you answer "O" for overwrite, the program will write over (and destroy) the existing disk file.

List And Select Patients (4)

The names of patients in the currently open file are listed on the screen one page at a time (see table 6.4). The light bar at the top of the page can be moved using the cursor control keys and, if more than 40 patients exist in the file, the next page can be viewed by pressing the PageDown and PageUp keys. The HOME and END keys move the light bar to the beginning and end of the list, respectively. A subset of patient names may be selected using the space bar or ENTER key. The ENTER key also advances the light bar to the next patient to facilitate quick selection of an adjacent series of patients. Pressing the space bar or ENTER key again deselects the patient. To select all the patients, press F10; to deselect all patients press F9. The selected subset of patients is used by the main menu functions listed below:

Menu item	Function
3)	Disk output file
5)	Change patient data
6)	Alter dialysis prescription
7)	Sort patients
8)	Patient reports

Selected patients are designated both by highlighting the name and by a small arrow preceding the name. If the patient list has been sorted using function #7, the names will appear in sorted order only if all patients have been selected. Otherwise the order of display is the order of entry or retrieval from the disk. The sort will remain in effect, however, for the functions listed above. As soon as another patient or group of patients is selected, the sorted sequence is lost. Function #4 serves as the master control of the sort done by function #7. It selects the patients for sorting and cancels the sort if another subset is selected. To indicate whether or not a sort remains in effect, the word *sorted* will appear at the top of the page.

Table 6.4 Listing of patients, function #4 screen display

PATIENTS IN CURRENTLY OPEN FILE: C:\UREA\EXAMPLE.KIN

1	TEST1, 1ST	21	TEST21, 21ST
2	TEST2, 2ND	22	TEST22, 22ND
3	TEST3, 3RD	23	TEST23, 23RD
4	TEST4, 4TH	24	TEST24, 24TH
5	TEST5, 5TH	25	TEST25, 25TH
6	TEST6, 6TH	26	TEST26, 26TH
7	TEST7, 7TH	27	TEST27, 27TH
8	TEST8, 8TH	28	TEST28, 28TH
9	TEST9, 9TH	29	TEST29, 29TH
10	TEST10, 10TH	30	TEST30, 30TH
11	TEST11, 11TH	31	TEST31, 31ST
12	TEST12, 12TH	32	TEST32, 32ND
13	TEST13, 13TH	33	TEST33, 33RD
14	TEST14, 14TH	34	TEST34, 34TH
15	TEST15, 15TH	35	TEST35, 35TH
16	TEST16, 16TH	36	TEST36, 36TH
17	TEST17, 17TH		
18	TEST18, 18TH		
19	TEST19, 19TH		
20	TEST20, 20TH		

PgUp PgDn Home End ESC F9=None F10=All CR,SP=select/deselect

View/Edit Patient Data (5)

Stored patient data in the currently open file may be edited with this routine. If kinetic variables are affected by the change, they will be recalculated automatically. Recalculation occurs when ESC is pressed to exit or when PageUp or PageDown is pressed to move to another patient. Recalculation can be requested manually at any time by pressing F10 while the current patient's data is displayed on the screen (table 6.5). The edit function is especially helpful if erroneous data have been entered from main menu function #1. The data are listed in tabular form and the cursor is active, allowing direct replacement of values on the screen (table 6.5). Calculated variables are shown on the right. All variables are highlighted when changes are made to them whether changes are made by direct keyboard entry or by the program after recalculation. Explanations for the symbols displayed on the screen are found in table 6.6.

Table 6.5 Edit Function #5 (screen display)

```
                    EDIT DATA FOR SELECTED PT # 1    ( 1 )
  1-POOL file: LASTNAME
                                                   calculated variables
patient name ....................LASTNAME, FIRST   ______________________
unit number .....................ID223344                 G .........6.49 mg/min
date of study ...................03/12/90              PCRn .........0.99 g/kg BWn/day
  dialysis schedule ...........246                       V .........43.2 liters
  weekday of study ..........2 (Mon=2)                     .........( 57% body wt )
  dialysis duration ............4 hrs                  Kdp .........246 ml/min
  usual dialysis duration ...4 hrs                    Kt/V .........1.34
dialyzer model ..................FR-F80               Av Pre .........60 mg/dl
whole blood flow ............300 ml/min                TAC .........40 mg/dl
dialysate flow ..................500 ml/min       Ideal TAC .........49 mg/dl
residual clearance ............0 ml/min           Ideal time .........3.08 hrs
dialyzer clearance ............241 ml/min
  BUN before dialysis .......70 mg/dl
  BUN after dialysis ..........22 mg/dl
weight before dialysis ......78.6 kg
weight after dialysis .........76.1 kg
  estimated V ....................58 % body wt = 38.0 liters
  patient height..................67 inches

 F1=Help   F2=Pool   F3=KR-calc   F4=Dialyzers   F9=Graph
   F10=Calc   ESC=Menu   HOME=Select pt   END=Restore
```

The number displayed at the top center of each screen represents the index position of the patient in the list of selected patients. Selections are made with main menu function #4. The number in parentheses represents the position of the patient in the file on disk. If all patients have been selected and no sorting has occurred, these two numbers will be identical. Sorting can be detected by selecting function #4 from the main menu. The word *sorted* will appear at the top of the list of patient last names.

It is important to emphasize that changes made here are permanently saved and stored on disk. You may wish to manipulate certain entry variables to see the effect on calculated variables, but before going to the next patient or exiting to the menu, you should restore the original values. It is better to use function #6 if you wish to tinker with the prescription to answer "what if" questions. Function #6 keeps V constant and nothing is saved on disk so you do not have to restore original values. The Edit function (#5) looks at the values for predialysis BUN and postdialysis BUN and recalculates V and G to fit these values. Function #6 keeps V and G constant and recalculates time on dialysis, altering values for predialysis BUN and postdialysis BUN to fit the new dialysis prescription.

In addition to the four cursor directional keys, the following control keys are active:

PageUp = previous patient	HOME= prompt for patient
PageDn = next patient	END = restore entery
INS = toggle insert mode	^Y = erase line
DEL = delete character	ESC = return to main menu

Active Function Keys (prompts at bottom of screen):

F1 = Help (this text file).
F2 = Toggle one- or two-pool model (see below).
F3 = K_r calculation. Calculates patient's residual urea clearance from timed urine volume and urea nitrogen concentration (explained in detail below).
F4 = List dialyzers and their mass transfer coefficients
F9 = Graph. A graphic image of BUN versus time for an entire week is displayed. On this graph, time-averaged BUN is shown as a dotted line labeled TAC, and ideal time-averaged BUN is shown as a dashed line labeled ITAC. If the two-pool model has been selected, intracellular concentrations are shown as a dotted line. An option for printing the graph is available if the printer installed is an Epson (TM) compatible or HP-Laserjet (TM).
F10 = Recalculation (see below).

Whole blood versus blood water

Because blood has an effective water content of 90%, and plasma or serum averages 93% water, adjustments must be made in blood flow, urea concentrations and clearance to obtain accurate modeled estimates of *V* and *G* (see chapter 7). This program always expresses BUN as the whole blood value. The adjustment to give true blood water concentration is hidden from the user but takes place before recalculation. Blood flow is also whole blood flow, adjusted before recalculation to true blood water flow. Dialyzer clearance, however, is depicted as true blood water clearance plus the contribution of ultrafiltration. Dialyzer clearance in the *Kt/V* expression is also blood water clearance and includes the usually minor contribution of ultrafiltration. The contribution of residual clearance is not included in this expression.

Dialyzer clearance

A "pop-up" list of dialyzer brands and their respective mass transfer coefficients appears after pressing F4. This list can be scrolled by pressing the PageUp or PageDown keys if it contains more than 20 dialyzers. The list can be edited by choosing #9 (utility programs) from the main menu and then #4 from the utility programs menu.

Table 6.6 Edit Function #5: meaning of symbols

Symbol	Unit	Meaning of symbol
Entered variables		
Patient name		Last, first
ID number		Identification number (alphanumeric)
Date of study		(mm/dd/yy) format
Dialysis schedule		Numbers correspond to days in week (e.g., 246 = MWF)
Weekday of study		(e.g., 3 = Tuesday)
Dialysis duration	hours	Time from start to end of study dialysis
Usual dial.duration	hours	Usually same as above
Dialyzer model		Code for dialyzer model or brand
Whole blood flow	ml/min	Averaged pumped flow rate
Dialysate flow	ml/min	Flow of dialysate through dialyzer
Residual clearance	ml/min	Residual (native kidney) urea clearance
Dialyzer clearance	ml/min	Urea clearance adjusted for blood water content and ultrafiltration
BUN before dialysis	mg/dl	Patient BUN at the start of dialysis
BUN after dialysis	mg/dl	Patient BUN at the end of dialysis
Weight before dialysis	kg	Predialysis weight
Weight after dialysis	kg	Postdialysis weight
Estimated V	% body wt.	Your estimate (or program average) of V
Patient height	inches	Height of patient, used for surface area
Calculated variables		
G	mg/min	Urea nitrogen generation rate
PCRn	g/kg BWn/day	Protein catabolic rate BWn = (normalized body weight)
V	liters	Volume of urea distribution
	% body wt.	Same, expressed as % actual body weight
K_{dp}	ml/min	Projected dialyzer urea clearance when V_p is substituted for V
Kt/V	/dialysis	Dialyzer urea clearance x time / V
Av Pre	mg/dl	Average predialysis BUN
TAC	mg/dl	Time-averaged BUN from area under curve
Ideal TAC	mg/dl	Calculated ideal TAC (see utilities)
Ideal time	hours	Duration of dialysis required to achieve ideal TAC

Dialyzer clearance is calculated from blood flow, dialysate flow, and the dialyzer's mass transfer coefficient for urea. Whenever the dialyzer name is changed, a search for a matching name is made. This search occurs as soon as the cursor is moved off the entry line (it does not wait for the F10 key). If a match is found, the corresponding mass transfer coefficient (*KA*) is used to calculate a new clearance. The new clearance is displayed on the screen in highlighted format. If no match is found, the clearance does not change, but a new value for *KA* is calculated based on the displayed values for blood flow, dialysate flow, and urea clearance. An asterisk (*) appears after the dialyzer name if no match is found in the list of dialyzers.

Recalculation of clearance occurs if blood flow, dialysate flow, or patient weight is changed. The new clearance is displayed in highlighted format as soon as the cursor moves off the line of entry. Predialysis and postdialysis patient weights affect urea clearance because they determine the ultrafiltration rate required during dialysis to bring the patient's weight back to its dry value. Ultrafiltration during dialysis adds a small component to urea clearance.

It is possible to change the dialyzer clearance directly by entering an arbitrary value on the *dialyzer clearance* line. Direct entry of dialyzer clearance in this manner causes an asterisk (*) to appear after the clearance value if the dialyzer name matches a stored name (i.e., if a stored mass transfer coefficient is available). The asterisk here reminds you that a stored *KA* is available but is not being used. The asterisk will disappear when blood flow, dialysate flow, or dialyzer type is changed. If any of these three variables is changed, the stored value for *KA* is used to calculate clearance. If the dialyzer type has no match in the stored list of dialyzers, no asterisk will appear when clearance is changed directly. Subsequent changes in blood flow, dialysate flow, or patient weight will cause changes in clearance based on the calculated value for *KA*, since there is no stored value.

The displayed value for dialyzer clearance is the actual clearance achieved during the study dialysis. Since the schedule is asymmetric (unless only one dialysis occurs/week) the clearance will differ slightly as more or less weight is gained between dialyses and lost during dialysis to achieve the same postdialysis dry weight. The extra ultrafiltration required after a longer interdialysis time interval will add slightly to the dialyzer clearance. The clearance displayed here will sometimes differ slightly from the clearance displayed using function #6 (alter prescription) that shows the average of all clearances during the week. Although the studied dialysis clearance is displayed here and the average clearance is displayed with function #6, the actual clearance for each dialysis is used during the modeling process.

Ideal time-averaged BUN

Ideal time-averaged BUN (ITAC) is determined by the program and printed on the right side of the screen. The user need not be concerned about its value here, but it can be altered with function #6 to see the effect on ideal dialysis time. If for any reason the value appearing on the screen does not agree with the program's calculated value, an asterisk (*) will appear after it. The asterisk will disappear when you press F10 to recalculate variables. Ordinarily this asterisk should not appear, but it may be seen after importing other files or after adjustment for dialysis time outside the range of 30 minutes to ten hours.

The last parameter listed on the right, *Ideal time*, is the time on dialysis (duration in hours) required to achieve the above desired time-averaged BUN.

Error traps

Numerous error traps are built into this function. For example, the entry routine for the dialysis schedule traps duplicate entries, will not allow schedules not in ascending order, and will not allow numbers less than one or greater than seven. If you change the value for *Day of dialysis* to a value that is not included in the schedule, the cursor will not move until a valid number is entered. A bell will sound and a prompt appears at the bottom of the screen if you attempt to move the cursor. Function keys are also turned off until a valid number appears. If you enter a variable value that causes the recalculation of dependent variables beyond limits set for them, or if the program fails to converge, a message appears at the bottom asking you to check for errors, and control returns to the main menu. Function keys are turned off if V or G are zero. This can occur if an error is encountered during the middle of keyboard entry (function #1). The data you have entered will be preserved, but the remaining variables will be set to zero or blank. When this patient's data is called up with function #5, only part of the data will appear and calculation of dependent variables will not be possible. If you complete the entry using function #5 (Edit), the recalculation key (F10) will not work, but recalculation will occur when you move to another patient or escape to the main menu. Under ordinary circumstances, the above errors should not occur.

If the patient's postdialysis weight is changed, V_p (your projected estimate of V) is automatically changed to maintain the estimated percentage. The patient's postdialysis weight is always used when weight is required for calculations. This weight is presumed to be closer to the patient's true (dry) weight.

Care must be taken to enter valid numbers where numeric values are required. If characters are entered where numbers are required, the value of the entry is zero.

More about the F2 key (Pool selection)

This program offers both a single-compartment (one-pool) and a two-compartment (two-pool) model for urea kinetic analysis. In general the two-pool model is more accurate, but because urea diffuses rapidly during dialysis, results obtained from the two models are nearly identical. For this reason and because the single-pool model also considers volume changes during dialysis, we prefer the single-pool model. Calculation of variables is also considerably faster using the single-pool model. Pressing the F2 key toggles the recalculation mode from single-pool to two-pool or vice versa, as indicated in reverse video at the upper left corner of the screen. Pressing F2 also signals the program to recalculate. Recalculation will occur when you exit or move to another patient. The model used for recalculation is that shown in the upper left corner of the screen.

A comparison of the results obtained from the two models can be obtained by viewing the screen or by pressing F9 for graphics. For the two-pool model, concentrations in the rapid-equilibrating (e.g., extracellular) pool and slow-equilibrating (e.g., intracellular) pool are depicted during dialyses and between dialyses. Differences are more marked when dialyzers with high clearance are used in patients with small urea volumes. The two-pool model uses the intercompartment mass transfer coefficient for urea (*KC*) to determine the rate of diffusion from one compartment to the other. Provision is made in the utilities menu function #12 to alter *KC* from its default value of 800 ml/min/1.76 m2. Regardless of the value entered, the program automatically normalizes it to the surface area of the patient (calculated from patient height and weight).

More about the F3 Key (KR-Calc)

The F3 key provides a simplified method of measuring and calculating patient residual urea clearance. Residual clearance refers to the remaining native kidney function that is not adequate to sustain life but may contribute substantially to overall urea removal. When present, it is usually in the 1 to 5 ml/min range and slowly diminishes over several months after starting dialysis therapy. After six months, few hemodialyzed patients have any significant residual clearance (0.5 ml/min or less). Urine output also falls during this initial phase. During the early months of dialysis therapy, residual clearance must be taken into consideration and included in the kinetic analyses. It must also be considered if the patient's urine volume persists at a high level (>1 liter/day).

The advantage of using this routine, rather than calculating clearance in the traditional way, is that no blood samples are required. The blood urea concentration, a necessary factor for the urea clearance calculation, is provided by the program. The operator enters the start and end times of collection and the program predicts the concentrations from the time of the last dialysis. While this may not be as accurate as measuring BUN at the start and at the end of the urine collection, it is a reasonable substitute for measurements that are practically impossible to obtain. Tests of the reliability of the model's estimation of BUN found that the average deviation from the true BUN was 7% after two days and 10% after one week. These are tolerable errors, since a low residual clearance is a minor factor in the kinetic analysis. If the residual clearance is closer to 5 ml/min or, the patient strays outside the limits of his dietary or dialysis prescription (i.e., is no longer in steady-state), more accurate analysis can be obtained by measuring BUN at the start and again at the end of the collection.

This module uses previously determined values for *V* and *G* to estimate BUN at any given time during the weekly cycles. This means that the patient must have had a previous kinetic analysis and must have remained in steady state since the last analysis (no change in prescription or status). Because the K_r-calc function is accessible only from main menu function #5, a kinetic analysis has obviously been done, but the results may be inaccurate if the correct value for residual clearance (K_r) was not entered. This routine calculates the value for K_r and inserts it into the data base. It then recalculates *V*, *G* and K_r a second time to ensure their accuracy.

More about the F10 Key (Recalc)

The F10 key provides a manual option for recalculation. Recalculation occurs automatically after changes are made when ESC is pressed to exit or when PageUp or PageDown is pressed to see the next patient's data. However, if you would like to see the results of recalculation, you can do so, without moving from the present patient screen, by pressing F10. All parameters on the right are recalculated and highlighted. No change in these parameters will be seen unless the model has been altered (e.g., by pressing F2) or a variable in the left-hand column has changed. Any changes on the left will also be highlighted.

A ceiling of 70 mg/dl is set for ideal time-averaged BUN (ITAC) in the right column. If the program's calculation of ITAC returns a higher value, ITAC is set to 70 mg/dl during recalculation. If the program determines that ITAC is not achievable within a dialysis time span of 30 minutes to ten hours, it will reset ITAC to a value 5 mg/dl higher or lower and then recalculate. This process continues until an achievable ITAC is reached. Each time ITAC is reset, a message appears at the bottom of the screen that pauses the program and directs attention to the change. An asterisk (*) appears after Ideal TAC when its value has been changed and no longer conforms to the ideal calculated value based on PCRn.

The value of ideal time-averaged BUN (ITAC) determines the ideal duration of dialysis (Ideal time) in the next column. Another variable that may be confused with ITAC is TAC (time-averaged BUN), shown above ITAC in the right column. TAC is the actual time-averaged BUN achieved with the present prescription and is shown on the graph as a dotted line. ITAC has a ceiling of 70 mg/dl, but there is no ceiling for TAC.

Meaning of the asterisk (*)

An asterisk may sometimes appear after dialyzer type, dialyzer clearance, or ideal time. Under ordinary circumstances, these should not appear. They are intended to direct attention to unusual conditions that can affect interpretation of the modeled parameters. For each parameter that precedes the asterisk, an explanation of its meaning is given above. Following is a brief review of the conditions that cause appearance of the asterisk:

Following dialyzer type:

An asterisk here means that the dialyzer name does not match one found in the list of dialyzers. This usually means that the dialyzer name is misspelled.

Following dialyzer clearance:

An asterisk here means that the dialyzer name matches one found in the list and that the corresponding dialyzer mass transfer coefficient has not been used to calculate clearance. This usually means that the dialyzer clearance has been changed by direct entry on the dialyzer clearance line.

Following Ideal TAC:

An asterisk here means that the displayed value for ideal time-averaged BUN does not match the calculated value based on PCRn. This will occur if ideal time is less than 30 minutes or greater than ten hours. It will also occur if a "foreign" file is loaded that contains erroneous values for Ideal TAC. In most cases, the asterisk will disappear when F10 is pressed and a new value for Ideal TAC is calculated

This function (#5, Edit Data) may be used to perform a new kinetic study on a patient previously studied. If the dialysis prescription and dialyzer are unchanged, one need only enter the new date, weights, and concentrations. A better way to do this, however, is to use menu item #1, as explained in the main menu help file.

A separate help function is provided (press F1) that explains each parameter in the View/Edit routine as well as the operation of the function keys.

ADJUST DIALYSIS VARIABLES (6)

This routine has a similar format to the VIEW/EDIT PATIENT DATA routine but has an entirely different application. The patient's name, unit number, and urea volume (V) are displayed in the left upper corner. These parameters are held constant by this function and cannot be changed. When you change the remaining parameters, only the target values for time-averaged BUN (ITAC), dialysis duration, average predialysis BUN, and Kt/V are affected. These are recalculated and redisplayed when the F10 key is pressed. This provides a "what if" approach to individual patient prescriptions by permitting temporary changes in the dialysis variables while setting the patient's urea volume constant. You may change dialyzer brand, blood flow, dialysate flow, dialysis schedule, residual clearance, dialyzer clearance, protein catabolic rate, and average weight loss during dialysis to see their effect on the ideal prescription. Table 6.7 displays the data for the same patient shown in table 6.5. Explanations for the symbols displayed on the screen are found in table 6.8.

For example, if you would like to see the consequences of changing dialysis from three times/week to two times/week, change the dialysis schedule (e.g., from "246" to "25") and press F10. New values for ideal time on dialysis and ideal average predialysis BUN will appear. The increase in time reflects the prescription change that is necessary to achieve the same time-averaged BUN (TAC) when the patient is dialyzed two rather than three times/week. Similar experiments can be done by changing the dialyzer model, dialyzer clearance, blood flow, dialysate flow, or residual clearance. Ideal or desired time-averaged BUN can be set to any value here, in contrast to Edit mode (function #5), and should be set to zero for automatic recalculation when the F10 key is pressed. Recalculation takes place only when you request it by pressing the F10 key. This contrasts with the Edit mode (function #5), where recalculation takes place automatically to avoid writing erroneous data to the disk.

The effect of weight gain on dialysis efficiency can also be demonstrated. Note that the average weight loss during dialysis is not the difference between predialysis and postdialysis weights. The rate of fluid accumulation and postdialysis weight are assumed to be constant. This means that more weight will accumulate during longer interdialysis intervals than during the shorter intervals. To bring postdialysis weight back to its constant dry value, more fluid must be removed during dialyses that follow the longer intervals. The program takes this into consideration when it calculates average weight loss. The result is only a slight change in kinetic parameters, but more importantly, the patient is not penalized if his weight gain (same as loss) occurs after a long interdialysis interval and is compared to other patients after a short interval. The reports issued by UREAKIN (see main menu function #8) all print the average weekly value, allowing comparisons among individual patients and a true average for the dialysis center as a whole.

Table 6.7 Adjust prescription variables (screen display)

```
              PRESCRIPTION VARIABLES FOR PT # 1  ( 1 )
     1-POOL                         file: LASTNAME
LASTNAME, FIRST
   I.D. NUMBER: ID223344
   STUDY DATE: 03/12/90
                      V = 43.2 liters (56.7 % body weight)

   dialyzer model ............................ FR-F80
   whole blood flow .......................... 300     ml/min
   dialysate flow............................. 500     ml/min
   dialysis schedule (2=Monday).. ........ 246
   residual clearance ........................ 0       ml/min
   dialyzer clearance ........................ 241     ml/min
   protein catabolic rate .................... 0.99    g/kg/day
   av. wt. loss during dialysis............... 1.9     kg
   desired time-averaged BUN ................. 49      mg/dl (set to 0 for ideal)
Target:
   dialysis duration ......................... 3.08    hrs
   av. predialysis BUN........................ 69      mg/dl
   Kt/V ...................................... 1.05

   (changes made here are temporary, data files are unaltered)

   F1=Help   F2=Pool   F4=Dialyzers   F9=Graph
   F10=Calc  ESC=Menu  HOME=Select pt  END=Restore
```

It is important to emphasize that any changes made within this subroutine will not be stored in memory or on disk. A reminder to this effect is posted on the screen. Even though recalculation of variables occurs, none of the revisions are stored. This differs from function #5, where all changes are stored in the disk file. You can demonstrate this by using the PageUp or PageDown keys after you have changed several variables. To restore the original values, simply press PageDown to go to the next patient and then PageUp to return to the current patient. All variables are restored from the disk file to the screen.

A pop-up list of dialyzer brands and their respective mass transfer coefficients appears after pressing F4. This list can be scrolled by pressing the PageUp or PageDown keys if it contains more than 20 dialyzers. The list can be edited by choosing #9 (utility programs) from the main menu and then #4 from the utility programs menu.

If the dialyzer brand is not found in the list of stored dialyzers, an asterisk (*) will appear after its name. If the dialyzer is stored in the list and you have evidence that its clearance is different from the value automatically provided, you may change it by moving the cursor to the dialyzer clearance line. An asterisk (*) will appear after the dialyzer clearance if its value is changed to anything other than the value determined by the dialyzer's stored mass transfer coefficient. The changed value will remain in effect during recalculation; its value will be changed by the program only when blood flow, dialysate flow or weights are changed.

The displayed value for dialyzer clearance is the average of all clearances during the week. Since the schedule is asymmetric (unless only one dialysis occurs per week) the clearance will differ slightly as more or less weight is gained between dialyses and lost during dialysis. The extra ultrafiltration required after a longer interdialysis time interval will add slightly to the dialyzer clearance. Although the average of these clearances is displayed here, the actual clearance for each dialysis is used during the modeling process.

To view a graphic display of BUN versus time for an entire week, press F9. This graph differs from the graph shown in Edit mode (function #5). All predialysis and postdialysis BUN values depicted are ideal values derived from the ideal time-averaged BUN shown by the dotted line as ITAC. The graph shows the steady-state profile of oscillating BUN levels that would be achieved if the patient were dialyzed using the ideal prescription variables shown on the text screen. Ideal or target time-averaged BUN (ITAC) is shown as a dotted line. Actual TAC is shown as a dashed line. ITAC represents the area under the curve of BUN versus time divided by a week's time. If ITAC equals TAC, the graph depicted by function #6 will be identical to that depicted in Edit mode (function #5), and the two lines representing TAC and ITAC will overlap.

Pressing F10 for recalculation in this mode causes recalculation of ideal values only. Ideal TAC, however will not change unless its value is set to zero. If set to zero, its value will change to the ideal value established by mechanistic interpretation of data provided by the U.S. National Cooperative Dialysis Study. None of the recalculated values are stored.

In addition to the four cursor directional keys, the following control keys are active:

PageUp	= previous patient	HOME	= prompt for patient
PageDn	= next patient	END	= restore entry
INS	= toggle insert mode	^Y	= erase line
DEL	= delete character	ESC	= return to main menu

Active Function Keys (prompts at bottom of screen):

F1 = Help (this text file).

F2 = Toggle one-pool versus two-pool model

F9 = Graph. A graphic image of ideal BUN versus time for an entire week is displayed. On this graph, ideal time-averaged BUN is shown as a dotted line labeled ITAC and actual time-averaged BUN is shown as a broken line labeled TAC. An option for hardcopy printing of the graph is available if the printer installed is an Epson compatible or HP-laserjet.

F10 = Recalculation (see above).

Table 6.8 Prescription variables: meaning of symbols

Symbol	Unit	Meaning of symbol
Dialyzer type		Code for dialyzer
Whole-blood flow	ml/min	Average pumped flow rate
Dialysate flow	ml/min	Average flow of dialysate
Dialysis schedule		Numbers correspond to days in week (e.g., 246 = MWF)
Residual clearance	ml/min	Residual (native kidney) urea clearance
Dialyzer clearance	ml/min	Urea clearance adjusted for blood water content and ultrafiltration
Av. weight loss	kg	Average weight loss during dialysis
Desired time-averaged BUN	mg/dl	Calculated ideal TAC (see utilities)
Target:		
Dialysis duration	hours	Time required to achieve ideal TAC
Av. predial BUN	mg/dl	Ideal average predialysis BUN
Kt/V	/dialysis	Dialyzer urea clearance x ideal time / *V*

Again, it is important to note the status of *desired time-averaged BUN* before you press the F10 key to recalculate. If its value does not match the program's calculated value for ideal TAC, an asterisk (*) will appear following the displayed number. If you wish to have the program recalculate this variable, you must set it to zero before pressing F10. If you forget to change it before pressing F10, just set it to zero and press F10 again. The value of this variable determines the ideal duration of dialysis (target time on dialysis).

A separate Help function is provided that explains each variable in the Adjust Dialysis Variables routine. This help file also explains the operation of the function keys.

Sort Selected Patients (7)

A virtual sort of selected patients occurs only in computer memory, not on the disk. To preserve the selected records in sorted order on disk, choose menu option #3 to rewrite the file after sorting. If the same file name is entered, the old file will be overwritten. The virtual sort uses a hidden index variable that calls up records from the file in sorted order for subsequent functions. Nineteen variables are listed for sorting. If the selected patients represent all patients in the file, then subsequent listing of patients using function #4 will be in the sorted order. If another group of patients is selected from the same file with menu option #4, the sort is erased and patient records will appear in the order they were entered on the disk. You will know that a sort has taken place and is currently active when the word "sorted" appears at the top of the screen when patient names are listed with menu function #4.

Two parameters may be used as a basis for the sort. If only one is desired, press ENTER when asked for the second patient parameter. The second defaults to the date of study.

Patient Reports (8)

This function provides a menu of reports that can be printed (hard copy) or displayed on screen (see table 6.9). Help is obtained by pressing F1 while the menu is on the screen.

Table 6.9 Reports Menu

1) SHORT REPORT
* 2) COMPLETE REPORT
* 3) INDIVIDUAL PATIENT REPORTS

4) PREPARE SUMMARY REPORT (means and SDs)
* 5) PRINT SUMMARY REPORT (means and SDs)

6) MAP IDEAL TREATMENT DOMAINS
7) GENERATE ASCII FILE

(* hardcopy only)

Report subroutines are selected from the above menu by entering the number at the left corresponding to the desired routine. Function keys 2 through 7 also enable entry of numbers; function key 1 calls up this help file.

All reports are designed to be printed with an 80-column dot matrix printer. The complete summary report and mean summary reports require a compressed format (130 columns). Alternatively, a printer with wide carriage may be used.

Only the short report can be viewed on the screen. All the information in these reports can be displayed on the screen using main menu function #5. The reports provided by function #8 make use of the larger display capacity of the printer to enhance the output and to show variables for multiple patients on the same page. This function cannot be duplicated on standard 25 line by 80 column screens.

SHORT REPORT (8-1)

This option provides an abbreviated report of all selected patient records. The output for each patient is a single line consisting of the patient's name, identification number, dialysis schedule, dialyzer, average blood flow, dialysate flow, and ideal dialysis duration. The patient records appear in the order established by the most recent sort (main menu item #7). The intent of this short report is to provide the nurse/technician staff with a synopsis, showing the dialysis prescription for each patient. Screen output is paged; hardcopy is optional.

COMPLETE REPORT (8-2)

This is a report of nearly all variables in a columnar format. There is no screen output. The page of variables is printed using the printer's compressed print font. Table 6.10 contains a brief description of each column heading.

Table 6.10 Complete report: meaning of heading labels

PATIENT NAME		Last name, first name
ID #		Patient identification number
STUDY DATE		Date blood was drawn
DIALYZER TYPE		Dialyzer label or brand abbreviation
Sch-dle		Weekly dialysis schedule (e.g., MWF)
Wt	(kg)	Postdialysis weight in kilograms
dW	"	Average weight loss during dialysis
T_{da}	(hrs)	Average duration of each dialysis
Q_b	(ml/min)	Whole-blood flow
Q_d	"	Dialysate flow
K_r	"	Residual (native kidney) urea clearance
K_d	"	Dialyzer urea clearance, adjusted for water and ultrafiltration
K_{dp}	"	Projected dialyzer clearance when $V = V_p$
PRE	(mg/dl)	BUN prior to dialysis
POST	"	BUN following dialysis
TAC	"	Time-averaged BUN from area under the curve
PREa	"	Average predialysis BUN
Kt/V	(per dialysis)	Dialyzer urea clearance x T_{da} / V
PCRn	(g/kg BWn/day)	Protein catabolic rate (expressed here as g protein/kg BWn (normalized Body Weight)
V	(% body weight)	Volume of urea distribution
V_p	(% body weight)	Your projected estimate of V
—IDEAL—		
TAC	(mg/dl)	Calculated ideal TAC
PRE	(mg/dl)	Ideal average predialysis BUN
Ideal hrs		Duration of dialysis necessary to achieve ideal TAC

The term *dW* requires further explanation. The average weight lost during dialysis is equivalent to the average weight gained between dialyses. This value is not simply the difference between predialysis and postdialysis weights. The rate of fluid accumulation is assumed to be constant, so more weight will accumulate during longer interdialysis intervals than during the shorter intervals. Consequently more fluid will be removed during dialysis after the longer intervals to achieve dry weight postdialysis. UREAKIN takes this into consideration when it calculates average weight loss.

INDIVIDUAL PATIENT REPORTS (8-3)

This routine prepares a report suitable for filing in the patient's chart. After the first report is printed a prompt appears for another report; if the answer is "Y"es, the question "stop after each report?" will appear. Entering "N"o answer here provides a convenient method for printing individual hardcopy reports of all selected patients without further input.

When protein catabolic rate (PCRn) is displayed, the units are in grams/gram normalized body weight (BWn). BWn is defined as *V*/.58 (assuming 58% of ideal body weight is the normal urea distribution space). PCR expressed per unit of normalized body weight (PCRn) is a more useful expression, because protein catabolic rate is determined more by lean body mass than by total body weight.

PREPARE SUMMARY REPORT (means and SD's) (8-4)

A separate file for each patient allows averaging previous kinetic data to improve accuracy and to observe trends. This part of the program facilitates development and maintenance of individual patient files. The routine is designed to be run following each month's kinetic study as follows:

Load the latest monthly unit file
Select patients for summary report (usually all)
At the main menu, select item (8)
At the reports menu, select item (4)
Follow instructions for summary report

Be certain that all files to be accessed are on the same disk drive: \ directory and that sufficient disk space remains to store a short file for each selected patient. Each study occupies 128 bytes of disk space, so 8 records can be stored per kilobyte of space. However, most floppy disks require a minimum of one kilobyte of disk space per file, and most hard disks require a minimum of two kilobytes of space per file. The program will prompt for the number *N* of previous records to include in the summary file (the default is 5, maximum 9). It will then prompt for an update of each patient's file. If the response is "N", each patient file will be left undisturbed, but the summary file will include data from the latest file. The second and final question provides an option to substitute an average of all previous values of *V* for projected V (V_p) in the current (unit) file. This eliminates the need to estimate *V* when the file is edited, substituting the next set of data. The average value of V generated by this routine becomes the default for V_p.

The default for both of the above questions is "N", i.e., you must deliberately enter "Y" to allow rewriting of files. As the program begins to work, it will address each selected patient's last name, look for a file on the disk with name corresponding to the first eight letters of the last name (extension .KIN) and update that file with information contained in the current file or create a new patient file if none is found. If no file corresponding to the patient's last name is found on the disk and you have responded "Y" to the update question above, the message "new file" appears as the new file is created. All of this is done in a few seconds and the results are displayed on the screen without user intervention.

While the program is creating, examining and updating patient files as described above, a new summary file is created and given the extension .SUM. This is a file of records, one for each selected patient, containing the mean and standard deviation for most of the patient's dialysis kinetic variables. The variables are selected from the last *N* records you previously specified. If less than three records are found for a given patient, only the mean value is stored. The number of records found for each patient is stored, the maximum determined by your response to earlier questioning (see above).

The report is provided in hard copy form only because the standard IBM screen cannot display all of the variables. First and last dates for the *N* records selected are listed; if there are more than two, the number of records used to generate the mean and SD are also listed.

Two safeguards are built into the file identification part of this program:

1. After finding the patient file (last names match) the ID number is checked for identity. If these do not match, the message "mismatched" appears and the file is skipped. This message should not appear, but when it does, check for patients with identical last names (or identical first eight letters in the last name). The program requires that each patient have a unique last name and that the first eight letters of that name be unique. For patients with identical last names it is suggested that a number be given (e.g., SMITH1, SMITH2, etc.).
2. While it is updating each patient file, the program looks at the date of each record in the file and compares it with the date in the current file. If any of these match, the message *no new update* appears and the patient file is left undisturbed.

PRINT SUMMARY REPORT (means and SD's) (8-5)

This routine prints previously prepared summary reports identified by the .SUM extension on disk. No further input is requested after this selection is made. The report is provided in hardcopy form only, so the printer must be ready.

MAP IDEAL TREATMENT DOMAINS (8-6)

The domain map shows all selected patients as points on a graph of normalized protein catabolic rate (PCRn) versus time-averaged BUN (TAC). This map is patterned after the mechanistic analysis of data collected by the National Cooperative Dialysis Study (NCDS), 1978-1980 (4). It differs from previously published graphs only on the *y*-axis, which is TAC instead of midweek predialysis BUN. By using TAC, all patients can be shown, regardless of their dialysis schedule, on one map; this allows direct comparisons among patients treated on different schedules.

The shaded area is the safe domain as determined by mechanistic analysis of the NCDS data. On the *x*-axis, it has an upper boundary of 1.4 g/kg/day and a lower boundary of 0.8 g/kg/day. These are arbitrary limits on either side of 1.1 g/kg/day, the accepted target protein intake for hemodialyzed patients. If PCRn falls below 0.7 g/kg/day or exceeds 1.5 g/kg/day, an asterisk appears on the report form listing all patients' variables (function #2, Reports menu) and a message appears under Comments on the single patient printout (function # 3, Reports menu). The domain map extends from 0.5 to 2.0 g/kg/day on the *x*-axis. For the occasional patient with PCRn outside of this range, no data point will appear.

On the *y*-axis, the upper boundary is a *Kt/V* of 0.9/dialysis and the lower boundary is a *Kt/V* of 1.5/dialysis for the first half of the graph (up to PCRn = 1.1 g/kg/day). *Kt/V* represents a constant dose of dialysis. The target therapy modeling line follows the *Kt/V* isopleth of 1.05/dialysis. *Kt/V* isopleths are linear on this type of map; three of these are shown as dotted lines extending to the upper right-hand corner of the graph. Although there is disagreement about how much more to administer, there is universal agreement that patients with higher protein catabolic rates require more dialysis. The mechanistic analysis suggests that the target TAC should stay within the bounds of NCDS patients who had favorable outcomes. This accounts for the new slope of the target line starting at a PCRn of 1.1 g/kg/day

Several features of this map require further explanation. The plotted data are PCRn on the *x*-axis and TAC on the *y*-axis. *Kt/V* isopleths are drawn for convenience only as a reminder of the linear relationship between *Kt/V* and the other two variables. The value for *Kt/V* shown for each patient on this graph may differ from that calculated by simple multiplication of *K/V* times *t*. For patients with residual kidney function, the graphed value of *Kt/V* includes the residual function component. Because residual urea clearance and dialyzer urea clearance occur at different times in the weekly cycle, they cannot be simply added. The graphed value for *Kt/V* includes a derived contribution from residual function that is explained in chapter 7. For patients dialyzed on weekly schedules other than Monday-Wednesday-Friday or Tuesday-Thursday-Saturday, the graphed value for *Kt/V* will also differ considerably from the calculated value. Although the *x*-axis and *y*-axis of the map are applicable to all patients regardless of their weekly schedule, the *Kt/V* isopleths assume a 3x/week schedule. The graphed *Kt/V* value represents the intensity of dialysis that would be administered three times/week to maintain TAC at the indicated level in a patient with the indicated protein catabolic rate. It correlates with the simple calculated value only if the patient is actually dialyzed three times/week (and has no residual function). This confusion between calculated *Kt/V* and modeled *Kt/V* underscores a pitfall from relying only on *Kt/V* for assessment of dialysis adequacy. It is much better to look at time-averaged BUN as a direct measure of dialysis outcome, compare it to protein intake, and ignore *Kt/V*. The latter parameter is best viewed as a measure of dialysis intensity, not adequacy.

An option is provided just before the graph is plotted to use Numbers or Points to plot the graph. If "N" is chosen, only the first nine points will be plotted to avoid hopelessly cluttering the graph with unreadable numbers. The Numbers option allows interpretation of each point on the graph as a specific patient. This feature can be used in conjunction with the sort routine (function 7) to examine groups of patients that fall outside the ideal domain (e.g., to evaluate patients with *Kt/V* < 1.0/dialysis or PCRn < 0.8 and > 1.4 g/kg/day.

GENERATE ASCII FILE (8-7)

This option provides an easy way to use the data generated by this program with Lotus 123 (TM Lotus Corporation) or in other programs, spreadsheets, data bases, and text processors. Twenty-nine variables are dumped to a comma-separated standard ASCII (American Standard Code for Information Interchange) file. String variables are bracketed with quotation marks. The file is given the same name as the open parent file but with the extension .PRN. To help identify the data, a heading line is sent first followed by the data corresponding to the headings. Patients are listed in selected order. If the .PRN file is loaded by a spreadsheet such as Lotus 123 or SuperCalc (TM Sorcim Corporation), the headings will occupy the first row of cells, followed by patient data in columns under the appropriate headings. To load the file with Lotus 123, first load the template UREA123.WK1. Then load the .PRN file with the File Import function and designate the data as Numbers. The template sets the appropriate column widths so that the headings line up with the columns of data.

Each variable, its unit designation, and a brief description is listed in table 6.11 in order of appearance in the file:

Table 6.11 Variables dumped to ASCII file

Variable	Unit	Description
Patient Name		Patient name
ID number		Patient identification number
Date		Date of study
Dialyzer		Model of dialyzer (your code)
Schedule		Days of week (e.g., MWF)
Day		The weekday of study (must be contained in Schedule)
#/wk		Number of dialyses per week
T_d	hours	Duration of dialysis for kinetic study
T_{da}	hours	Average time on dialysis
Q_b	ml/min	Dialyzer whole-blood flow (average)
Q_d	ml/min	Dialysate flow
K_d	ml/min	Dialyzer whole-blood urea clearance
K_r	ml/min	Patient native kidney clearance of urea
BUN-1	mg/dl	Predialysis blood urea nitrogen (BUN)
BUN-2	mg/dl	Postdialysis BUN
Wt-1	kg	Predialysis weight
Wt-2	kg	Postdialysis weight
TAC	mg/dl	Time-averaged BUN
PREa	mg/dl	Average predialysis BUN
Kt/V	per dialysis	Dialyzer clearance multiplied by time on dialysis divided by V
V	liters	Calculated volume of urea distribution
V_p	liters V	projected (estimated) by user
G	mg/min	Calculated urea generation rate
PCRn	g/kg/day	Protein catabolic rate, normalized
PCRp	g/kg/day	PCRn when $V = V_p$
K_{dp}	ml/min	Kd when $V = V_p$
—ideal——		
TAC	mg/dl	Target TAC calculated from area under the curve
PREa	mg/dl	Target average predialysis BUN
T_d	hours	Target duration (time) of dialysis

UTILITIES (9)

A menu of utility functions is displayed by pressing #9 from the main menu. These functions allow installation of printers, establishment of default entry parameters, and a variety of other useful support functions for UREAKIN. The stand-alone program that provides these services can be loaded from outside UREAKIN by entering UK-UTIL at the DOS prompt. Default parameters that are modified by these utility functions are stored in a file called UREAKIN.DTA discussed at the beginning of this chapter. These parameters include printer control codes, the default disk drive, the default subdirectory for data files, and the codes for dialyzers with their corresponding mass transfer coefficients (*KA*).

Subroutines are selected from the Utilities menu by entering the number at the left corresponding to the desired function and then pressing ENTER. Function keys 2 through 10 also substitute for numbered entries.

A HELP function is available by pressing F1 from the UTILITIES menu.

Table 6.12 UTILITIES MENU

1) Set default disk drive\directory for data files
2) Select printer driver
3) Calculate dialyzer mass transfer coefficient
4) List dialyzers/enter new dialyzers
5) Print kinetics input/instruction sheet
6) Print instructions for clearance and reflow analysis
7) Print residual clearance entry form
8) Print heading abbreviations for complete report
9) Calculate *V*, nitrogen removal during a single dialysis
10) Calculate average blood flow
11) Estimate initial dialysis time
12) Change intercompartment mass transfer coefficient (2-pool)
13) Toggle sound on/off
14) Change printer port (LPT1 or LPT2)

Following are descriptions of each function listed in the main Utilities menu:

Set default disk drive\directory for data files (U-1)

The MSDOS disk\directory for data files can be changed from UREAKIN's main menu by pressing "L," but the change is only temporary. Each time you load UREAKIN.EXE, the disk\directory will revert to its default configuration. This utility function changes the default drive\directory permanently. When a new drive\directory is entered, a check is made for proper format: a colon ":" after the disk drive and a backslash "\" before the subdirectory. If these are not found, you will be asked to reenter the drive\directory.

Select printer driver (U-2).

A list of available printer drivers appears on the screen with corresponding numbers to the left. These drivers are stored in a file called UREAKIN.DRV. After the list is printed you are asked to select one of the drivers by number.

The last entry in the list is ENTER YOUR OWN PRINTER CODES. If you select this entry, a series of questions about your printer codes will appear. You are asked to enter the decimal equivalent of the codes for turning enlarged print on and off, turning compressed print on and off, bidirectional and unidirectional printing, and codes for setting the printer's left margin. It may be necessary to seek expert advice if customized printer codes must be entered. For most users, one of the prepackaged drivers such as the Epson or HP-Laserjet can be selected by pressing "1" or "2," and nothing else need be entered.

Calculate dialyzer mass transfer coefficient (U-3)

A place (field) is provided in the patient data base for the patient's brand or model of dialyzer. An array of dialyzers models with their respective mass transfer area coefficients (*KA*) is stored in UREAKIN.DTA. When a parameter that affects clearance is changed within UREAKIN.EXE, the dialyzer array is searched for a match. If a match is found, the stored (*KA*) is used to calculate a new clearance. If no match is found, a calculated value for (*KA*) is used to modify clearance.

Since most dialyzer manufacturers do not provide values for *KA*, this utility routine is provided to calculate *KA* from blood and dialysate flows and clearance measurement. The formula for *KA* is given below:

$$KA = \frac{Q_b \cdot Q_d}{Q_b - Q_d} \ln \left[\frac{1 - K_d/Q_b}{1 - K_d/Q_d}\right] \qquad 6.1$$

K_d is urea clearance (ml/min)
Q_b is blood flow (ml/min)
Q_d is dialysate flow (ml/min)
ln is the natural logarithm

The mass transfer area coefficient (*KA*) is a measure of the dialyzer's intrinsic capacity to transfer urea from the blood to the dialysate compartment (or vise versa) and is independent of blood and dialysate flow. Conceptually, it can be considered the rate of urea clearance at infinite blood and dialysate flow. Under such purely theoretical conditions, the only factor limiting diffusion is the membrane itself, i.e., transport of urea is purely membrane- limited. Since the dialyzer membrane thickness and surface area are part of *KA*, its unit of measurement is ml/min rather than cm^2/min, the typical unit for a diffusion coefficient. The latter is a fundamental property of the membrane material, familiar to physiologists and students of membrane transport.

We strongly advise that each dialysis center measure *KA* for individual dialyzers and equipment employed in the treatment center during actual patient dialyses. The manufacturer's (package insert) data for dialyzer clearance is often obtained under in vitro conditions where the dialyzer is tested by pumping a saline solution containing urea through the blood compartment. Urea clearances for whole blood will differ from these in vitro clearances because of the higher viscosity, lower water content, and the presence of red cells. To help with these measurements, this utility function also allows input of predialyzer and postdialyzer BUN in place of clearance measurements for calculation of mass transfer coefficients.

Because blood pumps measure whole-blood flow and dialysis personnel are not accustomed to converting blood flow to blood water flow, only whole blood flow is requested by UK-UTIL.EXE. Likewise, blood urea nitrogen concentrations need no correction for plasma or serum water content prior to calculating *KA* or urea clearance. Conversions of blood flow to blood water flow and whole-plasma concentration to plasma water concentration are made by UREAKIN.EXE before modeling, but these conversions are transparent to the user. Similar conversions are made by UK-UTIL.EXE before calculation of *KA* and urea clearance. Urea clearance is always reported as blood water clearance, not whole-blood clearance. In the absence of ultrafiltration, blood water urea clearance = 0.9 (whole-blood clearance).

Because *KA* is the mass transfer area coefficient for blood water in the absence of ultrafiltration, the effect of ultrafiltration is subtracted.

Note also that nothing is stored in memory or on disk. This utility is simply a calculation aid, so the results should be jotted on paper, to ensure accurate transfer to utility function #4, described below.

List dialyzers/enter new dialyzers (U-4)

This utility function lists dialyzer codes with their individual mass transfer coefficient (*KA*) and allows addition, deletion or editing of existing codes (refer to the above discussion of *KA*). Each code may have a maximum of eight characters. If more than eight are entered, the additional characters on the right will be omitted.

A maximum of 50 dialyzers may be entered.

Print kinetics input/instruction sheet (U-5)

A single page of instructions to the nurse/technician is printed with blank spaces for the patient's name, identification number, dialyzer, etc. You may elect to print more than one copy if you have a fast printer; otherwise print one sheet and copy the remainder. This form offers a convenient method for requesting kinetic studies and helps to ensure that all the pertinent data are properly collected during the dialysis. The printer's compressed print font is used to print the form.

The instruction sheet is stored in file INSTRUC.TXT and can be modified to suit your needs with any word processor or text editor that produces standard ASCII text.

Print instructions for clearance and reflow analysis (U-6)

Dialyzer urea clearance is easily measured by sampling arterial and venous blood simultaneously. By adding a third blood sampling, recirculation or *reflow* in the blood access device can be measured. Reflow is defined as the recirculation of blood from the venous access device (needle) to the arterial access device. This is an undesirable phenomenon that diminishes the removal of small-molecular-weight, easily dialyzable solutes. The high dialyzer ex traction ratio of these solutes produces a low venous concentration that accentuates the effect of reflow. Measurement of reflow in blood access devices should be done routinely during suspect dialyses. Suspicious signs include excessively high venous pressures, dark blood, collapsing arterial lines, inordinately high predialysis BUN, and a urea volume (V), measured by urea modeling, that is much higher than predicted. The latter results from overestimation of dialyzer clearance. When actual dialyzer urea clearance is lower than the value used for modeling, the BUN does not fall as much as it would if the clearance were accurate. To allow for the small decline in BUN at a higher clearance, the mathematical analysis concludes that the urea volume must be large.

The output here is similar to that for utility function #5: a single sheet of instructions with blank spaces for filling in three BUN values and blood flow. Space is also provided for the patient's identification, nurse/technician initials, date, and time. The peripheral blood sample is most conveniently obtained by slowing the blood pump momentarily, occluding the venous line and sampling from the arterial line. Unless blood flow through the access device is markedly reduced, this technique provides a sample representative of peripheral blood.

Like the previous instruction sheet, this sheet is stored in the file CLEARANC.TXT and can be modified with any word processor or text editor capable of ASCII output.

Print residual clearance entry form (U-7)

When this single sheet is placed at the patient's bedside, it signals the nurse/technician that a residual clearance measurement is requested while providing instructions and an input form. This form is meant to be used in conjunction with UREAKIN's main menu function #5, subroutine KR-Calc. The advantage of using this routine, rather than calculating clearance in the traditional way, is that no blood samples are required. The blood urea concentration, a necessary factor in the urea clearance calculation, is provided by the program. The patient records the starting and ending times of urine collection, and the program predicts the midcollection BUN. This prediction is based on the weekly BUN profile depicted by the graph shown under function #5 and the time of the last dialysis (see further discussion in the help routine for function #5, Edit Data). This entry form can also be modified by the user.

Print heading abbreviations for complete report (U-8)

The complete report, in Reports menu item #2, is printed with compressed figures (132/line) so that all data will fit within a single line on 8 1/2-inch-wide paper. To fit all the data in this way, the headings for each column of data must necessarily be abbreviated. Those familiar with urea modeling parameters may be able to interpret these abbreviations, but their meaning is not obvious to most. UK-

UTIL provides a hard copy of the abbreviations on the left with a detailed explanation of their interpretation on the right. The list can be posted for quick reference; a copy is best stored with the reports generated by main menu function #8.

Calculate V, nitrogen removal during a single dialysis (U-9)

This function provides support for teaching and research purposes so that students of dialysis may gain a better feel for the effects of change in certain variables on the outcome of dialysis and modeling in general. Several variables, such as protein catabolic rate (PCRn), that are requested can be approximated. The single-compartment model is used.

To calculate total urea nitrogen removed during a dialysis, only the predialysis BUN need be measured. If the urea volume (*V*) is not known, the postdialysis BUN can be substituted and *V* will be calculated from it. The curve for BUN versus time is integrated over the dialysis time interval to give total urea nitrogen removed. Since urea generation is small compared to urea removal during dialysis, only a rough estimate of PCRn is necessary. Relatively large changes in PCRn have only a minor effect on the calculation of total nitrogen removed. Likewise, the change in weight or fluid removed during dialysis has only a minor effect and can be approximated if patient weight is not monitored.

Calculate average blood flow (U-10)

Because blood flow often changes during dialysis, the rates must be averaged for each time interval to give a more accurate estimate of clearance. This routine will quickly average different flow rates of varying duration.

Estimate initial dialysis time (U-11)

When the patient with end-stage renal disease is first started on dialysis therapy, urea concentrations are not in a steady state, so UREAKIN cannot be used to prepare the maintenance dialysis prescription. Regardless of the prescription followed, steady-state urea kinetics will eventually be reached, but without a means of estimating the ideal dialysis time, this state may be far from what is desired. Initially, it is wise to dialyze frequently for short intervals to avoid symptoms of disequilibrium, but at some time early in the course of therapy, the question of ideal dialysis duration will arise.

This routine will give a reasonable estimate of ideal time on dialysis by making several assumptions about the patient. These assumptions may cause significant errors in some patients, especially for those with abnormal distributions of body fat or water or excessively low or high PCR. Therefore, as soon as the patient is established on a regular dialysis regimen (after three dialyses at full blood flow rate and time), predialysis and postdialysis BUN should be measured and formal urea kinetic analysis should be undertaken.

Change intercompartment mass transfer coefficient (U-12)

With addition of the two-compartment model, another variable, *KC*, is introduced that may require adjustment for certain patients and for kinetic studies of other solutes. The cellular mass transfer coefficient (*KC*), like the dialyzer mass transfer coefficient (*KA*), has units of ml/min. It is considered by most as a measure of the diffusion rate across the intracellular/extracellular compartments. The principal barrier to diffusion is the cell wall, a heterogeneous structure that includes virtually all cells in the body. Only one diffusion constant is included in the equations that describe movement of urea as a two-compartment phenomenon. The default value for *KC*, an average coefficient of diffusion for all cells, is 800 ml/min. This is the value obtained from analysis of rebound postdialysis and from extrapolation of in vitro studies.

This function permits the user to change value of *KC*. No testing of values outside the range of 100-1000 has been done, so unpredictable results may occur if the value is drastically altered.

Toggle sound on/off (U-13)

In a quiet environment, the beeping sound that occurs at the end of lines, with error messages, and as patient data is changed on the screen may be annoying. This routine offers an option to turn off the sound. Under ordinary circumstances, sound will be helpful and should not be turned off.

Change printer port (U-14)

If your computer has two printers or another device connected to the standard printer port (LPT1), you may divert all printer output from UREAKIN to the second printer port (LPT2) using this function.

Quit (U-Q)

This function exits the utility menu and returns to UREAKIN, to DOS, or to the calling program.

DELETE (D)

You will be asked twice to confirm deletion of the file name you have entered. Be careful with this function; once deleted, the file cannot be retrieved.

LOG (SET) DISK\DIRECTORY FOR DATA FILES (L)

This utility changes the default drive or subdirectory for patient files that is displayed at the main menu. It is possible to store different patient shifts on separate disks or subdirectories, but this will create problems with the summary function (8-4). It is better to store all patient files on the same disk\ directory. This disk\ directory may be separate from the program files. The summary files with extension .SUM and monthly .KIN files may be moved to another location after they have been created, since only the individual patient files are used by the summary routine.

QUIT (Q)

This utility function exits to DOS or to the calling program.

REFINEMENTS TO UREAKIN

UREAKIN includes many practical refinements and adaptations that have been added after several years of experience and experimenting with urea modeling. For instance, we assume that the rate of weight gain between dialyses is constant and that each treatment is designed to bring the patient's weight back to a stable dry weight. Since more fluid accumulates during a long interdialysis interval, the rate of weight loss (*dV*) during dialysis is higher after a weekend off dialysis. This adaptation adds a small degree of precision to the analysis that assumes more importance for patients with larger fluid accumulations between dialyses. It also means that the theoretical effect of weight loss or gain between and during dialyses can be explored and accurate predictions can be made above the range that is seen clinically.

A similar refinement concerns the average weight loss during dialysis shown on the Alter Prescription screen (main menu item #6). The weight loss shown is not the weight change recorded for the study dialysis but is the average value computed from the interdialysis weight gain. The fluid removal required to bring the patient's weight back to its dry value is computed, summed, and averaged for the week. Thus if a patient who is dialyzed three times/week is studied on the first day after the weekend, the weight loss during dialysis is assumed to be greater than that lost during subsequent dialyses, where shorter interdialysis intervals are associated with less weight gain. Thus, for similar weight changes, the average weight loss and modeling results will differ modestly when the study is done on different days of the week.

Urea kinetic modeling can be fun. The computer takes the drudgery out of mathematical computations and makes an enjoyable exercise out of an otherwise tedious, repetitive task. The ability to quickly change input variables and recompute *V*, *G*, TAC, and ideal time on dialysis provide an especially satisfying solution to questions that, in the past, may not have been asked because they would have taken several months to resolve. "What if" adventures can be taken by the physician, dietitian, or anyone, including the patient, to determine the optimum prescription that will fit the individual's needs and the dialysis center's schedules, equipment, and time. The instrument that makes all of this possible, the computer, fulfills the dream of Von Leibnitz, whose quote at the beginning of this chapter reflects his need and respect for mathematics tempered by a realistic perspective. The computer allows the intellect to extend itself even beyond the range of this seventeenth century mathematician's gifted imagination.

REFERENCES

1. Sanfellipo ML, Hall DA, Walker WE, Swenson RS: Quantitative evaluation of

hemodialysis therapy using a simple mathematical model and a programmable pocket calculator. Trans Am Soc Artif Intern Organs 21:125-131, 1975.

2. Walker WE, Hall DA, Sanfelippo ML, Swenson RS: Application of a programmable pocket calculator to a single compartment mathematical model of solute kinetics. Comput Programs Biomed 75:99-104, 1975.
3. Guthke R, Gunther K, Stein G, Knorre WA: Two-pool model analysis of data in hemodialysis by means of programmable pocket calculator TI 59. Comput Prog Biomed 19:189-195, 1985.
4. Gotch FA, Sargent JA: A mechanistic analysis of the National Cooperative Dialysis Study (NCDS). Kidney Int 28:526-534, 1985.

Chapter 7

REFINEMENTS AND APPLICATION OF UREA MODELING

MEASURING BLOOD UREA CONCENTRATIONS

Urea versus urea nitrogen

Nonprotein nitrogenous compounds (NPN) are the major excretory constituents of urine, so serum levels are a measure of renal function or clearance. The most abundant NPN compound in both serum and urine is urea. Urea constitutes approximately 50% of serum NPN in people with normal renal function and a higher percentage in those with renal failure (1). Traditionally, clinical laboratories report the serum urea concentration as urea nitrogen in milligrams per deciliter. This unusual practice of expressing urea concentration as urea nitrogen instead of whole urea is better understood after reviewing the history behind blood testing of renal function. Specific tests for urea have replaced the older clinical tests for nonprotein nitrogen. These older assays for total nitrogen in deproteinized serum measure a variety of nitrogenous compounds, including creatine, creatinine, uric acid, and amino acids, so the sum of their concentrations had to be expressed in terms of their common element, nitrogen. Specific urea assays replaced the more tedious NPN test, but the results continued to be reported as milligrams of nitrogen per deciliter, a more familiar term to the clinicians of old. The two nitrogen atoms in each urea molecule make up nearly half (28/60) of its molecular weight, so urea nitrogen concentrations are slightly less than half the value of total urea concentrations. Today, the assays for blood urea are more specific than they were in the past, but the results are often expressed in the same units as the older tests for nitrogen. North American laboratories continue to report urea concentrations as urea nitrogen while clinical laboratories in other countries report whole urea concentrations. It is important to take note of the units of measurement, especially when laboratory results are received from foreign sources.

Blood urea nitrogen versus serum urea nitrogen

Blood urea nitrogen concentration or BUN is measured in samples of clotted whole blood sent to the clinical laboratory, where the serum is separated and assayed for urea. Red cells are separated from the serum before the sample is analyzed for urea, so BUN is really not ***blood urea nitrogen*** but ***serum urea nitrogen***. The term BUN was popularized in the distant past, when whole-blood urea nitrogen was routinely analyzed, avoiding the red cell separation step. This expression has been entrenched in medical parlance and refuses to let go, despite ongoing attempts to change BUN to the more accurate term, SUN.

The difference between whole-blood urea concentration and serum urea concentration is 3% to 5% for patients with normal hematocrits (1). SUN is slightly higher than BUN because urea is distributed only in the water compartment and erythrocytes have a lower water content than plasma: 72% for erythrocytes versus 93% for plasma (2). But the distribution coefficient for red cell urea is closer to 86% (2). The discrepancy between red cell water distribution and urea distribution is at least partially explained by studies that show binding of urea to red cells, principally hemoglobin. Considerable evidence for approximately 20% red cell binding of urea has accumulated over the past eight decades (3,4,5,6,7,8). Figure 7.1 shows a diagram of red cell urea distribution. The additional urea bound within the red cell increases the apparent urea distribution to 86% of cell volume (2). Noncovalent binding of this type has a very short time constant, usually measured in milliseconds (9,10). Because of the rapid equilibration at the binding interface and the relatively slow movement of red cells through the dialyzer (transit time = 10-30 seconds), intracellular binding has the effect of increasing the total (bound + unbound) amount of urea available for diffusion within the erythrocyte.

Clinical laboratories may differ slightly in BUN measurements for the same sample of blood. Also, two different instruments in the same laboratory may give somewhat different results. This can lead to errors in modeling urea kinetics. Better accuracy is obtained when all blood samples from a single patient are analyzed at the same time with one instrument in one laboratory.

Compensation for blood and plasma water content

Effect of whole blood on kinetic measurements

The artificial kidney extracts less urea from whole blood than from equimolar aqueous solutions of urea, though the concentration in plasma water is higher under these conditions. The cause of this reduction in urea clearance is multifactorial and not completely understood. Table 7.1 lists the causes that have been invoked, some on purely theoretical grounds without experimental evidence.

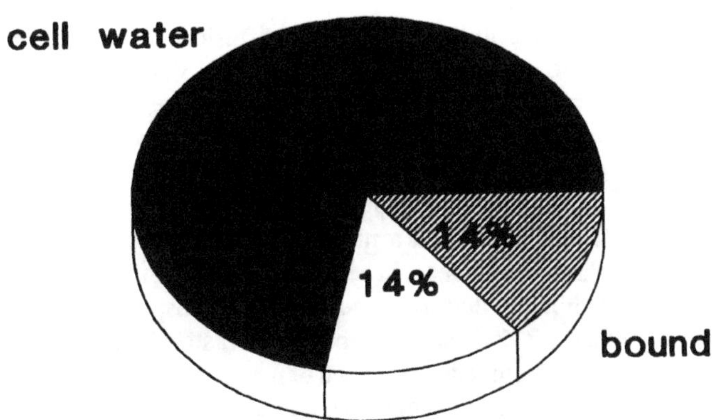

Figure 7.1. Erythrocyte urea content as percent of cell volume. Total urea consists of freely diffusible urea in the cell water compartment (solid area) plus 14% bound to intracellular hemoglobin (striped area). The distribution coefficient is the sum of these (86%). The clear area represents nonaqueous red cell volume that contains no urea.

Table 7.1 Potential causes of lower urea clearance in whole blood versus pure aqueous solutions

1. Reduced red cell water content
2. Reduced plasma water content
3. Impaired diffusion of urea through red cells
4. Increased depth of unstirred layers at the dialyzer membrane
5. Reduced diffusion of urea in plasma
6. Red cell or plasma binding of urea
7. Uneven blood flow due to clotting or cellular clumping
8. Loss of surface area for diffusion due to fibrin formation or "coating" of the membrane

The first two causes listed in table 7.1 probably contribute most of the effect (11,12). The fourth and fifth potential causes may explain small differences in clearance measured in plasma versus pure aqueous solutions. The last two causes can affect clearance in dialyzers that promote thrombus formation or when patients are inadequately anticoagulated. Clotting and fibrin formation should not be a factor in properly constructed and adequately heparinized dialyzers.

There is little evidence for significant plasma binding of urea, but intracellularly, urea may bind to hemoglobin. Because of its low affinity for urea and the rapid time constant for equilibration, hemoglobin binding actually increases urea transport, as discussed above.

Erythrocyte urea transport

Concern about the possibility of impaired diffusion of urea in red cells has heightened since the arrival of erythropoietin therapy (13). Genetically engineered erythropoietin replaces the natural kidney-derived hormone that serves to maintain red cell production in normal individuals. Erythropoietin is probably even more important in the uremic environment where inhibitors of bone marrow erythropoiesis are found (14). The hematocrit response to pharmacologic doses of erythropoietin in patients with renal failure has a marked beneficial effect on oxygen delivery and energy metabolism. The increase in hematocrit also has the potential to decrease solute transport within the dialyzer. The diffusive transport of solutes such as phosphate and vitamin B12 is reduced when the hematocrit rises, but most of the evidence supports free unimpeded diffusion of urea from red cells during the relatively slow transit of these cells through modern dialyzers (15,16,17). The red cell may be unique in its capacity to transport urea rapidly across its cell membrane.

Other cells and tissues do not enjoy such free and rapid diffusion of urea. The postdialysis rebound in urea concentration, as discussed in chapter 5, clearly shows that the rate of urea diffusion between body compartments is finite. Resistance to urea diffusion during dialysis causes a gradient to develop between the blood and other more slowly equilibrating compartments. The rate of decline in urea concentration, especially at the end of dialysis, determines the severity of urea disequilibrium between body pools at the end of dialysis and hence the magnitude of rebound. A steeper fall in urea concentration with time causes more disequilibrium and a larger postdialysis rebound.

The steepest decline in urea concentration occurs inside the dialyzer itself, where within a time span of approximately 20 seconds, the concentration falls from blood inlet levels, often 50 to 100 mg/dl, to outlet levels of 10 to 30 mg/dl. Consequently, blood cells traversing the dialyzer are subject to the greatest disequilibrium, and a lag in urea diffusion should be demonstrable. Despite early reports of red cell urea disequilibrium (11), no lag in urea diffusion has been convincingly shown in recent years using modern techniques (15). Erythrocytes have been studied more extensively than perhaps any other cell in the body, and the transport of urea across the red cell membrane is no exception. Studies of urea diffusion in red cells have concluded that the mammalian red cell behaves differently from other cells in the body (18,19,20).

Urea transport across red cell membranes appears to occur by facilitated diffusion (18,19,20,21,22). Facilitated diffusion is a carrier-mediated, often energy-requiring transport that is more efficient than passive diffusion across inert membranes. This efficient pathway for movement of urea may have evolved to protect red cells from osmotic damage during their passage through the renal medulla (21). In the normal renal medulla, red cells are subject to very high urea concentrations within a time span measured in seconds; they then reappear within seconds in the cortex, where they are subject to low osmotic pressures. Osmotic damage to red cells passing through the renal medulla can be prevented by a rapidly equilibrating mechanism for urea transport that maintains osmotic stability (21). Teleological reasoning suggests that urea was selected for its role in the countercurrent concentrating mechanism because of its rapid diffusibility, or conversely that facilitated diffusion developed to allow medullary concentration gradients to enlarge. Transit of red cells through hollow fiber dialyzers is analogous to movement of these same cells through the capillaries of the renal medulla. In the dialyzer they encounter marked fluctuations in urea concentration within a time span also measured in seconds. Hemodialyzers take advantage of the facilitated transport pathway in red cells to increase the blood clearance of urea. Facilitated diffusion of urea partially explains the disparity between dialyzer creatinine clearance and dialyzer urea clearance. No facilitated diffusion has been shown for creatinine.

For urea modeling, it is important to measure urea clearance in vivo, preferably using equipment within each dialysis treatment center. Corrections for plasma or blood water content are usually necessary when calculations call for blood water or plasma water clearance. Many calculations used in modeling urea kinetics involve ratios of urea concentrations, e.g., $(C_{in} - C_o)/C_{in}$ in the equation for calculation of dialyzer clearance (equations 7.6, 7.11). Where ratios are used, errors are canceled and no correction for plasma water is necessary. However, if dialysate urea is measured and used to calculate clearances, blood flow must be corrected for water content. Compensation for blood water content is also necessary if whole-blood clearance is used to calculate urea volume. Following are more detailed analyses of these corrections.

Effect of hematocrit on urea removal

When blood traverses the dialyzer, urea diffuses out of red cells at a rapid rate, apparently unimpeded by the cell membrane. The additional urea loosely bound to and in rapid equilibration with hemoglobin within the cell helps to compensate for the low water content of the red cell compared to plasma. The water fraction of red cells is 0.72% whereas the urea distribution coefficient for these cells is 0.86 (cell/plasma water) compared to a plasma coefficient of 0.93 (23). The flux of urea from whole blood across the dialyzer membrane is only slightly less than urea flux when the dialyzer is perfused with cell-free plasma solutions. Since the red cell behaves like plasma, only a small additional compensation for hematocrit is necessary even for patients with relatively high hematocrits undergoing high-flux dialysis (16,24,25).

When it is necessary to measure precisely the effects of hematocrit and plasma water on urea delivery to the dialyzer, a quantitative expression of the above relationships is helpful. The following equation gives the appropriate correction of blood flow rate (Q_{bi}) for hematocrit (hct) and plasma water:

$$Q_{biw} = Q_{bi}\left[0.86(\text{hct}) + 0.93(1 - \text{hct})\right] \qquad 7.1$$

Q_{biw} is effective blood water flow through the dialyzer

Equation 7.1 can be used to show that over the range of hematocrits observed in dialyzed patients (_15% to 40%), the fractional adjustment varies from 0.902 to 0.920. This variation is not considered significant enough to warrant inclusion of hematocrit as a variable in some programs that describe urea kinetics.

Urea distribution volume: compensation for plasma and blood water

Plasma or serum volume is made up of 93% water and 7% nonaqueous bulk, most of which is plasma protein. Since urea dissolves only in the water fraction and is not significantly bound to plasma proteins, plasma or serum urea concentrations measured by the lab are about 7% less than true concentrations in plasma water.

For reasons discussed above, the urea distribution fraction or apparent water fraction of whole blood is approximately 90%. This means that blood water flow through the dialyzer appears to be 90% of whole-blood flow, and blood water clearance will be 90% of whole-blood urea clearance. Dialyzer urea clearance can be measured in vivo using serum urea or urea nitrogen concentrations and whole-blood flow rates (equation 7.6). The resulting whole-blood clearance must be used cautiously. If whole-blood urea clearance is used to compute the amount of urea removed from the blood side of the dialyzer, no compensation for blood water content is necessary. If urea volume is modeled from whole-blood clearances and whole-plasma or serum concentrations, a significant error is introduced unless adjustments are made for blood water. Figure 7.2 illustrates the source of this error. The calculated urea volume is presumed to behave like blood, i.e., approximately 90% is water. More than 80% of urea volume consists of interstitial and intracellular fluid that contains no plasma or red cells. A correction factor for the inflated volume shown by the dotted line in figure 7.2, must be made to bring the urea volume down to its true value, represented by the solid lines in figure 7.2.

To obtain the true urea distribution volume, two methods of correction have been applied. Before calculating V (e.g., using equation 4.17), dialyzer urea clearance can be corrected to blood water clearance, reducing it to approximately 90% of whole-blood clearance. The second technique applies a similar 90% correction factor to the urea volume after its value is determined from the model calculations using whole-blood urea clearance (equation 4.17). These adjustments provide more accurate estimates of V. It is also important to note that corrections for plasma water content must be included to obtain accurate values for G. The BUN before and after dialysis must be expressed as plasma water concentration, an adjustment that increases the value in whole serum by approximately 7%. If no corrections are made for blood and plasma water content with either of these techniques, both V and G will be overestimated. These methods of correction are valid both for one-compartment and multicompartment models.

In summary, to avoid the pitfalls associated with differing water content of blood and plasma, include a correction factor that converts urea clearance to average blood water clearance (~90% of whole-blood clearance) and another correction factor to

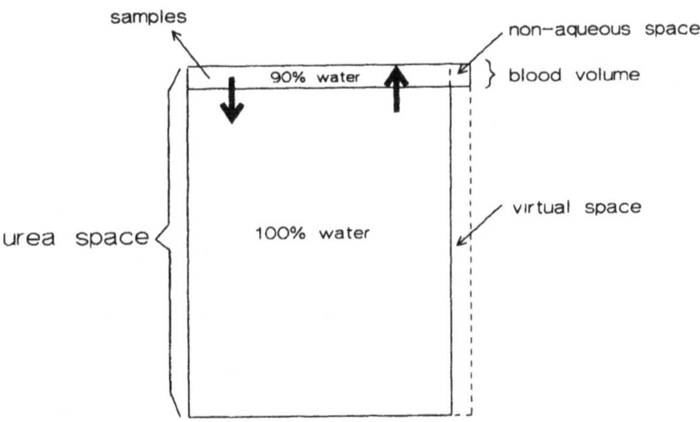

Figure 7.2. Calculating urea distribution volume (V). Urea volume is depicted as area. Ten percent of blood volume is nonaqueous space, shown in the upper right corner. Only the aqueous space equilibrates with interstitial and intracellular spaces. If no correction is made for blood water, urea volume will be inflated by 10% due to addition of the area labeled *virtual space*.

correct urea or urea nitrogen concentrations for average plasma water content (~93%). Blood water clearance is estimated by multiplying whole-blood clearance by 0.90; plasma water concentrations are obtained by dividing each BUN by 0.93. It then becomes important to distinguish these corrected values when reports are issued. When blood flow and clearance are reported, they should always be distinguished as blood water flow/clearance or whole-blood flow/clearance. Modeling programs should convert all serum water concentrations back to whole serum values before reporting results, because clinicians are familiar with whole-blood values.

The correction factors for water content of blood and plasma are average values that are reasonably constant from patient to patient. Significant errors will occur if these factors are applied to patients with severe lipemia, sometimes observed during hemodialysis (26).

Dialyzer urea clearance

How dialyzer clearance is measured

Clearance instead of *dialysance* is used as a measure of dialyzer function because most modern dialysate delivery systems use the *single-pass* technique. For single-pass systems, fresh dialysate is pumped into the dialyzer and all outflow from the dialyzer is discarded to waste. Because the dialysate inflow concentration of urea for such systems is always zero, the expressions of clearance and dialysance are equivalent (equation 3.1). Refer to the discussion of dialysance and clearance in chapter 3.

The major variables that affect urea clearance are blood flow, dialysate flow, membrane permeability, and membrane surface area. The contributions of transmembrane pressure and convective filtration to clearance are discussed below. Other variables such as blood and dialysate channeling have additional minor effects that are usually ignored. Each dialyzer model has a different but constant membrane permeability and surface area. So the capacity of each dialyzer to remove urea can be condensed to a single expression, its *mass transfer area coefficient* (*KA*) for urea. *KA* is a solute-specific property of the dialyzer that is independent of blood or dialysate flow. Expressed in units of ml/min, *KA* may be considered the maximum urea clearance attainable, i.e., the clearance at infinite blood and dialysate flow rates. This concept is graphically illustrated in figure 7.3. The value for *KA* depends on the relative direction of dialysate and blood flow and can be computed from blood flow (Q_b), dialysate flow (Q_d), and dialyzer urea clearance (K_d). Equation 7.2 shows the relationship between *KA* and these three variables for countercurrent blood/dialysate flow used in nearly all modern systems (16):

$$KA = \frac{Q_b \cdot Q_d}{Q_b - Q_d} \ln \left[\frac{1 - K_d/Q_b}{1 - K_d/Q_d} \right] \qquad 7.2$$

KA is the mass transfer area coefficient (ml/min)

Q_b is blood flow (ml/min)
Q_d is dialysate flow (ml/min)
ln is the natural logarithm
K_d is dialyzer urea clearance (ml/min)

Equation 7.3 gives the expression for the less efficient co-current flow (dialysate and blood flow are parallel) (16):

$$KA = \frac{Q_b \cdot Q_d}{Q_b + Q_d} \ln \left[\frac{Q_b \cdot Q_d}{Q_b \cdot Q_d - K_d(Q_b + Q_d)} \right] \qquad 7.3$$

The only unknowns in these equations are blood flow, dialysate flow, and urea clearance. Therefore, a single measurement of urea clearance at fixed dialysate and blood flow rates allows calculation of *KA*. Selection of blood and dialysate flow in the therapeutic range will ensure accuracy of *KA*. Once determined, *KA* can be used to predict clearance at other dialysate and blood flows using the following equations, obtained from rearrangement of equation 7.2.

$$z = (KA/Q_b) \cdot (1 - Q_b/Q_d) \qquad 7.4$$

$$K_d = Q_b \frac{(e^z - 1)}{(e^z - Q_b/Q_d)} \qquad 7.5$$

Blood flow and clearance should be corrected for blood water because it is

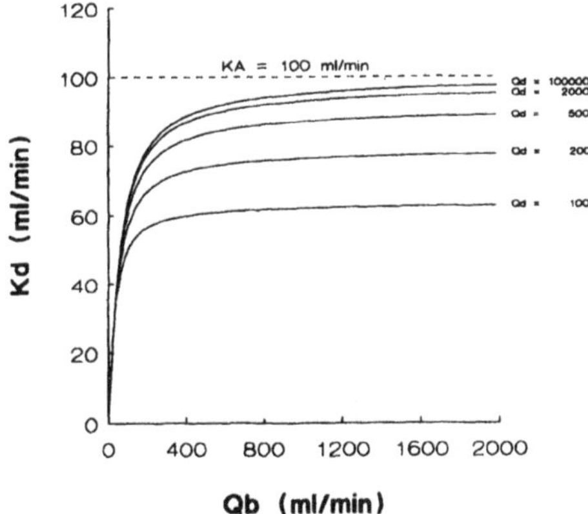

Figure 7.3. $KA = K_d$ at infinite Q_b and Q_d. Dialyzer urea clearance (K_d) is enhanced by raising blood flow (Q_b) and by raising dialysate flow (Q_d). Both effects show a ceiling or plateau effect where further increases no longer increase K_d. The dialyzer mass transfer area coefficient (*KA*) can be represented as the maximum clearance achievable at infinite blood and dialysate flow (dashed line).

inappropriate to add whole-blood flow to dialysate flow or indeed to combine any whole-blood and blood water values in the same equation. When *KA* is properly determined using effective blood water flow (equation 7.2), clearances calculated from *KA* (equation 7.5) will be blood water clearances. These are approximately 10% lower than whole-blood clearances, as noted above.

Figure 7.4 shows the relationships among the three variables, K_d, Q_b, and Q_d. Changing blood flow (bottom graph) or dialysate flow (top graph) establishes a new curve of flow versus clearance that has the same effect as changing *KA*. The steep upwardly rising portion of each of these curves represents the *flow dependent* range of clearance while the flat, more horizontal portion approaches the purely *membrane-dependent* clearance. Figure 7.3 shows that increasing either blood or dialysate flow increases urea clearance but there are diminishing returns as each of these variables is further increased.

Traditionally, nephrologists have considered dialysate cheaper than blood; that is, dialysate flow has always exceeded blood flow. Theoretically, the two solutions are equivalent, i.e., Q_b and Q_d are interchangeable in equations 7.2 and 7.3 (see figure 7.4), so to minimize total flow, Q_b and Q_d should be equal. But this conclusion ignores other costs and benefits of increasing dialysate flow versus increasing blood flow. Blood flow costs little to increase, but it is limited by the blood access device as prepump pressure falls and postdialyzer pressure rises. Dialysate flow is

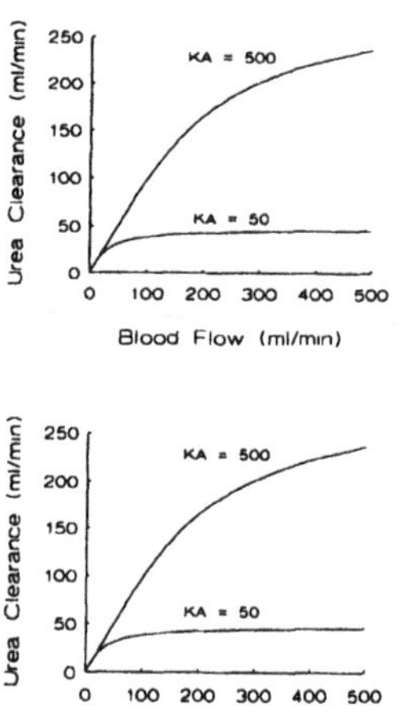

Figure 7.4. Blood and dialysate flow have equivalent but additive effects on dialyzer urea clearance. For solutes with low mass transfer coefficients (e.g., *KA* = 50 ml/min) clearance quickly reaches a plateau, the so-called *membrane limited* region. For solutes with high mass transfer coefficients (e.g., *KA* = 500 ml/min), clearance is linearly dependent on flow over a wide range of typical blood or dialysate flows, the *flow-dependent* region.

unlimited and can be increased without adverse effects on pressure, but it incurs progressively higher costs as the flow increases. The concern about costs has caused some nephrologists to reexamine this practice, especially when the newer high-flux dialyzers are used. Blood flows have approached dialysate flows in some dialysis centers where high-flux techniques are in place.

Measured versus modeled clearance

Two methods are used to estimate urea clearance on the blood side of the dialyzer:

1. Direct Method:

Measure inlet blood flow and both inlet and outlet urea concentrations.

$$K_d = Q_{bi} \frac{C_{in} - C_o}{C_{in}} \qquad 7.6$$

2. Modeled Method:

Measure the patient's blood urea concentration before dialysis and after dialysis and fit K_d to a mathematical model using a predetermined value for urea volume.

The first method measures clearance directly, using one of the techniques outlined above that gives an "instantaneous" clearance at any moment. This method tests the dialyzer itself and is commonly used to assess equipment function. Also, it can be done in vitro, without connection to a patient, although care must be taken to distinguish between whole-blood clearance and the higher clearances obtained from pure aqueous solutions.

The second method uses a mathematical model that determines the clearance required to cause the measured change in concentration during a timed dialysis interval. This is a patient or whole-body clearance, an average effective dialyzer clearance during the studied dialysis. If we ignore residual clearance, generation rate and weight loss during dialysis, effective dialyzer clearance of urea can be shown to be proportional to the log of the ratio of predialysis to postdialysis BUN. From the single-compartment model, equation 4.14 gives

$$K_d \cdot t = V \cdot \ln(C_1/C_2) \qquad 7.6a$$

The simplified equation 7.6a shows that the volume of urea distribution (V) is required to calculate urea clearance using the second method. Usually V is obtained from the average of previous modeling results. The modeled clearance is a more indirect method for measuring K_d but it provides a "whole patient" urea clearance. The modeled clearance is preferred by the clinician because it considers variables such as blood access recirculation and changes in clearance during dialysis that are not included with the more direct method.

The two techniques for measuring clearance are totally independent of one another. Thus, one can be used to check the other. A significant benefit of urea modeling comes from the assessment of patient urea clearance during dialysis.

Clearance computed using method 2 can be compared with expected clearance promoted by the manufacturer of the dialyzer or with direct measurements of urea clearance during the same or other dialyses. When significant differences are detected, attention is directed to potential flaws in the system. Efforts can then be concentrated on those few patients whose modeling results suggest problems. This approach forms the basis of an efficient quality assurance program.

EFFECT OF FLUID BALANCE ON KINETIC MEASUREMENTS

Effect of ultrafiltration on urea clearance

Does ultrafiltration during dialysis add to the clearance of urea? It seems that convective removal of urea from ultrafiltration during dialysis should add to diffusive transport, augmenting urea removal by the dialyzer. The increment in urea removal should correlate with ultrafiltrate volume. The intuitive answer to the above question is "yes," but the additional urea clearance from ultrafiltration is much less than one might hope. A consideration of mass balance for urea across the dialyzer gives the quantitative answer. The rate of urea removal by the dialyzer, J_u, is the difference between the entry rate, $C_{in} \cdot Q_{bi}$, and the exit rate, $C_o \cdot Q_{bo}$ (27).

$$J_u = C_{in} \cdot Q_{bi} - C_o \cdot Q_{bo} \qquad 7.7$$

Ju is urea removal rate by the dialyzer (flux)
C_{in} is inlet (arterial) urea concentration
C_o is outlet (venous) urea concentration
Q_{bi} is blood flow into the dialyzer
Q_{bo} is blood flow out of the dialyzer

Dialyzer urea clearance, K_d, is the removal rate divided by the inlet concentration, C_{in}.

$$K_d = J_u/C_{in} \qquad 7.8$$

$$= \frac{C_{in} \cdot Q_{bi} - C_o \cdot Q_{bo}}{C_{in}} \qquad 7.9$$

Any unit of measurement for Cin and Co can be used provided that the same unit is used for both. When ultrafiltration occurs during dialysis, the inlet blod flow rate can be expressed as a function of the outlet flow rate and Qf:

$$Q_{bi} = Q_{bo} + Q_f \qquad 7.10$$

Qf is the ultrafiltration rate.

Combining equations 7.10 and 7.9 and rearranging:

$$K_d = Q_{bo}(C_{in} - C_o)/C_{in} + Q_f \qquad 7.11$$

Equation 7.11 describes total clearance as the sum of ultrafiltration rate (Qf) and clearance calculated using outlet blood flow instead of inlet blood flow.

If we substitute Q_{bo} from equation 7.10 into equation 7.9:

$$K_d = Q_{bi}(C_{in} - C_o)/C_{in} + Q_f(C_o/C_{in}) \qquad 7.12$$

Equation 7.12 describes total clearance as the sum of conventional clearance (assuming no ultrafiltration) plus a fraction of the ultrafiltration rate. The fraction is the ratio of outlet to inlet solute concentration.

Equations 7.11 and 7.12 are equivalent and apply to any solute, because they are derived from consideration of mass balance alone. For urea and other small-molecular-weight solutes, additional expressions can be derived because the concentration in ultrafiltrate is essentially the same as in blood water (sieving coefficient of 1.0). These will be shown in the next section.

Equation 7.12 is the most practical expression, since blood pumps are usually placed upstream from the dialyzer and flow (Q_{bi}) is determined from blood pump calibration. Equation 7.12 shows that dialyzer urea clearance is the sum of the urea clearance assuming no ultrafiltration ($Q_{bi} = Q_{bo}$) plus the ultrafiltration rate factored by C_o/C_{in}. The latter term is best understood by examining extremes. If $C_o = 0$, i.e., all the urea is removed by the dialyzer, there is no additional contribution from ultrafiltration. This is intuitively obvious, since one cannot expect to improve on a 100% removal rate. Conversely, if $C_o = C_{in}$, i.e., no urea is removed by diffusion, then the additional clearance afforded by ultrafiltration is simply the ultrafiltration rate. This is the case for isolated ultrafiltration where there is no dialysate. For the usual dialysis, the actual contribution of ultrafiltration to urea clearance is small. The reduction in urea concentration across a standard dialyzer is usually 60 to 70%. From equation 7.12 we can anticipate that the contribution of ultrafiltration to urea clearance will average 30% to 40% of the ultrafiltration rate for the average dialysis. Since ultrafiltration rates during hemodialysis, even when high, are a small fraction of the dialyzer urea clearance, the additional boost in urea clearance is usually small. For example, consider a dialyzer with urea clearance of 200 ml/min. At an ultrafiltration rate of 2 liters/hour (33 ml/min), the contribution to urea clearance is 10 ml/min or an additional 5%.

For purposes of illustration, it is instructive to separate ultrafiltration (convective clearance) from dialysis (diffusive clearance) by placing a filter proximal to the dialyzer or by placing a filter distal to the dialyzer. Dialysate pressure can be adjusted to prevent ultrafiltration through the dialyzer. If filtration is considered to occur at the venous (distal) end of the dialyzer and the sieving coefficient is 1.0, C_o will not be affected by the filtration rate. The first term in equation 7.12 can then be considered to be the urea clearance when Q_f is zero:

$$K_d = K_{d0} + Q_f(C_o/C_{in}) \qquad 7.12a$$

K_{d0} is dialyzer clearance with no ultrafiltration

If filtration is considered to occur at the arterial proximal end of the dialyzer and the sieving coefficient is 1.0, the contribution of Q_f can be estimated by calculating the clearance at a reduced blood flow (Q_{bi} - Q_f) and adding Q_f (from equation 7.11):

$$K_d = K_{d2} + Q_f \qquad 7.12b$$

K_{d2} is dialyzer clearance when $Q_b = Q_{bi} - Q_f$ with no ultrafiltration

Although equations 7.11 and 7.12 are equivalent, equations 7.12a and 7.12b are not. Equation 7.12b will yield a slightly higher clearance than equation 7.12a because filtration at the arterial end of the dialyzer provides slightly improved urea removal. The difference can be appreciated only at high filtration rates and when clearance is considerably less than blood flow. If filtration occurs through the dialyzer along its entire length, then the actual contribution of filtration rate to clearance might be considered an average of equations 7.12a and 7.12b. But this is not the case, because filtration interferes with dialysis when both occur simultaneously across the same membrane (28). Thus even equation 7.12a slightly overestimates the contribution of ultrafiltration to urea clearance.

It is possible to estimate the contribution of ultrafiltration to urea clearance without measuring outflow urea concentration as required by equations 7.11, 7.12 and 7.12a. When there is no ultrafiltration, outflow urea concentration is obtained from a rearrangement of equation 7.9:

$$C_o = C_{in}(1 - K_{d0}/Q_{bi}) \qquad 7.13$$

When the above expression for C_o is replaced in equation 7.12a, we obtain:

$$K_d = K_{d0} + Q_f (1 - K_{d0}/Q_{bi}) \qquad 7.14$$

K_d is dialyzer clearance that includes the ultrafiltration component
K_{d0} is dialyzer clearance with no ultrafiltration

Equation 7.14 is useful when Q_f is known and clearance at zero ultrafiltration can be calculated from measured blood flow (Q_{bi}) and the dialyzer urea mass transfer coefficient. The ultrafiltration rate (Q_f) can be estimated from changes in weight that occur during dialysis or by directly measuring ultrafiltrate flow, a task made easy by volumetric dialysate delivery systems of recent vintage.

Finally, it is important to note that the urea clearance term in *Kt/V* discussed in chapter 3 and in more detail below, must include the contribution of ultrafiltration as calculated above.

Consequences of weight gain between dialyses

Fluid gained between dialyses must be matched by that removed during dialysis if the patient is to remain in fluid balance. Between dialyses, fluid accumulation causes a significant reduction in the predialysis BUN. This effect results from both expansion of V that diminishes the effect of urea generation and from dilution of V from exogenous fluid that contains no urea. Theoretically it would be possible to maintain postdialysis BUN constant if enough fluid were retained between dialyses. If this massive expansion could be achieved, simple ultrafiltration without dialysis would be sufficient to sustain patients with end-stage renal disease. Unlike maintenance hemofiltration currently used as an alternate treatment to hemodialysis, filtration of patients with massive weight gains would not require administration of intravenous fluids during treatments. For example, if postdialysis BUN is 50 mg/dl and urea generation rate is 3 mg/min, then 6 ml/min of fluid accumulation between dialyses would dilute the added urea sufficiently to maintain BUN constant. Filtration treatments would then be required to bring V back to dry weight as BUN remained constant. This would amount to over eight liters of fluid accumulation per day, a rate that even the most noncompliant patients do not achieve and that few patients would tolerate. But lesser quantities of fluid retained between dialyses can have a marked lowering effect on predialysis BUN. This emphasizes the importance of the variable-volume model for evaluating urea kinetics in patients with significant weight gain between dialyses. The constant-volume model cannot fully incorporate fluctuations in V during or between dialyses and can cause significant errors in estimates of predialysis and postdialysis BUN, as discussed in chapter 4. Figure 7.5 shows the BUN/time profile of a patient whose weight is constant. Superimposed is the same patient's profile when fluid accumulation

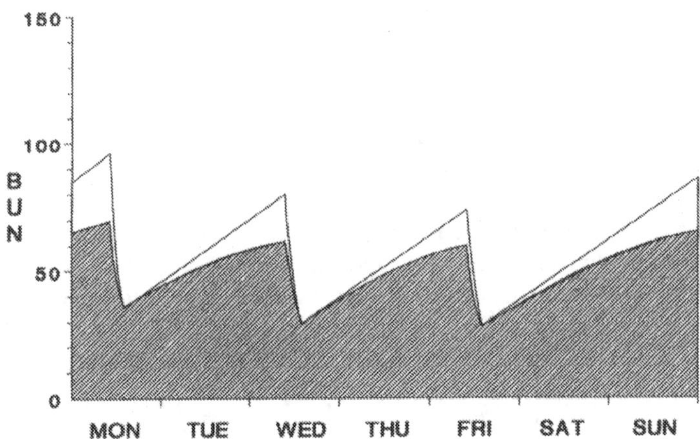

Figure 7.5. Effect of weight gain between dialyses. Steady-state BUN profiles are shown for a patient with no residual clearance dialyzed Monday, Wednesday, and Friday. The upper profile shows BUN levels when no fluid is gained between dialyses. The lower profile (shaded area) represents the same patient dialyzed under the same conditions except that fluid accumulates at 3 ml/min between dialyses.

between dialyses is 3 ml/min (shaded area of graph). The interdialysis portion of the curve is a straight line when no fluid accumulation occurs because the patient lacks residual clearance. When fluid accumulation occurs between dialyses, the plot becomes curvilinear, an effect similar to the effect of residual clearance. For the patient who gains a large volume of fluid between dialyses, both predialysis and time-averaged BUN are lower due to the effect of dilution. As discussed above, the urea clearance also improves, but the effect is offset by the lower urea removal rate at lower blood concentration during dialysis; the net effect is no change in postdialysis BUN. Positive fluid balance between dialyses can markedly reduce predialysis BUN, as figure 7.5 shows, but will have little or no effect on postdialysis BUN.

Effect of saline or blood infusion during dialysis

If weight is maintained during dialysis by infusing normal saline at a rate equal to ultrafiltration losses, the mathematics of urea kinetic modeling are simplified. The dV term in equation 4.27 becomes zero. This does not interfere with the overall kinetic analysis, so the resulting V and G are valid provided an accurate value for K_d is used. If the change in patient weight is used to calculate ultrafiltration rate, a small error occurs in the calculation of K_d because the ultrafiltration rate is not zero. The ultrafiltration component must be added to the dialyzer clearance to obtain the true net urea clearance (see effect of ultrafiltration discussed above). Instead of using weight change as a measure of ultrafiltration, the volume of normal saline infused in the above example can be used to obtain the true rate of ultrafiltration and clearance. This simple correction is valid even if the fluid infused is hypertonic or hypotonic, blood or plasma. If diffusive transport of urea decreases, due for example to poor membrane permeability, convective transport of urea (from ultrafiltration) increases and becomes a more important pathway of urea removal. At high rates of filtration with little or no dialysis, the necessity for correction becomes easier to appreciate. If weight change during treatment is the measure of filtration or convective transport, we would conclude that no urea is removed during pure hemofiltration treatments that maintain the patient's weight constant. So as larger volumes of fluid are replaced during dialysis, it becomes increasingly more important to distinguish dV/dt in equation 4.27 from Q_f in equation 7.14. The term dV/dt represents the actual rate of change in volume or patient weight, while Q_f is the true ultrafiltration rate. Q_f is obtained by subtracting dV/dt from the fluid infusion rate. The value for dV/dt is usually negative and Q_f is usually positive. When no fluid is administered during dialysis, $-dV/dt = Q_f$.

Just as ultrafiltration itself usually has only minor effects on urea removal during the usual hemodialysis treatment, the correction for fluid supplements during dialysis usually has a minor effect on the results of kinetic analysis even in patients with large requirements for supplemental fluids. So most modeling programs ignore saline or blood infusions during dialysis to keep to a minimum the amount of input data required of the user. To avoid errors of this type, it is best to do the study of urea kinetics during a stable dialysis, avoiding fluid supplements. This policy also will prevent errors from irregularities in blood flow, since many centers routinely reduce blood flow when patients experience symptoms during dialysis that require saline infusions.

Recirculation of Dialyzer Venous Blood

Blood returning to the patient from the dialyzer venous line can find its way back into the dialyzer arterial line when certain conditions exist. This unhappy event occurs when blood flow in the access device is compromised due to a proximal or distal stenotic lesion or when venous and arterial needle positions are reversed. Venous reflow is shown diagrammatically as Qr in figure 7.6. At its extreme, all venous blood would recirculate and no blood would return to the patient as the dialyzer continues to clear the same small volume of blood in a constant recirculating mode. More commonly, a 5%-20% venous admixture occurs when access problems arise. The higher urea concentration in peripheral arterial blood (Q_b in figure 7.6) is diluted by blood returning from the dialyzer that has a very low urea concentration. This tends to defeat the purpose of dialysis by diminishing the urea concentration in blood entering the dialyzer, lowering the gradient or driving force across the membrane. The result is a reduction in the rate of urea removal. Measurement of dialyzer urea clearance using conventional arterial and venous port sampling will not detect this aberrancy, but urea kinetic analysis will (see discussion of measured versus modeled clearance above).

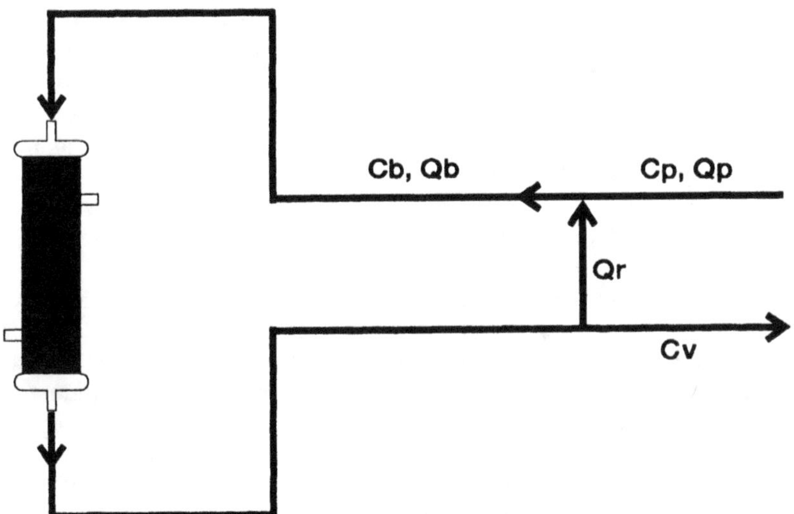

Figure 7.6. Dialyzer venous reflow. Recirculated venous blood is shown as *Qr*. *Qp* is peripheral inflow to the dialyzer line. Q_b is arterial inflow to the dialyzer (*Qp* + *Qr*). *Cp*, *Cb*, and *Cv* are urea concentrations.

The fraction of arterial flow that consists of recirculating blood from the venous limb can be calculated from simple mass balance considerations (see figure 7.6):

If Q_b is arterial blood inflow (measured at the blood pump)
Q_p is peripheral blood inflow
Q_r is flow of recirculating blood
C_b is arterial BUN
C_p is peripheral BUN
C_v is venous BUN

$$Q_p = Q_b - Q_r \tag{7.15}$$

$$C_b \cdot Q_b = C_p \cdot Q_p + C_v \cdot Q_r \tag{7.16}$$

arterial urea delivery	=	peripheral urea influx	+	reflow urea

Substituting Q_p from equation 7.15,

$$\frac{Q_r}{Q_b} = \frac{C_p - C_b}{C_p - C_v} = F_r \quad \text{(reflow fraction)} \tag{7.17}$$

Reflow can be measured directly by obtaining a peripheral blood sample simultaneous with samplings from the arterial and venous blood lines during hemodialysis. If significant recirculation exists, blood from the arterial line will have a lower urea nitrogen concentration than the peripheral blood sample. In actual practice it is not necessary to make a separate venipuncture. After the arterial and venous blood samples are drawn, the blood pump can be momentarily stopped and a sample withdrawn from the arterial needle. Care must be taken to aspirate the dead volume from the arterial sampling site to the site of recirculation before taking the second sample. This is best done when BUN levels are high at the beginning of dialysis and reflow is more easily detected. However, increased reflow can occur during later hours of dialysis, especially if blood pressure falls, compromising flow in the blood access device.

Clearances measured by standard arterial-venous difference across the dialyzer will be normal despite significant recirculation of dialyzer venous blood. Urea modeling uncovers reflow problems because it examines the overall effect of dialysis from beginning to end rather than an instantaneous clearance (refer to the discussion above of modeled versus measured clearance). Urea modeling techniques do not directly measure urea clearance; instead they fit a clearance to the change in urea nitrogen concentration from start to end of dialysis. To model clearance this way, the user must provide an estimate of the urea distribution volume. Urea volume is most conveniently obtained from averaging the values of V calculated from previous kinetic analyses. If the modeled urea clearance is significantly lower than clearance calculated from the dialyzer mass transfer coefficient and flow rates, recirculation must be suspected. Other causes of low effective clearance also must be investigated, but a frequent causes of discrepancy between the expected and the modeled urea clearance is reduced blood flow in the arteriovenous access device that leads to reflow of dialyzer venous blood.

Although dialyzer clearance is not affected by venous reflow, whole-body clearance is reduced. The decrease in whole body clearance can be quantified by comparing clearance measured using a peripheral BUN measurement with clearance measured using an arterial BUN measurement (equation 7.6). The effect of reflow on whole-body urea clearance can be calculated from further consideration of urea mass balance in figure 7.6.

Rearranging equation 7.17,

$$C_p = C_v + (C_b - C_v)/(1 - F_r) \qquad 7.18$$

The standard equation for diffusive clearance is

$$K_d = Q_b\left[\frac{C_b - C_v}{C_b}\right] \qquad 7.19$$

and for effective clearance,

$$K_{de} = Q_p\left[\frac{C_p - C_v}{C_p}\right] \qquad 7.19a$$

Substituting equation 7.18 into equation 7.19a we have

$$K_{de} = \frac{K_d(1 - F_r)}{1 - F_r + K_d \cdot F_r/Q_b} \qquad 7.20$$

Equation 7.20 shows that reflow will diminish the diffusive urea clearance by a factor that is a complex function of the reflow fraction (F_r), dialyzer clearance (K_d), and dialyzer inlet blood flow (Q_b).

Conversely, the reflow fraction can be computed from the difference between expected or measured dialyzer clearance and true patient clearance either measured or modeled:

$$F_r = \frac{K_d - K_{de}}{K_d - K_{de} + K_d \cdot K_{de}/Q_b} \qquad 7.21$$

When K_{de} is equal to K_d, equation 7.21 appropriately returns a reflow fraction of zero. When K_{de} is zero, reflow fraction is 1.0 or 100%. Figure 7.7 shows the effect of K_d/Q_b on the severity of reflow. As K_d approaches Q_b, the fractional recirculation rate approaches the fractional reduction in clearance. Conversely, for poorly functioning dialyzers or for solutes with low clearance relative to Q_b, reflow will have less impact. Equation 7.20 shows that the effect of reflow is potentially higher at low blood flow rates. This is reasonable since dialyzer outlet (venous) BUN is lowest when blood flow is diminished and will have a greater diluting effect when mixed with incoming arterial blood. But because flow in the access device is usually the cause of reflow, the recirculation fraction itself often increases with increasing blood flow and diminishes with decreasing flow. This means that there is no escape from the effects of venous recirculation. When detected, efforts must be taken to correct the problem.

As an example, let us consider a patient whose modeled urea clearance is 50% of expected clearance. When blood flow is 300 ml/min, if the expected dialyzer clearance is 200 ml/min the reflow fraction is 0.6 (equation 7.21). This means that 60% of dialyzer arterial blood flow is recirculated blood from the venous return line. Conversely, when reflow is 20%, equation 7.20 shows that urea clearance will be reduced by 14%. These relationships between reflow and effective clearance help to quantify the negative impact of reflow on the effectiveness of dialysis. After discovering significant reflow, the clinician has the option of attempting to improve clearance by choosing a larger dialyzer or repairing the access site to eliminate reflow. The clearance from a larger dialyzer will be less than expected because the resulting lowered postdialyzer (outlet) BUN will magnify the effect of reflow on clearance. Again, it should be noted that increasing the blood pump rate may serve only to increase the reflow fraction.

Recirculation in the blood access device is a more significant problem for urea clearance than for less easily dialyzed solutes. This can be appreciated from examination of equation 7.17. If *Cv* is high due to poor removal by the dialyzer, the

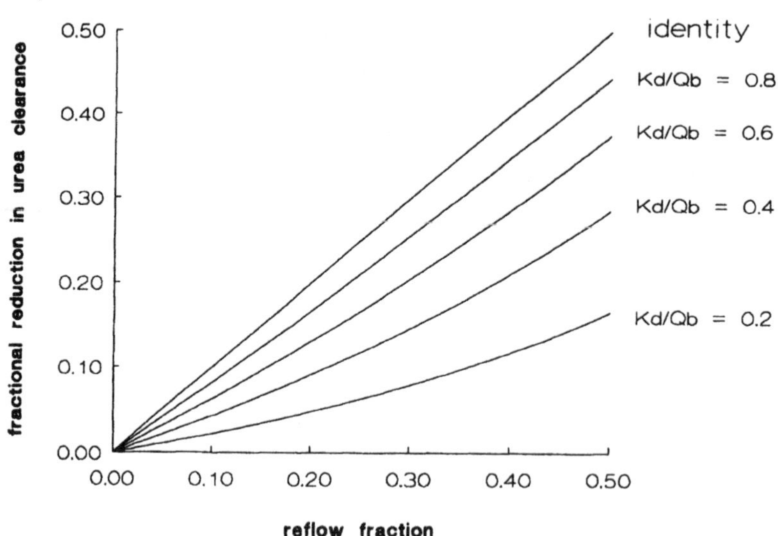

Figure 7.7. Effect of venous reflow on whole-body urea clearance depends on K_d/Q_b. As dialyzer urea clearance (K_d) approaches blood flow (Q_b), the effect of access recirculation is maximized. When K_d equals Q_b, venous blood coming from the dialyzer is free of urea and has a maximal diluting effect on incoming arterial BUN.

denominator will be larger and the numerator will be smaller. Since other more toxic and presumably less diffusible solutes are thought to be responsible for uremic toxicity, we might conclude that recirculation will raise the BUN but have less effect on the outcome of dialysis. The correlation of BUN with outcome uncovered by the National Cooperative Dialysis Study tends to refute this position (see chapter 8) (29). It suggests that urea and other easily dialyzable solutes are the primary determinants of dialysis outcome (see chapter 8). Until more light is shed on the long-standing question of uremic toxicity, we must pay attention to recirculation and its consequent elevation of the BUN. An additional reason for monitoring recirculation is that it warns of impending failure of the blood access device.

Residual (native kidney) urea clearance: its significance

Measuring residual clearance

Residual clearance is the urea clearance contributed by what remains of the patient's own native kidneys. It is often ignored in urea modeling calculations, largely because its value is seldom measured. Measurements of residual clearance require extra time and effort. Also, most patients lose all significant residual function an average of six months after starting hemodialysis for reasons that are not clear (27). Patients treated with peritoneal dialysis seem to retain residual function longer than hemodialyzed patients (30).

If the low level of accuracy required to measure residual urea clearance is appreciated, there might be less reluctance to measure residual function routinely in the dialysis center. Accuracy of urine collection, for instance, is less important because the difference between residual clearances of 1.0 and 1.5 ml/min will have little effect on the dialysis prescription. From the clinical standpoint, we are interested only in an approximate figure when clearance is very low. Another reason for the reluctance to measure residual function is the problem with oscillating BUN values. When the patient turns in a timed collection of urine, unless a precisely timed serum sample was drawn at midcollection, the value to use for serum urea nitrogen is partly a guess. Clinical labs demand a serum urea nitrogen concentration to calculate clearance (they usually do not tolerate guesses), so the data are often withheld or returned in partial form for dialysis personnel to complete. Urea modeling programs can provide an easy solution to this problem by estimating the midcollection BUN and calculating the clearance. This means that the physician need only order a timed urine for urea nitrogen, which is easier and less expensive than a urea clearance. Alternatively, to speed the process further, dialysis personnel can simply request a urine urea nitrogen concentration from an aliquot of the timed urine collection, recording the urine volume before doing so. Slight to moderate deviations from the routine schedule or clearance will be well tolerated because the model is insensitive to variations in residual clearance.

Because creatinine clearance correlates better with true glomerular filtration rate

(GFR), many clinics have protocols for creatinine clearance determinations but not urea clearance. For urea modeling, the urea clearance is necessary; we are not looking for a true measure of GFR. But if creatinine clearance is measured, it may be converted to urea clearance when multiplied by 2/3 (31).

Effect of residual renal function on Kt/V

We discussed *Kt/V* as a measure of the "dose" of dialysis in chapter 3. This parameter, with units of dialysis^{-1}, is a measure of the fraction of total body water that is cleared of urea during a single dialysis. A value of 1.0/dialysis or greater is considered a minimum for patients dialyzed three times/week. This does not mean that all urea is removed during a single dialysis, because the removal of urea follows an exponential decay as the dialysis progresses. A *Kt/V* of 1.0 translates to approximately 60% removal of urea during the dialysis. How does residual urea clearance contribute to this value?

At first glance it might be expected that the dialyzer and native kidney clearances could be added. But simple addition cannot be applied because of the intermittent or discontinuous nature of hemodialysis. Residual clearance (K_r) can be added to dialyzer clearance only during the dialysis when it contributes an insignificant effect. Its major contribution comes during the long interval between dialyses when the BUN is rising and the mathematical relationship between K_d and K_r is complex. Note also that it is not possible to simply multiply the time between dialyses by K_r and add it to the dialyzer clearance multiplied by time on dialysis. If this is done, the contribution of K_r is underestimated because the residual nephrons are more effective in removing urea as the BUN rises between dialyses. Removal of urea decreases progressively with time during dialysis, whereas removal of urea increases with time between dialyses. Because the driving force for urea removal, the urea concentration, is fluctuating rapidly and over a wide range, these two clearances cannot be added. One way to measure the contribution of K_r is to calculate *V* and *G* when K_r is set to zero, and then to observe the change in *Kt/V* when target outcome (TAC) remains fixed and K_r is included in the calculations. This can be done quickly with urea modeling programs that can set *V*, *G*, and time-averaged BUN constant while changing other variables.

For example, consider a 70 kg patient studied during a four-hour dialysis on Monday of a Monday-Wednesday-Fridau schedule with K_d = 180 ml/min, K_r = 0. If the predialysis BUN is 80 and postdialysis BUN is 30, the generation rate (*G*) is 6.1 mg/min, urea volume (*V*) is 41.0 liters, and *Kt/V* approximates ideal *Kt/V* at 1.05. If K_r is changed to 3 ml/min, *Kt/V* falls to 0.75. The change in *Kt/V* that occurs when residual clearance is added to the prescription variables reflects the contribution of K_r to *Kt/V*. To quantify this contribution, simply subtract the new *Kt/V* from *Kt/V* when residual clearance is zero. In this example, the contribution of 3 ml/min residual urea clearance to *Kt/V* is 1.05 - 0.75 = 0.30.

A formula that relates residual clearance contributions to *Kt/V* is given in equation 7.22 for dialysis three times/week (32):

$$\frac{K't}{V} = \frac{K_d \cdot t + K_r \cdot f}{V} \qquad 7.22$$

K_d is dialyzer urea clearance (ml/min)
K_r is patient's residual urea clearance (ml/min)
K' is total effective urea clearance (dialyzer + residual)
t is time on dialysis (min)
V is volume of urea distribution (ml)
f is a coefficient that converts residual clearance into units that allow reduction of K_d while maintaining the same time-averaged BUN when $K_r = 0$.

Residual clearance effect on time-averaged versus midweek predialysis BUN

The formula in equation 7.22 is an approximation that works well for residual clearances in the 0 to 5 ml/min range. The value of f is 4000 min for a three dialyses/week schedule. For dialyses two times/week, the value for f is 6500 min. Note that these values are lower than those published by others who seek to maintain predialysis BUN constant instead of controlling time-averaged BUN (32). We prefer to control time-averaged BUN (TAC) instead of predialysis BUN, since the former does not depend on dialysis schedule and since it was the outcome variable used by the National Cooperative Dialysis Study (NCDS, see chapter 8) (29). A comparison of weekly BUN profiles when predialysis BUN is kept constant

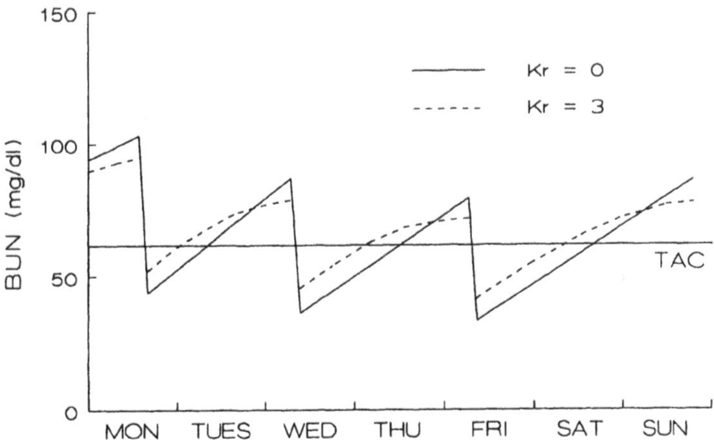

Figure 7.8. Constant time-averaged BUN. Steady-state BUN profiles are shown for a patient whose time-averaged BUN (TAC) is maintained at 60 mg/dl before (dashed line) and after (solid line) losing native kidney clearance of 3 ml/min. Dialyzer clearance and time on dialysis are increased (solid line) to keep TAC constant after the loss of residual function. Predialysis BUN is lower and postdialysis BUN is higher when native kidney function contributes to urea clearance.

compared to the same patient's BUN profile when TAC is kept constant is shown in figures 7.8 and 7.9. Figure 7.9 shows that if the goal is to maintain predialysis BUN constant, TAC is higher when significant residual function exists. This means that in patients with significant residual function, more dialysis is required to maintain TAC constant than is required to maintain predialysis BUN constant.

SIMPLIFIED METHODS FOR UREA MODELING

Predialysis and postdialysis BUN measurements can be used to estimate Kt/V quickly. One method expresses the change in BUN as a fraction or percent of the predialysis value (33,34,35,36). Jindal et al. have shown that Kt/V varies linearly with percent reduction in urea concentration over the range of clinical interest, i.e., for values of Kt/V between 0.7 and 1.5 (34). In a larger group of our own patients, we have shown that the relationship is approximately linear in the clinically applicable range of Kt/V between 0.9 and 1.5/dialysis. These data are shown in figure 7.10. For the linear range

$$Kt/V = 0.026 \cdot P - 0.49 \qquad 7.23$$

P is percent reduction in BUN

If P is 57%, Kt/V is 1.0. Kt/V changes by 0.13 for each 5% change in P. This provides a quick estimate of the adequacy of dialysis if the goal is to maintain $Kt/$

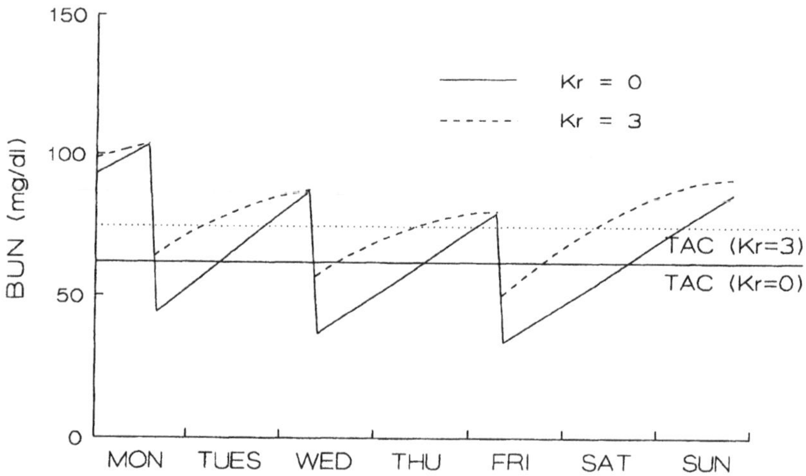

Figure 7.9. Constant predialysis BUN. To maintain the same predialysis BUN after loss of residual function, dialysis intensity must increase more than required to maintain the same time-averaged concentration (TAC) (figure 7.8). TAC is lower when predialysis BUN is kept constant after residual clearance is lost. Conversely, when TAC is kept constant, more dialysis is dictated for patients with residual function.

V above 1.0. When no computer and software are available, this is probably the best technique for estimating the adequacy of any particular dialysis treatment. A 55%-60% fall in BUN is usually associated with an adequate dialysis. However, data from the NCDS has been used to show that ideal *Kt*/*V* varies with protein catabolic rate (PCRn) and that a *Kt*/*V* of 1.0/dialysis was inadequate to maintain some patients with high urea generation rates. It is therefore hazardous to use *Kt*/*V* as the sole determinant of ideal dialysis time without simultaneously assessing the patient's protein intake. In addition, figure 7.10 shows considerable scatter due to individual differences in PCRn, weight gained and lost during dialysis, residual clearances, and dialysis schedules. Since *Kt*/*V* is an estimated value, *Kt*/*V* determined by percent reduction in BUN is an ***estimate of the estimate***.

The so-called quick "bedside" analyses of dialysis urea kinetics or simplified determinations of ideal dialysis time are largely an illusion (32). A pocket calculator or pad and pencil are required to do the arithmetic after data, including predialysis BUN, postdialysis BUN, weight, and other variables, have been entered. Calculators are generally slower than desktop computers and have no permanent storage or hard copy capability. In the end, there is no time saved for a solution that is an approximation. Formal kinetic modeling, using a computer, gives a more precise

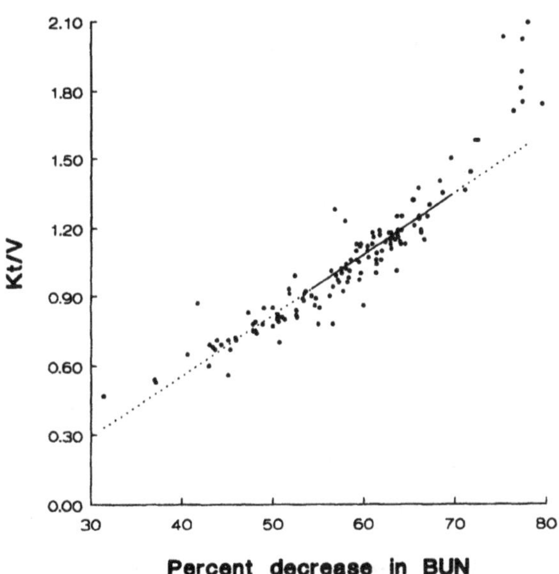

Figure 7.10. Reduction in BUN expressed as a percent of the predialysis value correlates with *Kt*/*V* in a curvilinear pattern. Data are shown for 136 dialyses in 44 patients. The solid line is the least squares linear regression for *Kt*/*V* in the clinically significant range of 0.9-1.5/dialysis: *Kt*/*V* = 0.026_P - 0.49.
P is percent reduction in BUN during dialysis.

measurement. The computer also provides permanent storage for data, hard copies for filing in the patients' charts, and statistical comparisons within and among patients.

Simplified formulas make assumptions that are correct for the average patient but ignore certain individual differences that can lead to erroneous results in patients whose size or water and fat content stray from the norm. This is not the intent of urea modeling or modeling of other biological systems. A major purpose of urea modeling is to detect outliers, patients who do not fit the model because of unforeseen problems with the dialyzer itself, the dialysate delivery system, or the access device. If the patient has significant residual clearance, the bedside formulas will not work. If the patient is dialyzed twice a week or has unusually large weight gains, large errors will appear. Simplified formulas are sometimes useful for "eyeballing" the BUN values, but the final analysis is more accurately done with formal kinetic modeling.

The concept of bedside analysis is also misleading. This suggests that one need only measure BUN predialysis and again postdialysis to measure the adequacy of that dialysis instantly, without leaving the patient's bedside. Ignored is the lag time between blood sampling and return of BUN data. Ordinarily, blood or serum samples are sent to a laboratory and the results return later, usually the next day. Formal kinetic analysis of the BUN using a urea model can be interjected during the lag time to provide results from formal kinetic modeling as fast or faster than the bedside method. It is possible to obtain immediate BUN values by installing and maintaining a modern BUN analyzer in the dialysis center. If such an effort and expense is felt to be justified, then addition of a microcomputer that can interpret the results should be considered. In the latter case, true bedside analysis of formal urea kinetics would be possible. Chapter 10 includes a discussion of the future feasibility of real-time BUN modeling to obtain on-line assessment of urea mass balance during dialysis.

References

1. Sunderman W, Boerner F: *Normal Values in Clinical Medicine*, Saunders, Philadelphia, p110, 1950.
2. Colton CK, Smith KA, Merrill EW, Reece JM: Diffusion of organic solutes in stagnant plasma and erythrocyte suspensions. Chem Eng Progr Symp Ser 66:85-100, 1970.
3. Marshall EK, Davis DM: Urea: its distribution in and elimination from the body. J Biol Chem 18:53, 1914.
4. Wu H: Separate analyses of the corpuscles and the plasma. J Biol Chem 51:21, 1922.
5. Berglund H: Nitrogen retention in chronic interstitial nephritis. JAMA 79:1375, 1922.

6. Ralls JO: Urea is not equally distributed between the water of the blood cells and that of plasma. J Biol Chem 151:529, 1943.
7. Pasynskii AG, Chernyak RS: Sorption of molecules of nonelectrolytes by proteins. Doklady Akad nauk SSSR 73:771, 1950.
8. Murdaugh HV, Doyle EM: Effect of hemoglobin on erythrocyte urea concentration. J Lab Clin Med 57:759-762, 1961.
9. Maes V, Engelborghs Y, Hoebeke J, Maras Y, Vercruysse A: Fluorimetric analysis of the binding of warfarin to human serum albumin: equilibrium and kinetic study. Mol Pharmacol 21:100-107, 1982.
10. Scheider W: The rate of access to the organic ligand-binding region of serum albumin is entropy controlled. Proc Natl Acad Sci USA 76:2283-2287, 1979.
11. Nolph KD, Bass OE, Maher JF: Acute effects of hemodialysis on removal of intracellular solutes. Trans Am Soc Artif Intern Organs 20:622-627, 1974.
12. Ilstrup K, Hanson G, Shapiro W, Keshaviah P: Examining the foundations of urea kinetics. Trans Am Soc Artif Intern Organs 31:164-168, 1985.
13. Eschbach JW, Egrie JC, Downing MR, Browne JK, Adamson JW: Correction of the anemia of end-stage renal disease with recombinant human erythropoietin: Results of a combined phase I and II clinical trial. N Engl J Med 316:73-78, 1987.
14. Taniguchi S, Tsunefumi S, Harada M, Niho Y: Prostaglandin-mediated suppression of in vitro growth of erythroid progenitor cells. Kidney Int 36:712-718, 1989.
15. Cheung AK, Alford MF, Wilson MM, Leypoldt JK, Henderson LW: Urea movement across erythrocyte membrane during artificial kidney treatment. Kidney Int 23:866-869, 1983.
16. Colton CK, Lowrie EG: Hemodialysis: Physical principles and technical considerations, in *The Kidney*, Brenner BM, Rector FC Jr (eds), Philadelphia, Saunders, pp 2425-2489, 1981.
17. Bosch JP, Barlee V, von Albertini B: Impact of hematocrit on solute transport in hemodialysis. Kidney Int 35:241, 1989.
18. Mayrand RR, Levitt DG: Urea and ethylene glycol facilitated transport system in the human red cell membrane. J Gen Physiol 81:221-237, 1983.
19. Levitt DG, Mlekoday HJ: Reflection coefficient and permeability of urea and ethylene glycol in the human red cell membrane. J Gen Physiol 82:239-253, 1983.
20. Kaplan MA, Hays L, Hays RM: Evolution of a facilitated diffusion pathway for amides in the erythrocyte. Am J Physiol 226:1327-1332, 1974.
21. Macey RI, Yousef LW: Osmotic stability of red cells in renal circulation requires rapid urea transport. Am J Physiol 254:C669-674, 1988.
22. Brahm J: Urea permeability of human red cells. J Gen Physiol 82:1-23, 1983.
23. Colton CK, Henderson LW, Ford CA, Lysaght MJ: Kinetics of hemodiafiltration I. In vitro transport characteristics of a hollow fiber ultrafilter. J Lab Clin Med 355-371, 1975.

24. Besarab A, Gaughan W, Anzalone D, Medina F: Effect of erythropoietin-induced changes in hematocrit on dialysis solute clearances. Kidney Int 35:240, 1989.
25. Collins A, Keshaviah P, Berkseth R, Opsahl J, Abraham P: Impact of erythropoietin therapy on rapid high efficiency hemodialysis. Kidney Int 35:243, 1989.
26. Bagdade JD, Porte D, Bierman EL: Hypertriglyceridemia: A metabolic consequence of chronic renal failure. N Engl J Med 279:181, 1968.
27. Gotch FA, Keen ML: Care of the patient on hemodialysis, in *Introduction to Dialysis*, Cogan MG, Garovoy MR (eds), New York, Churchill Livingstone, pp 73-143, 1985.
28. Husted FC, Nolph KD, Vitale FC, Maher JF: Detrimental effects of ultrafiltration on diffusion in coils. J Lab Clin Med 87:435-442, 1976.
29. Lowrie EG, Laird NM, Parker TF, Sargent JA: Effect of the hemodialysis prescription on patient morbidity: Report from the National Cooperative Dialysis Study. N Engl J Med 305:1176-1181, 1981.
30. Lysaght M, Pollock C, Schindhelm K, Ibels L, Farrell P: The relevance of urea kinetic modeling (UKM) to CAPD. Trans Am Soc Artif Intern Organs, Abstracts, p 84, 1988.
31. Milutinovic J, Cutler RE, Hoover P, Meijsen, B, Scribner BH: Measurement of residual glomerular filtration rate in the patient receiving repetitive hemodialysis. Kidney Int 8:185-190, 1975.
32. Gotch FA: Kinetic modeling in hemodialysis, in *Clinical Dialysis* (2ed), Nissensen AR, Gentile DE, Fine RN (eds), Norwalk CT, Appleton and Lange, pp 118-146, 1989.
33. Lowrie EG, Teehan BP: Principles of prescribing dialysis therapy: Implementing recommendations from the National Cooperative Dialysis Study. Kidney Int 23 (Suppl 13):S113-122, 1983.
34. Jindal KK, Manuel A, Goldstein MB: Percent reduction in blood urea concentration during hemodialysis (PRU): A simple and accurate method to estimate Kt/V urea. Trans Am Soc Artif Intern Organs 33:286-288, 1987.
35. Daugirdas JT: Bedside formulas for urea kinetic modeling. Contemp Dial Nephrol 10:23-25, 1989.
36. Barth RH: Direct calculation of KT/V: A simplified approach to monitoring of hemodialysis. Nephron 50:191-195, 1988.

Chapter 8

MEASURING DIALYSIS: HOW MUCH IS ENOUGH?

These fundamental questions about each patient's dialysis prescription should be frequently asked:

1. *How much dialysis is prescribed?*
2. *What is the outcome?*
3. *Is the outcome adequate?*
4. *Is the outcome commensurate with the amount prescribed, i.e., is the equipment functioning properly?*

Historical methods

In the early years of dialysis therapy, when treatment was barely adequate, most patients received the same dialysis prescription. The duration of each dialysis was long and was dictated more by schedules and equipment than by patient need. Larger patients or those with high protein catabolic rates were "difficult to dialyze" and were prone to suffer from symptoms of chronic underdialysis. Technical improvements over the past 30 years have allowed the average time spent on dialysis, for patients dialyzed three times/week, to diminish from eight to six to four hours and more recently to two and three hours, while patient well-being has improved. These technical improvements come from changes in membrane composition and permeability, changes in dialysate components and concentrations, increases in blood and dialysate flow rates, and improved filtration control. The progressive reduction in average dialysis duration evolved as a consensus, from empiric trial and error, that equal benefit could be obtained from newer equipment with shorter dialysis exposure. When time on dialysis is shortened, equipment performance and time itself become much more important to monitor to guarantee adequate dialysis for the individual patient. A more quantitative and individualized approach is needed. To be effective, a quality assurance program should quantify dialysis as it addresses each question listed above. Before attempting to devise a strategy that will guarantee adequate dialysis for each patient, it is important to know how to measure a "dose" of dialysis.

Kt/V: A yardstick for dialysis therapy

The intensity or "dose" of dialysis seems easy to quantify. The dose should include a measure of the effectiveness of the dialyzer, i.e., its urea clearance and the time spent on dialysis. In chapters 3 and 4, we discussed the mathematical origins of the term Kt/V. K is the dialyzer urea clearance, t is time on dialysis and V is the patient's urea volume. Since patient size, or more precisely, urea pool size (V), is a determinant of need for dialysis, the dose is factored by V. This allows the dose of dialysis to be expressed much like the dose of an antibiotic like penicillin as mg drug/kg body weight. When expressed this way, both Kt/V and the antibiotic doses are dimensionless. Kt/V provides a more uniform parameter that can be applied to larger groups of patients. However, size or urea pool is not the only determinant of dialysis need, so it is inappropriate to apply the same Kt/V to all patients. As a component of the dialysis prescription, *Kt/V* offers the advantage of expressing dialysis intensity independent of patient size, blood flow, dialysate flow, time on dialysis, or dialyzer clearance, since these variables are factored in the expression.

Much of the confusion about the meaning of *Kt/V* stems from a misunderstanding about how it is calculated (1,2). There are two ways to calculate *Kt/V*:

1. by multiplying estimated *K* times t and dividing by *V*.
   This gives a measure of dialysis "dose" or amount prescribed factored for patient or urea pool size.
2. by kinetic modeling. This provides a measure of outcome, as described below.

These two methods of calculating *Kt/V* relate directly to the two methods for calculating *K* and the ratio *K/V*, discussed in chapter 7.

The first method gives a measure of how much dialysis is prescribed. Accurate computation of *Kt/V* using this method requires careful consideration of each of the component terms:

K: By convention, *K* is blood water clearance of urea by the dialyzer, not whole-blood clearance (see discussion in chapter 7 of corrections for water content of blood). *K* is the sum of diffusive and convective clearances. Therefore the contribution of ultrafiltration to urea clearance must be added to the clearance calculated from blood and dialysate flow rates when there is no ultrafiltration (see chapter 7, Effect of ultrafiltration on dialyzer urea clearance). Residual clearance is usually not added to *K*, since its con tribution during dialysis is insignificant. But if *Kt/V* is used as a measure of dialysis outcome, a component of residual urea clearance must be added (refer to the contribution of K_r to *Kt/V* discussed in chapter 7).

t: Since *Kt/V* is an expression of average dialysis intensity, the time component, *t*, is the average time spent with each dialysis. Dialyzer clearance (K_d) must be constant during this time interval.

V: The volume parameter (*V*) is the apparent urea distribution volume at dry weight, i.e., postdialysis.

This is the most tedious method of calculating *Kt/V*, and it gives only a measure of how much dialysis is prescribed. Similar to orally administered medications, the amount delivered (or absorbed) is often considerably less than the amount prescribed. We need a measure of the *bioavailability* of dialysis, i.e., a measure of outcome.

Measuring dialysis outcome

If outcome criteria for each organ system were easily delineated and measurable, the success of dialysis would be unquestioned. The heart rate given in the example above for treatment of atrial fibrillation with digoxin is the simplest example of an easily measurable outcome criterion. Unfortunately, there are no specific signs or symptoms of dialysis adequacy to follow, only threshold phenomena that signify gross underdialysis. We must rely on laboratory tests that measure the severity of uremia as a practical means of assessing dialysis need and completeness. Laboratory measurements are easy to apply to nondialyzed patients with renal disease, but once dialysis treatments have started, interpretations become more difficult because of the rapid perturbations in solute concentration induced by the dialyzer (figure 4.1). Which BUN is the best measure of dialysis outcome: predialysis BUN, postdialysis BUN, or some measure of average concentration? Two measures of average concentration introduced in chapter 3 are time-averaged BUN and average predialysis BUN. The latter is approximated for patients dialyzed three times/week as the midweek predialysis BUN. This value slightly underestimates the true mean predialysis value for most patients whose three-day interdialysis interval occurs over the weekend. For patients dialyzed twice per week, the mean predialysis BUN has no easily measured counterpart.

Time-averaged BUN (TAC) as a measure of outcome

Although more difficult to compute, the time-averaged BUN (TAC) has several advantages, as outlined in table 8.1. Time-averaged BUN is the area under the BUN-versus-time profile for an entire week divided by the number of minutes in the week (figure 8.1). It encompasses a measure of postdialysis as well as predialysis BUN. This gives a measure of average tissue exposure to urea concentrations during the week. Patients with the same average predialysis BUN may have very different TACs if their urea volumes or generation rates differ significantly. Figure 8.2 illustrates this point in two patients whose individual prescriptions are designed to keep predialysis urea concentrations at an ideal level but whose postdialysis concentrations differ because of differences in urea production.

Another advantage of TAC is its previous selection by the National Cooperative Dialysis Study (NCDS) as the major outcome parameter in their quest to quantify the effect of dialysis. In analyzing the study's results, Laird comments that time-averaged BUN was the "...strongest predictor of failure. ....Midweek predialysis BUN had predictive value, but TAC_{urea} was consistently better...." (3). The NCDS is discussed below in more detail.

Still another advantage of TAC over other outcome parameters is its simplicity. Use of time-averaged urea concentrations allows a comparison of urea levels in hemodialyzed patients with urea levels in patients before starting dialysis or in those treated with continuous peritoneal dialysis as discussed below. Once maintenance dialysis is started, TAC can be used to prescribe dialysis and to compare outcomes among patients regardless of their dialysis schedules. In contrast to midweek predialysis BUN, the target value for TAC is the same for patients dialyzed once/week, twice/week, or three times/week. For patients dialyzed twice weekly, there is no midweek measurable concentration that approximates the average predialysis BUN.

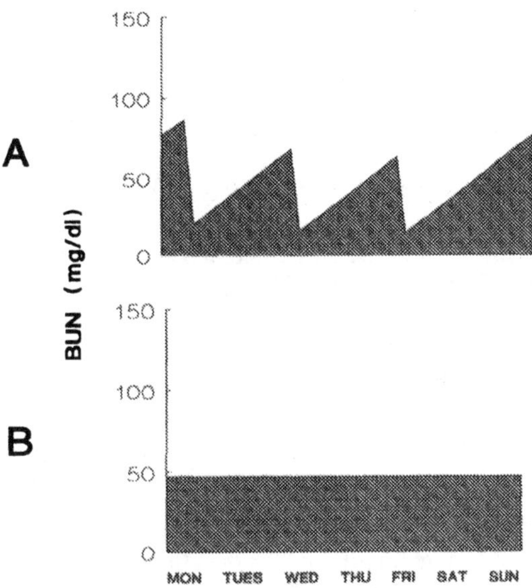

Figure 8.1. Time-averaged BUN is the area under the BUN versus time profile divided by the time. The average urea exposure for patient A is considered equal to patient B.

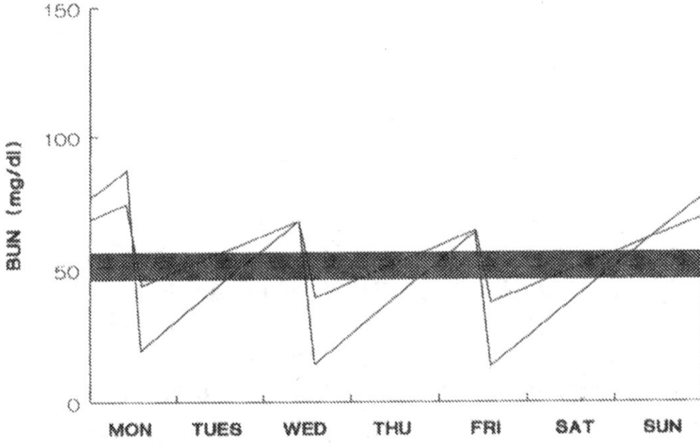

Figure 8.2. Two patients with the same midweek predialysis BUN. Despite constant midweek predialysis BUN levels, time-averaged values may differ significantly in different patients. The shaded area represents the difference between time-averaged BUN values in these two patients. According to the NCDS data, the patient with the lower urea generation rate (upper profile) with higher TAC is at higher risk for complications. Compare this to figure 7.8.

Table 8.1 Advantages of time-averaged BUN as a measure of dialysis outcome

1. Includes a measure of postdialysis BUN and predialysis BUN in the outcome criteria
2. Was the major outcome parameter used by the National Cooperative Dialysis Study (see text)
3. After starting maintenance dialysis, allows comparison of outcome with prior BUN levels
4. Allows comparison of outcome in hemodialyzed patients with outcome in patients maintained with peritoneal dialysis
5. Simplifies outcome criteria by applying the same measure of outcome to all patients, regardless of dialysis schedule or residual function
6. Allows uniform criteria to apply regardless of urea kinetic model used (no shifting of criteria for one-compartment versus two-compartment model)

Proponents of monitoring predialysis BUN may argue that the toxic effect at higher concentrations is more important than the effect of average concentration. This is a difficult question to test, since admittedly urea has little toxicity of its own and serves only as a marker for other easily dialyzed toxins that remain to be identified. For toxic drugs such as the aminoglycoside antibiotics, trough concentrations are felt to correlate better with toxicity than peak concentrations (4). For chemotherapeutic drugs such as methotrexate, levels greater than 100 times the mean toxic level are tolerated for brief periods, suggesting that mean levels are more important determinants of toxicity (5). The answer to this question will likely come only after specific uremic toxins are identified. If peak concentrations prove more important to control then hemodialysis should be applied at regularly spaced intervals, avoiding the three-day weekend stretch.

Kt/V as a measure of dialysis outcome

Kt/V also can be used to measure dialysis outcome because, as discussed below, it is highly correlated with the predialysis and postdialysis BUN. But only the modeled value for *Kt/V* can be used to judge dialysis outcome. The differences between measured and modeled *Kt/V* are listed below.

A closer inspection of its components suggests that true *Kt/V* is difficult to measure or at least should be highly prone to error. Although time (*t*) is easy to measure, *K*, the dialyzer clearance, is more of a challenge. Calculated values for *Kt/V* may have significant errors due to inaccuracies in measurement of blood flow, dialysate flow, or dialyzer mass transfer coefficient (see chapter 7). Direct measurement of clearance during the modeled dialysis will also fail to prevent errors unless it is measured several times during dialysis, taking into account modifying factors such as venous reflow in the blood access device, clotting in the dialyzer, and blood pump variance. The patient's urea volume (V) is even more difficult to measure and seems out of reach for most clinicians. Why then do we even bother with Kt/V?

To answer this question, it is important to appreciate that for purposes of modeling, it is not necessary to resolve K or V, only their ratio. K/V is recognized, from equations 4.3, 4.4, and 4.8 and the accompanying discussion in chapter 4, as the urea elimination constant. Like the elimination constant, K/V has units of time^{-1}, better understood as the fraction of distribution volume cleared per unit of time or the fractional removal rate. For urea and the single-compartment model, the distribution volume is total body water; so K/V is the fraction of body water that is cleared of urea per minute. When K/V is multiplied by total time on dialysis (t), it becomes the fraction of body water cleared per dialysis, often abbreviated as *fractional clearance* (6) or *fractional removal*. Kt/V, therefore, does have a dimension; the unit is dialysis^{-1}. When Kt/V for urea is 1.0/dialysis, the fractional urea clearance is 1 (100%) per dialysis or 3 (300%) per week for the patient dialyzed three times/week. This does *not* mean that the urea distribution space is entirely cleared of urea three times/week. Because of the diminishing effect of dialysis as time progresses and the BUN falls, much less urea is removed than is suggested by this interpretation of Kt/V.

When no weight is gained or lost during dialysis and G, the urea nitrogen generation rate, is ignored, Kt/V is simply determined by the ratio of postdialysis BUN to predialysis BUN (C_0/C). This relationship is seen in the single-compartment equation 4.14, applied to the intradialysis interval, and reproduced here with the assumption that G is zero:

$$C = C_0 \cdot e^{-K \cdot t/V} \qquad 8.1$$

$$Kt/V = \ln(C_0/C) \qquad 8.2$$

The second example in chapter 9 (table 9.2) also illustrates this point. For the simplified example patient, it makes no difference what value is given for dialyzer clearance during single-compartment modeling, because K/V remains constant (determined by C_0/C), and Kt/V, the measure of dialysis intensity, does not change. There is no need to resolve V or K when determining the elimination constant (K/V) from measured urea concentrations. When this modeling method is used, Kt/V reflects whole-body (fractional) clearance and is designated the *modeled* value of Kt/V.

Only the modeled value of Kt/V can be used as an indicator of dialysis outcome. The calculated value is difficult to obtain and limited in its usefulness; the modeled value is easier to calculate and can be used to judge outcome. So, when Kt/V is determined by modeling predialysis and postdialysis BUN, it becomes a measure not of the amount of dialysis prescribed but of the amount delivered. With this understanding, the modeled value for Kt/V truly correlates with dialysis outcome.

However, the correlation is tight only in the "ideal" patient who does not have large weight changes or residual urea clearance and who dialyzes three times/week. If adjustments are made for patient dialysis schedule, residual clearance, protein catabolic rate (PCRn), and ultrafiltration rate, modeled *Kt/V* can be used as a measure of outcome. Despite these shortcomings, *Kt/V* has gained popularity as an expression of the intensity of dialysis. We consider time-averaged BUN (TAC) a better measure of outcome because it is less abstract than *Kt/V*, it can be visualized better by the clinician, and it derives from direct measurements of predialysis and postdialysis BUN without adjustments. Other advantages of TAC are outlined in table 8.1.

Neither *Kt/V* nor time-averaged BUN can be used as the sole criterion to judge the adequacy of the total dialysis prescription because, protein intake is not included. The net protein catabolic rate (PCRn) or protein intake in metabolically stable patients is a major determinant of the need for dialysis.

The adequacy of dialysis

Limitations of simple methods for measuring dialysis adequacy

Estimates of dialyzer solute clearance, patient size, and dialysis duration give us a measure of the intensity or dosage of dialysis treatment. Modeling gives us measures of outcome. Other criteria must be considered to answer the question, "how much is enough?" For dialysis, it seems reasonable to establish some criteria of outcome success or failure that can be examined frequently and can provide feedback to the clinician and to the patient. Using the calculated value of *Kt/V* to decide how much dialysis to prescribe is much like prescribing a dose of digoxin for atrial fibrillation without following the heart rate. Another way to monitor the dosing of drugs is to follow serum levels. For drugs such as digoxin or the aminoglycoside antibiotics, this adjustment of dosage is critical because of the drug's low therapeutic index (ratio of toxic to therapeutic dose). The dosage is adjusted for each patient to achieve a target blood level. Conditions that affect absorption and elimination of drugs may change from time to time, requiring periodic measurement of blood levels and readjustment of dosage. For dialysis, it seems reasonable to establish a patient outcome parameter, such as a target concentration of urea above which dialysis is necessary and below which it is not. This would be analogous to the method described for the antibiotic in the example above, but the therapeutic effort for dialysis is intended to keep the level down rather than up.

This type of approach to dialysis is often taken just before initiating dialysis treatments in a new patient. When the BUN rises consistently above 80-100 mg/dl, we begin to prepare for dialysis. Each patient requires frequent reassessment and careful attention as renal function falls to inadequate levels. The physician looks for vague symptoms and signs of uremia, weighing each of these against the urea and creatinine concentrations and clearance as he or she attempts to select the most opportune time for commencing dialysis. Dialysis is sometimes initiated well before the BUN reaches 100 mg/dl if symptoms warrant it. This "seat of the pants" approach requires physician experience, skill, and diligence. But even in the best hands the time to start dialysis is often decided more by the common cold or seasonal influenza virus than by the toxins of uremia. The early symptoms are vague and nonspecific, often going unrecognized even by the patient. At the start of replacement therapy, it may not matter that dialysis is delayed until symptoms are evident. But once a dialysis regimen is established, the criteria for deciding dialysis dosage should include a guarantee that symptoms of uremia will never again appear. This means that monitoring for symptoms and signs of uremia is not sufficient to guarantee adequacy of dialysis.

The dose of dialysis necessary to keep patients alive could be considerably less than that necessary to keep them healthy. Many physicians continue to use the predialysis BUN as the only objective parameter to judge the adequacy of chronic maintenance dialysis. The selected cutoff level varies considerably. For the average patient with no residual function, average weight gain between dialyses, and a constant average protein intake, the predialysis BUN serves marginally as a control on dialysis outcome. For most patients who do not fit this description, the predialysis BUN can be severely misleading. For instance, if a constant midweek predialysis BUN of 70 mg/dl is selected as a target for all patients, those with low PCR may be markedly underdialyzed, and patients with high PCR values will spend an inordinately long time on dialysis (7).

Other clinical data have been and continue to be used to decide how much, how long, and how often to dialyze. These include predialysis serum creatinine and potassium concentrations; fluid balance and acid/base status; urine output and residual urea or creatinine clearance; and nutritional status and dietary assessment. Together, these major and other minor criteria help to formulate the clinician's decision about how much dialysis to prescribe. Lack of uniformity among physicians in their perception of dialysis adequacy was evident at our dialysis center before initiating urea modeling. We observed a cyclic readjustment of dialysis duration, blood flow rates, or even the schedule of two or three dialyses/week as staff physicians rotated onto and off the dialysis service. Even when one physician was in control, the dialysis prescription was altered when symptoms suggesting underdialysis appeared in the patient. If the patient was "doing well" and the predialysis BUN was high, it tended to be ignored. Sometimes this was appropriate (e.g., if PCRn was high), but often it was not. Also, responding to symptoms is somewhat like chasing the blood sugar with insulin. The response is always too late. A better approach would anticipate problems and prevent them with a prescription that is appropriately tailored to each patient's needs.

The National Cooperative Dialysis Study (NCDS)

In 1975 a conference was convened in Monterey, California to address the issue of dialysis adequacy (8). One important outcome of this meeting was the formulation of a National Cooperative Dialysis Study (NCDS) sponsored by the National Institutes of Health as described in chapter 1. The purpose of the study was to examine objective measures of dialysis outcome in controlled studies using varied dialysis prescriptions. At the time this investigation was formulated, the most popular theory that proposed to explain uremic toxicity was the *middle molecule* hypothesis (9). Accumulated evidence suggested that beyond a certain minimum, blood and dialysate flow had little effect on outcome because removal of compounds in the 350-5000 dalton range mitigated the success or failure of dialysis. These middle molecules are membrane-limited (see figure 3.9); their removal is almost entirely dependent on dialyzer size (surface area) and time on dialysis. The NCDS was designed to maximize the difference between low- and middle-molecular-weight solute clearances by varying time and urea clearance independently. One hundred sixty patients were studied for a minimum of six months over a period of three years. The patients were divided into two groups, one with time-averaged BUN held at 50 mg/dl and the other maintained at 100 mg/dl using a single-pool variable-volume kinetic model. Each of these groups was in turn divided into two groups of long- and short-time dialysis. Average duration for the short-time dialysis was 3.25 hours and for the long-time dialysis 4.50 hours. Dialyzer urea clearance was controlled by varying blood flow and dialysate flow or by altering relative flow direction, countercurrent or cocurrent. Clearance and levels of marker middle molecules were not measured, but urea kinetics were repeatedly measured to confirm that the differences in dialysis prescriptions were actually achieved. The experiment was not blinded to the physician or to the patient, but by March 1980 the study was halted by the patient safety committee. Differences among the groups were striking, as shown in figure 8.3. Prescriptions designed to keep the time-averaged BUN at 50 mg/dl were more successful than those designed to maintain the time-averaged BUN at 100 mg/dl regardless of the time spent on dialysis. Those patients who had high average BUN levels, independent of the duration of dialysis, had more frequent medical complications and a higher dropout rate than in those whose BUN levels were kept low. Within each of these groups there was a significant difference between failure rates in patients dialyzed for a long versus short duration, only when number of hospitalizations (F2) was included in the criteria for success (figure 8.4). The statistical correlation between time spent on dialysis and outcome was weak compared to the correlation with time-averaged BUN (3).

The NCDS was designed to keep TAC for the low BUN group at 50 mg/dl. Detailed analysis of the data, however, showed that the ideal level of urea varied with PCRn (10).

The importance of protein catabolic rate

All who have interpreted the NCDS data have concluded that ideal time-averaged BUN (ITAC) varies with protein catabolic rate (PCR) (10,11,12,13). Because urea is only a marker for uremia and has low intrinsic toxicity, the patient with higher protein intake should be allowed to have a slightly higher BUN. Those

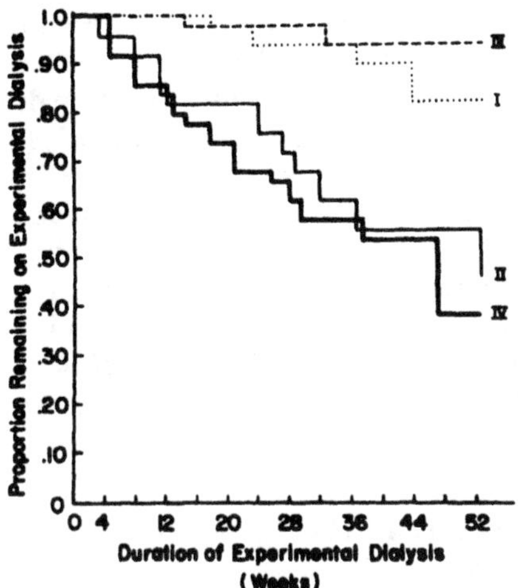

Figure 8.3. Actuarial plot from the National Cooperative Dialysis Study. The fraction of patients not withdrawn for reason of death or medical complications (F1) is shown on the *y*-axis. Groups I and III are the low-BUN group (TAC = 53); Groups II and IV are the high-BUN group (TAC = 89). Reprinted with permission from Lowrie EG et al., N Engl J Med 305:1178, 1981.

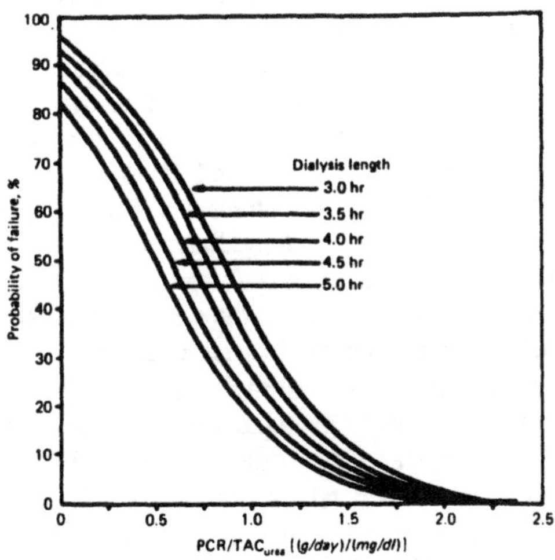

Figure 8.4. Patients treated for a longer duration per dialysis had slightly improved outcomes in the NCDS. The *y*-axis is the fraction of patients not withdrawn for reason of death, medical complications, or hospitalization (F2). Reprinted with permission from Lowrie EG, Teehan BP, Kidney Int 23 (Suppl 13):S120, 1983.

with low PCRs should be dialyzed to keep the BUN low to maintain the concentration of other more toxic solutes at a low level. The inference here is that patients with low protein intake must require more dialysis than is indicated by the level of blood urea. The additional dialysis is required to remove other unidentified toxins. Carried to an extreme, this approach would dictate the same dialysis dose for all patients, i.e., *Kt/V* would be the same for all patients, regardless of the BUN.

Opposing this viewpoint is the well-known observation that protein restriction improves the symptoms and signs of uremia and that serum urea is a good marker for uremia. This theory proposes that patients with lower PCRs should require less dialysis and patients with high protein intakes should require more dialysis. Carried to its extreme, this approach would dictate the same BUN for all patients, i.e., TAC would be maintained at 50 to 60 mg/dl in all patients regardless of the protein intake or other factors.

The mechanistic approach

Since the NCDS did not control PCRs, it is difficult to assert from their data that more dialysis for those with low PCRs will improve clinical outcome (3). Rigorous statisticians could effectively argue that such a conclusion would require a separate study and that no conclusion regarding outcome and intensity of dialysis can be gleaned from the NCDS. However, after reviewing the NCDS data, Gotch and Sargent found no difference in probability of failure as modeled *Kt/V* ranged from 0.8 to 1.5 (11). The probability of failure sharply increased as *Kt/V* fell below 0.8 for all patients in the study (figure 8.5). This is a surprising finding, one that has been challenged by others (12). But it is not without precedent in biological systems. The response to hormones, for example, is often sigmoidal; no response is seen until a critical concentration is reached, but slightly above the critical level the response is complete. The effect of dialysis may not be continuous over a wide range of BUN values. Lowering the BUN below a critical level may accomplish 90% or more of the task, beyond which benefits are diminished. These investigators reasoned that a constant dose of dialysis factored for patient size (*Kt/V*) is appropriate, provided the BUN, measured as TAC or midweek predialysis BUN, does not rise above the range found in the group of successfully dialyzed patients with lower BUN values shown in figure 8.6. This approach, which they termed a *mechanistic* strategy, is a combination of the two extreme viewpoints mentioned above. The term *mechanistic* simply means a mechanical or "eyeballing" interpretation of the NCDS data. Applications of the mechanistic strategy make use of the linear association between outcome and PCRn at a constant *Kt/V* (figures 8.7 and 8.8).

This convenient mathematical relationship can be shown using a simplified fixed-volume model, to apply to average predialysis BUN, average postdialysis BUN, and time-averaged BUN. If G is ignored during equally spaced dialyses, equation 4.14 simplifies to

$$C_2 = C_1 \cdot e^{-K \cdot t/V} \qquad 8.3$$

C_1 is predialysis BUN
C_2 is postdialysis BUN

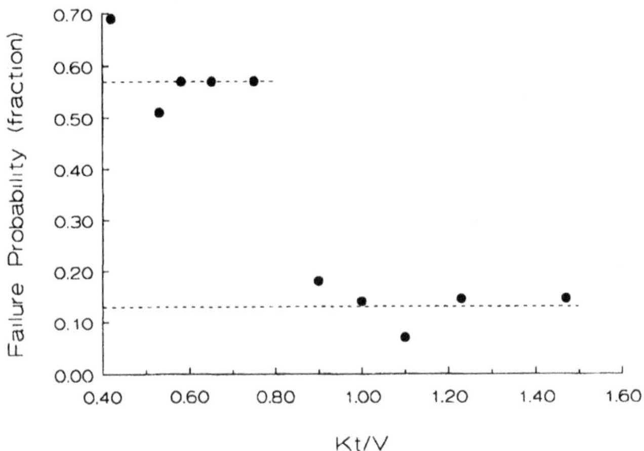

Figure 8.5. Mechanistic analysis of the NCDS data shows a discontinuous abrupt decrease in failure probability for the low-BUN group in the lower right-hand corner compared to the high-BUN group in the upper left-hand corner. Within each group there is no trend. Adapted and reprinted with permission from Gotch FA, Sargent JA, Kidney Int 28:529, 1985.

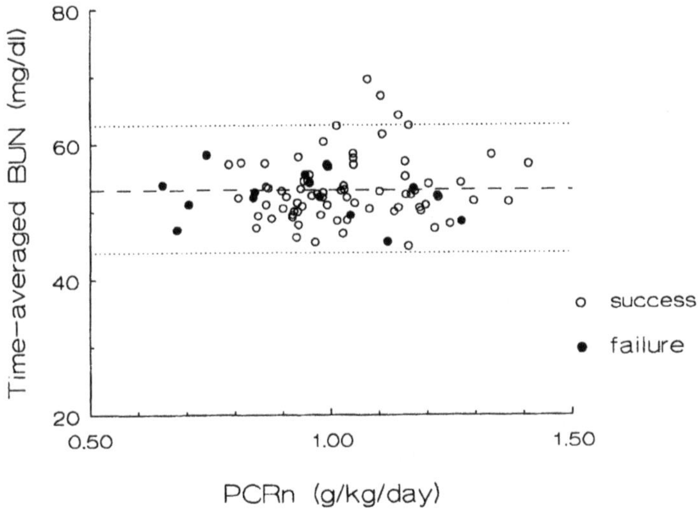

Figure 8.6. The low-BUN groups (1 and 3) in the NCDS. Mean time-averaged BUN was 53 mg/dl for the group of successfully treated patients enrolled in the NCDS. Dotted lines are two standard deviations from either side of the mean. Failures, shown for comparison, tended to have lower protein catabolic rates (PCRn).

When there is no residual clearance, G is simply described by equation 5.10, reproduced here as the ratio G/V:

$$G/V = (C_1 - C_2)/T_i \qquad 8.4$$

The ratio G/V is closely related to PCRn, as equation 3.9 shows. If we substitute C_2 from equation 8.3 into equation 8.4, we have

$$G/V = C_1(1 - e^{-K \cdot t/V})/T_i \qquad 8.5$$

Equation 8.5 shows that if Kt/V is kept constant, PCRn varies linearly with C_1. The linear relationship is shown graphically in figure 8.7. If instead we substitute C_1 from equation 8.3 into equation 8.4, we have

$$G/V = C_2 \cdot T_i(e^{K \cdot t/V} - 1) \qquad 8.6$$

Because both C_1 and C_2 are linearly related to PCRn at constant Kt/V, time-averaged BUN (TAC), the average of C_1 and C_2 in this simplified model, is also

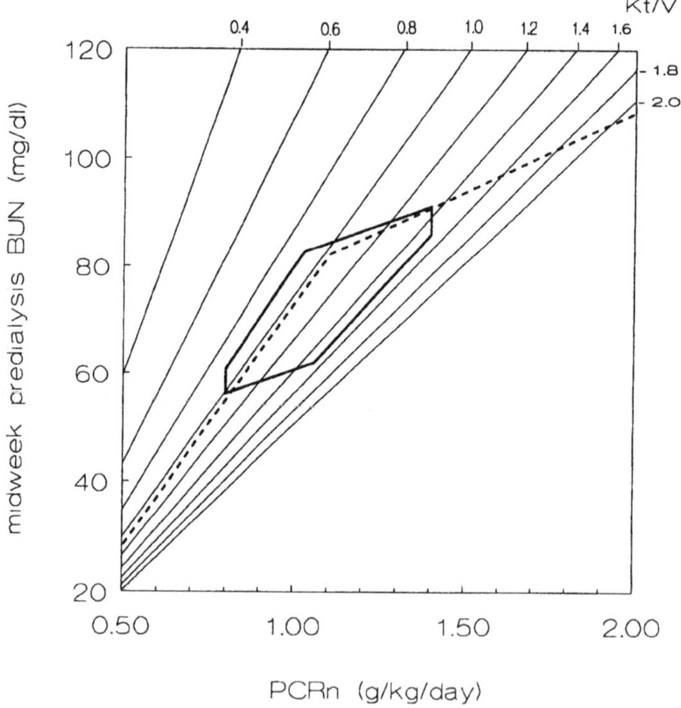

Figure 8.7. Map of ideal outcome zone. The left axis is midweek predialysis BUN. The right axis refers to the linear Kt/V isopleths. The dashed line is the ideal target predialysis BUN at varying PCRn values. Adapted and reprinted with permission from Gotch FA, Sargent JA, Kidney Int 28:529, 1985.

linearly related. This association between PCRn and TAC is shown in figure 8.8.

The linear relationship described above, between PCRn and outcome parameters, also applies to more complex single-compartment models that include isosmotic fluid shifts and residual clearance (7). When formal kinetic analysis is applied under widely varying conditions of urea flux, volume, and ultrafiltration, the relationship remains linear and the line relating TAC to PCRn (at $Kt/V = 1.0$/dialysis) is amazingly constant, as figure 8.9 shows.

According to the mechanistic analysis, target Kt/V should be maintained constant at 1.0/dialysis or slightly above until the PCRn rises above 1.1 g/kg/day, as shown in figure 8.7 (11). Above this level of protein intake, Kt/V must increase to keep midweek predialysis BUN from rising above levels observed in the successfully dialyzed patients studied by the NCDS (figure 8.7). An ideal or target therapy line (broken line in figure 8.7) shows this boundary of safety, designed to assure adequate dialysis while avoiding undue and possibly unnecessary time on dialysis.

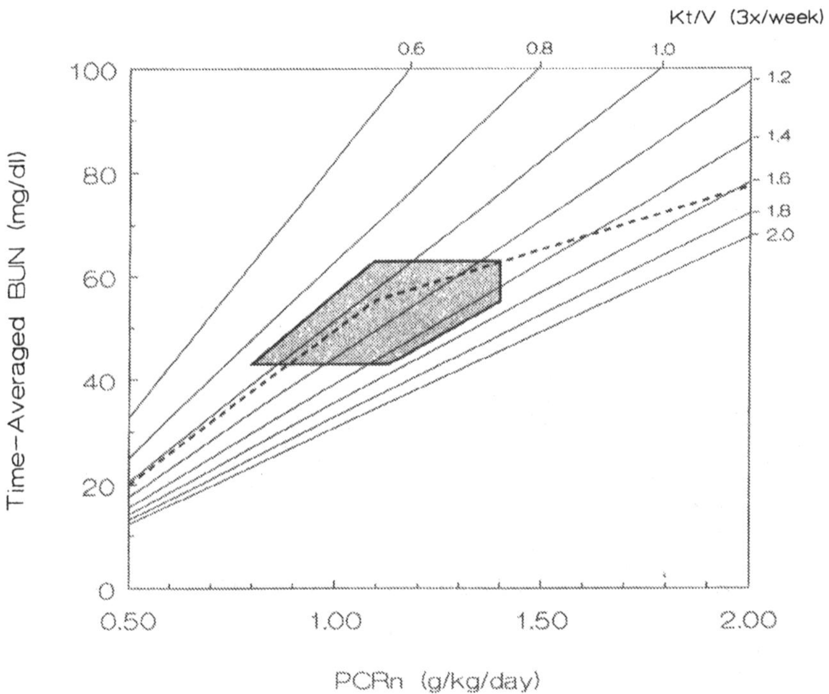

Figure 8.8. Ideal time-averaged BUN, shown as the dashed line (see text for source), depends on normalized protein catabolic rate (PCRn). Shaded area represents the "safe" zone. Extracted from NCDS data kindly provided by Dr. E.G. Lowrie (10).

Statistical analysis of the NCDS

When multivariate statistical analysis was applied to the NCDS data, analyzing probability of failure measured either as dropout for medical reasons or dropout plus number of hospitalizations, the major determinant of success or failure was the time-averaged BUN as defined in figure 8.1. Outcome success or failure also depended on PCRn, so statistical contour maps were generated, relating the probability of failure as the dependent variable to time-averaged BUN (TAC) and PCRn. Figure 8.10 shows one of these contour maps where probability of failure is assessed by death or dropout for medical reasons. Care was taken to caution the interpreter against applying the data directly to patient care situations (10). This is a wise approach for several reasons, but one that stands out is the inability of the study results to distinguish among various causes of failure. Since the patients were randomized, it was assumed that any differences in outcome among them would be due primarily to the deliberate modifications of their dialysis prescriptions. Because the diet part of the prescription was not controlled, any attempt to segregate the data according to PCR must consider other determinants of PCR that might influence outcome. The outcome variable TAC is dependent on both PCRn and dialysis

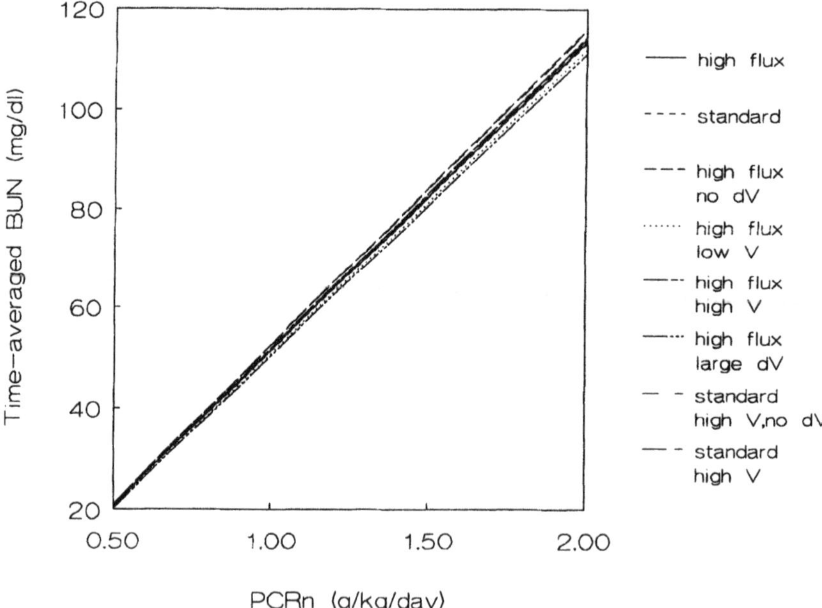

Figure 8.9. For hemodialysis three times/week, if *Kt/V* is maintained at 1.0/dialysis, PCRn is linearly related to time-averaged BUN under a variety of commonly encountered dialysis conditions. Nearly all *Kt/V* lines are superimposable.

intensity. It would appear from examining the contours in figure 8.10, especially in the region extrapolated to very low PCRn values, that intensive dialysis can reverse the ill effects of an inadequate protein diet. In other words, can dialysis in some way replace adequate nutrition? Consideration of what is known about body metabolism and the effects of dialysis render this conclusion untenable. This idea also opposes the well-known observation that a low protein diet improves uremic symptoms of anorexia, nausea, and sedation (14,15).

When judgment of outcome success or failure of the dialysis regimen included the number of hospitalizations, the low PCR group had a high failure rate. The sources of failure in this group may be twofold: 1) those related to poor nutrition, and 2) those related to underdialysis. The statistical method of analysis can be misinterpreted by placing too much emphasis on the second cause, attributing most of the morbidity to inadequate dialysis. In some cases this may be true, i.e., uremic symptoms of anorexia may cause low protein intake. But in other cases an independent underlying disease state, such as congestive heart failure or infection, may be responsible for the low PCRn and increased hospitalization rate. In such cases, increasing the dialysis intensity is unlikely to improve outcome.

Limitations of the NCDS

The three-BUN, single-pool variable-volume model was used to evaluate urea kinetics for enrollees in the NCDS. As discussed in chapter 5, this model provides a reasonable approximation of urea kinetics for most patients. No correction for serum water content was made for BUN determinations in this study. In addition, whole-blood clearances were used instead of blood water clearances, so the urea

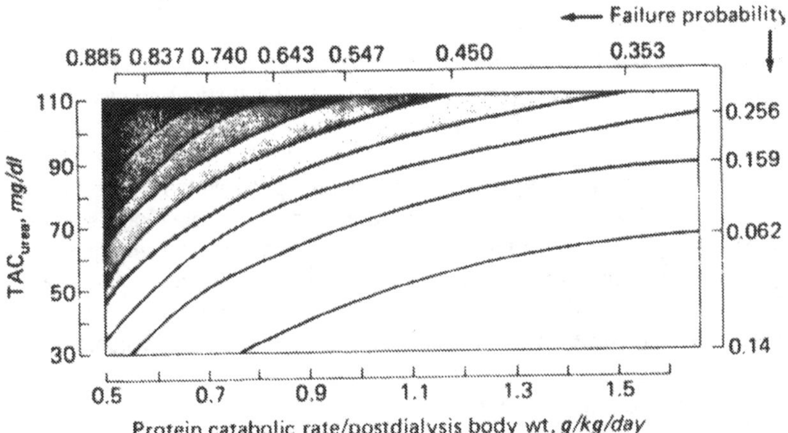

Figure 8.10. Failure probability shown at the top and right was based on death or withdrawal for medical reasons. Probability of failure increased with increasing TAC and decreasing PCRn. Data are from the National Cooperative Dialysis Study. Reproduced with permission from Laird NM et al., Kidney Int 23 (Suppl 13):S105, 1983.

volume calculated for all patients enrolled in the NCDS was falsely inflated by approximately 10%. This has no consequence for the derived parameter *Kt/V* because the errors in numerator and denominator cancel each other. Similarly, modeled values for *G* are nearly accurate because of offsetting errors in *V* and in (C_2 - C_1) in equation 5.10. Because urea concentrations were expressed as whole serum instead of serum water concentrations, the falsely low BUN concentration in whole serum partially cancels the error in *G* due to false inflation of *V*. PCR is normalized to body size by dividing by *V* to give PCRn. This leaves an uncompensated error in PCRn due to the use of whole serum urea concentrations instead of serum or plasma water values. Plasma is approximately 93% water, so the error in PCRn is about 7% on the low side. This is fortunate, because the single-compartment model used by the NCDS tends to overestimate *G* (see chapter 5). We would not expect a significant error in *G* due to application of single-pool kinetics, because conventional, low-flux dialyzers were used in this study; but if an error did occur, it would partially offset the error described above from failure to compensate for serum or plasma water. These errors should be kept in mind when attempts are made to establish guidelines for patient treatment using current patient information that does not contain these errors.

Limitation of the NCDS data are encountered when attempts are made to extrapolate the results. This study was not designed to measure the effectiveness or adequacy of dialysis over the spectrum of BUN values at a given PCRn. It was instead designed to find which of four discrete dialysis prescriptions produced the best outcome (16). This fundamental question is clearly answered by the study. Attempts to glean more from the data by correlating success or failure over the range of time-averaged or predialysis BUN values observed are fraught with hazard. As discussed above, the study lumped together successes and failures due to various causes and tended to attribute all failure to failure of dialysis. This is especially evident in the patients with low PCR values, where statistical analysis predicts that more dialysis will correct the higher morbidity rate in this group. This conclusion can only be reached if the PCRn is also controlled and randomized like the remainder of the dialysis prescription. An attempt to exclude patients with PCR values in the low range due to causes other than inadequate dialysis would be an impossible task that begs the question; that is, the answer to the question being sought by the study would have to be known before the results could be analyzed. The mechanistic analysis acknowledges these limitations and establishes a minimum prescription based on the data that will help ensure an adequate outcome.

Hemodialysis is a self-limiting treatment

Time-averaged BUN is logarithmically related to dialysis dose

Unfortunately, time on dialysis is logarithmically, not linearly, related to the commonly accepted dialysis outcome variables, TAC and midweek predialysis BUN. This means that as BUN falls, toward the end of dialysis, the decision to lower it further commits the patient to a disproportionately longer intradialysis time. For example, if the postdialysis BUN is 20, lowering it to 10 may require doubling the time on dialysis, as figure 8.11A shows. Figure 8.11B shows that the time-averaged BUN falls even less. For the small patient (V=20 liters) in figure 8.11, reducing TAC from 47 to 44 mg/dl requires doubling the dialysis time from four to eight hours. The dialyzer becomes less efficient as the BUN falls. This is not due to a flaw in the dialyzer but to the nature of the dialysis process itself, and is common to diffusion and all other first-order processes (see chapter 4). The driving force for diffusion across the dialyzing membrane, the concentration gradient, dissipates as time progresses, leading to less and less removal of solute per unit of time. Dialysis can be considered a self-limiting process for solutes that are easily removed. The logarithmic relationship between dialysis efficiency and time on dialysis must be considered when one proposes to set an outcome goal. As the goal becomes more stringent (i.e., lower BUN), the responsiveness to time on dialysis decreases. Because dialysis time is usually viewed by the patient as a negative factor, the establishment of a *minimum* goal is critical.

The NCDS, widely used as a basis for setting dialysis time, was not designed to

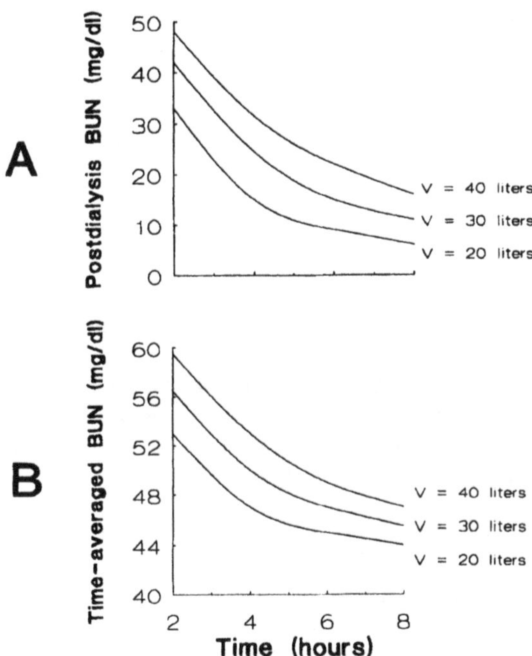

Figure 8.11. A. Logarithmic relationship between postdialysis BUN and dialysis time required to achieve the indicated postdialysis levels. B. Similar logarithmic relationship for time-averaged BUN.

establish minimum goals for dialysis. Data from this study can only be used to identify outcome regions (figures 8.7 and 8.8) to avoid. Within the broad range of acceptable dialysis prescriptions, it is impossible to say whether the lower BUN is better than the higher BUN. Because of the sensitive relationship between dialysis duration and dialysis goal as discussed above and because of the strongly negative psychological impact of time on dialysis, most nephrologists have opted for the short end of the spectrum, i.e., the minimum time required to keep the TAC within an acceptable range. This is illustrated as the target therapy lines in figures 8.7 and 8.8.

Kt/V *is logarithmically related to dialysis outcome*

The decreased sensitivity of dialysis outcome to time on dialysis as the BUN falls also applies to *Kt/V*. If *Kt/V* is the only criterion for the success or failure of dialysis, one might assume that a 20% increase in *Kt/V* (e.g., from 0.9 to 1.1) results in a 20% improvement in the effectiveness of dialysis and a proportionate improvement in outcome. If, as suggested by the NCDS, the efficiency of dialysis is best measured as the rate of decline in urea and other small-molecular-weight solute concentrations, the above reasoning is incorrect. When the postdialysis BUN is very low, there will be little or no improvement in outcome after a 20% increase in *Kt/V*. The logarithmic relationship between the dose of dialysis (*Kt/V*) and outcome, best illustrated by the lines in figures 8.7 and 8.8, helps to explain much of the controversy about where the limits of therapy should be set. If dialysis time is doubled, *Kt/V* is doubled but outcome is not doubly improved. This puts additional emphasis on our reluctance to use even the modeled value of *Kt/V* as a measure of outcome.

While attempting to balance the well-known psychological and economic benefits of shortened dialysis time against the potential ill consequences of inadequate dialysis, we note less and less change in outcome when dialysis time or clearance is increased beyond a certain point. Perhaps a goal of dialysis should be to reach that point and then not be concerned about admittedly slight changes in outcome as the prescribed dose of dialysis varies about it. The point of diminishing returns depends upon the postdialysis BUN. This value in turn depends on the dialysis dose (*Kt/V*) and the patient's protein intake (PCRn). Both must be considered in the determination of outcome. This reasoning is the basis for the mechanistic approach to establishment of target dialysis therapy.

The importance of dialysis membranes

Hemofiltration versus hemodialysis

While the NCDS was under way, improvements in membrane technology were successfully increasing convective clearance of middle molecules. Synthetic membranes were produced that had filtration coefficients at least an order of magnitude greater than standard cellulosic membranes used in the study (17). It was expected that the marked improvement in middle molecule clearances by these membranes would be matched by an equally marked improvement in the outcome of dialysis. Clinical application of the new membranes using the methods of hemofiltration or *hemodiafiltration* began to be eyed closely by the nephrology community as the expected next leap forward. Results, however, were disappointing. After several crossover studies lasting more than one year, the only significant differences in outcome were better vascular stability during treatment and better control of hypertension using the new membranes (18,19). Similar improvements have been observed after switching to noncellulosic membranes for hemodialysis. Middle molecule clearance was improved by hemofiltration but outcome was not. The major determinants of dialysis success remained the dialyzer urea clearance and the patient's urea concentration.

High flux dialysis

Following publication of the NCDS results in 1981, attention was turned to methods for improving urea clearance. The same membranes designed to improve middle molecule clearance could be used to improve urea clearance simply by increasing blood and dialysate flow rates. Increasing flow through standard cellulosic membrane dialyzers did not improve urea clearance because most patients had reached the membrane-limited portion of the clearance curve for these dialyzers (figure 3.9). The curve is shifted upwards for the more permeable dialyzers and reaches its plateau at a higher clearance level. This means that as blood flow increases from 200-300 ml/min to 400-500 ml/min, the clearance rises more steeply than it does with the older membranes. Because the membranes are more porous, the increase in urea clearance is accompanied by an increase in middle molecule clearance. The term *high-flux dialysis* was introduced to distinguish the improved clearance of a variety of compounds including urea. After resolution of several problems inherent in this method of dialysis (most troublesome were acetate toxicity and erratic control of extracellular volume), experiences with the high-flux technique were encouraging. Patients seemed to tolerate the new membranes better, and dialysis time could be reduced without apparent detriment to the patient (7,20). This experience contrasted sharply with previous attempts to reduce dialysis time with the older cellulosic membranes. To reduce time with the older membranes, an increase in surface area had to accompany the increase in blood flow through the dialyzer to increase the dialysis dose (K in Kt/V). Because cellulosic membranes have toxic effects, these efforts were poorly tolerated by the patients (21,22). The newer synthetic membranes allow small molecule clearances to increase without increasing surface area, and the membrane itself is less toxic.

Nephrologists continue to apply outcome parameters derived from the NCDS to prescriptions for patients on high-flux dialysis. This practice may not be appropriate because the Cooperative Study was done with older cellulose-derived dialyzers. Since no other standards exist — and in particular, standards for high-flux dialyzers — we accept extrapolation of the NCDS data as a reasonable compromise. Ideally, an additional study should be done to determine the optimum prescription for high-flux dialysis (see chapter 10).

A note about middle molecules

It might be argued that successful shortening of dialysis time disproves the middle molecule hypothesis, since removal of these compounds depends more on dialysis time and membrane surface area. This conclusion is not justified, because the newer membranes are also more permeable to middle molecules; despite shortening the duration of dialysis, the clearance of middle molecules increases.

The National Cooperative Dialysis Study accomplished its goal admirably. By demonstrating a strong correlation between dialysis outcome and average serum urea concentration, this study showed that low-molecular-weight solute removal is more important than middle-molecular-weight solute removal. When dropout from the study for medical reasons was the criterion for success or failure, urea removal was the sole determinant of outcome within the list of criteria considered. If the number of hospitalizations was added, time on dialysis also played a role but was a relatively weak determinant compared to the time-averaged BUN (3). Despite these findings, most seasoned nephrologists would not discount the importance of middle molecules as uremic toxins. Middle molecules should not be excluded from the search for uremic toxins, especially when long-term benefits of dialysis are considered, because the minimum NCDS follow-up was only six months.

Comparison of peritoneal dialysis dosage with hemodialysis dosage

The peritoneal dialysis membrane is considered much more permeable than standard hemodialysis membranes. Clearances are lower by an order of magnitude, presumably because of lower blood flow. The dose and outcome of peritoneal dialysis may be measured using *Kt/V* as well, but this exercise is often misinterpreted. Just as *Kt/V* derived from residual renal clearance cannot be compared or added directly to dialyzer *Kt/V*, the *Kt/V* generated by continuous peritoneal dialysis cannot be compared to *Kt/V* generated by intermittent hemodialysis when one is attempting to predict the steady-state average BUN. This lack of linear proportionality among *Kt/V* values generated by multiple treatment modalities underscores the futility of using *Kt/V* to predict ideal dialysis regimens.

We can illustrate this dilemma by considering extremes. To devise a treatment regimen that would achieve with one dialysis/week the same time-averaged or predialysis BUN that is usually achieved with three dialyses/week, we might suggest simply multiplying *Kt/V* by three and apply this goal to the single dialysis regimen. Very quickly it becomes apparent that this cannot be done. The reason for the failure of this simple extrapolation is that removal of urea, not clearance of urea, ultimately determines BUN. Adding more time to a single dialysis removes very little additional urea because the dialyzer is working at low efficiency during those final hours of dialysis when BUN is low. Conversely, if we were to dialyze daily, a *Kt/V* of 3/7 per dialysis would lower the time-averaged BUN below levels achieved with thrice-weekly dialysis and a *Kt/V* of 1.0 per dialysis.

Use of *Kt/V* to compare different dialysis modalities is not valid. Peritoneal dialysis operates at constant efficiency because BUN does not vary significantly from hour to hour. Hemodialysis, because of the intermittency, higher urea clearance, and consequent diminishing urea removal toward the end of dialysis, requires a higher *Kt/V* to achieve the same time-averaged BUN. Failure to appreciate this fact can lead to the false conclusion that peritoneal dialysis has additional hidden benefits or effects because it can keep the patient feeling well at a lower *Kt/V* than is possible with hemodialysis (23).

Ideal dialysis outcome from time-averaged BUN

Method for targeting ideal time-averaged BUN

Equations 8.5 and 8.6 show that *Kt/V* is linearly related to the ratio TAC/PCRn. An ideal target modeling line similar to that described in figure 8.7 for midweek predialysis BUN can also be developed for TAC. The following equations used to calculate ideal time-averaged BUN (ITAC) are represented by the dashed line in figure 8.8.

When PCRn is below 1.1 g/kg/day,

$$\text{ITAC} = 0.60\,(\text{PCRn}) - 0.10 \qquad 8.7$$

When PCRn is equal to or above 1.1 g/kg/day,

$$\text{ITAC} = 0.25(\text{PCRn}) + 0.28 \qquad 8.8$$

ITAC is ideal time-averaged BUN in mg/ml
PCRn is protein catabolic rate in mg/kg BWn/day
BWn is normalized body weight defined as: *Vl*/.58
Vl is volume of urea distribution in liters

When PCRn is 1.1 g/kg/day, both equations return the same value for ITAC (0.56 mg/ml).

Equations 8.7 and 8.8 were derived from the mechanistic model of urea kinetics based on the NCDS as discussed above. A diagram illustrating the relationships among TAC, PCRn, and *Kt/V* is shown in figure 8.8. The data used to construct this diagram differ from those published by the NCDS and others (3,7,11) in four respects:

1. The *y*-axis is time-averaged BUN instead of midweek predialysis BUN.
2. Dialyzer clearance is blood water clearance, about 10% lower than whole-blood clearance.
3. Urea volume is the true volume instead of the inflated value derived from whole-blood clearance.
4. PCRn is a true estimate of protein catabolic rate obtained from calculations using blood water concentrations of urea instead of the falsely low values derived from whole-blood concentrations.

When PCRn is below 1.1 g/kg/day, the equation that relates PCRn to ideal TAC is the line that generates a constant *Kt/V* of 1.05 (figure 8.8). When PCRn is above 1.1 g/kg/day, the new prescription line has a lower slope. This new slope is designed to keep the time-averaged BUN from rising above the upper limit (63 mg/dl), two standard deviations above the mean value of 53 mg/dl established for the successfully treated low-BUN group in the NCDS. This approach, in patients with higher PCRn values, is consistent with many previous studies showing that keeping the average BUN from rising too high will avoid protein-induced uremic toxicity. Yet it acknowledges that patients with higher PCRn values can be allowed to have somewhat higher average BUN levels without increased risk of toxicity.

Most urea modeling programs calculate *V*, *G*, and PCRn using the iterative methods described in chapters 4 and 5. Ideal time-averaged BUN (ITAC) can then be calculated from equation 8.7 or 8.8. A new value for time on dialysis (Tb) is established using the double-loop iterative process illustrated in figure 4.10. This calculation sequence begins by selecting an arbitrary time on dialysis and then adjusts predialysis BUN until a steady state is found. Either the one-compartment or two-compartment model can be used to generate the BUN versus time profile for an entire week. The weekly steady state is defined when the measured predialysis BUN matches (within 1%) calculated predialysis BUN one week later. The area under the curve of BUN versus time (AUC) is then measured for this set of BUN values, and TAC is calculated as AUC/10,080. Estimated time on dialysis (Tb) is then adjusted up or down to bring TAC closer to ideal TAC (ITAC), and the whole process repeats itself until TAC is within 1% of ITAC. For each iteration, the AUC is calculated for the BUN versus time plot over an entire week. Time on dialysis is adjusted with each iteration until predialysis BUN values match (weekly steady state).

Values for predialysis and postdialysis BUN are allowed to vary during these iterations, in contrast to the iterative process used to calculate *V* and *G* described in chapters 4 and 5, where the measured values of predialysis and postdialysis BUN are held constant. This explains the difference between target and actual BUN profiles. When ITAC differs from TAC, target values for predialysis and postdialysis BUN are significantly different from actual measured values. For example, to reduce the AUC when ITAC is significantly lower than measured TAC, predialysis and postdialysis BUN must be lower than the measured values. This means that time on dialysis must be increased above that currently prescribed if the remainder of the dialysis prescription is unchanged.

Of lesser importance is the difference between AUC for one-compartment versus two-compartment analysis. The two-pool model accounts for rebound in urea concentration postdialysis and this causes AUC to be slightly greater for the same predialysis and postdialysis BUN. Since time-averaged BUN is computed from AUC, two-compartment analysis dictates a slightly greater dialysis intensity to reach the same time-averaged BUN.

Rationale for the target modeling lines

Careful analysis of the NCDS data by experienced statisticians could not define precise guidelines for dialysis control of BUN but only regions to avoid (figures 8.7 and 8.8). Urea modeling provides a measure of PCR not otherwise available, as well as a measure of TAC. With these two measurements in hand, it is possible to redefine the dialysis prescription so that TAC falls within the "safe" zone defined by the successfully dialyzed patients in the NCDS. The range for this zone is relatively large. The ideal target lines (dashed lines in figure 8.8) are derived from the mechanistic interpretation of the NCDS data as described above. These lines are deliberately poised on the low-intensity side of the ideal zone because of the psychological and economic advantages of shortened dialysis time and because of the diminishing returns inherent in the dialysis process itself that limit the benefits obtained from more intensive dialysis (see discussion above). These simple modeling lines derived from the NCDS shortly after the data became available have been successfully applied in our unit over the past six years.

A plot of ideal predialysis BUN versus PCRn in a series of 103 patients is shown in figure 8.12 for patients dialyzed three times/week. All values are derived from the linear relationship between ideal time-averaged BUN and PCRn expressed by equations 8.7 or 8.8 and are well within the "adequate" range (11). The scatter observed is a consequence of weight changes during and between dialyses that have

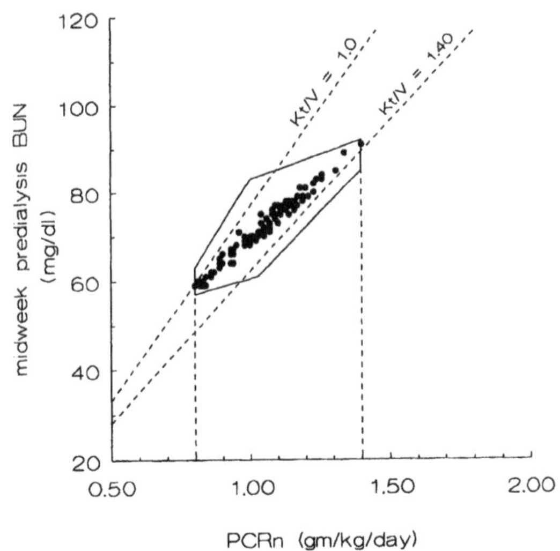

Figure 8.12. Ideal midweek predialysis BUN values versus PCRn in a series of 100 patients dialyzed three times/week where ideal TAC was computed using equations 8.7 and 8.8. Data are superimposed on the ideal therapy map (see figure 8.7). These data confirm the validity of the ideal TAC target therapy lines in figure 8.8.

a larger effect on predialysis BUN than on time-averaged BUN. A large weight gain between dialyses will lower the predialysis BUN more than the time-averaged BUN (see figure 7.5). Thus predialysis BUN can take on slightly different values at the same value of TAC.

A map of the adequate range for dialysis prescriptions is shown in figure 8.13. The ideal target line is identical to that in figure 8.8, as is the upper limit on TAC up to PCRn values of 1.1 g/kg/day. Above this point, the upper and lower excursions about the target line are fixed, since there is no reason to expect that tighter bounds should be applied to patients with PCRn values above 1.1 g/kg/day.

The real value of the NCDS is found in the proof that urea concentrations are important to control. The study served to place more emphasis on kinetic analysis, but the data analyzers warn against too rigorous interpretation of failure probability functions (3). Some reviewers of the study have demanded a precise adherence to the statistical probability curves generated by the study. We prefer the mechanistic approach because it acknowledges the limits of the study beyond which the data are not interpretable.

Prescriptions for patients with very high or very low protein catabolic rates

For patients with PCRn values less than 0.8 g/kg/day or greater than 1.4 g/kg/day, we cannot confidently judge the adequacy of dialysis from urea modeling, but can only suggest that protein intake is inadequate or excessive. Reviewers of the NCDS

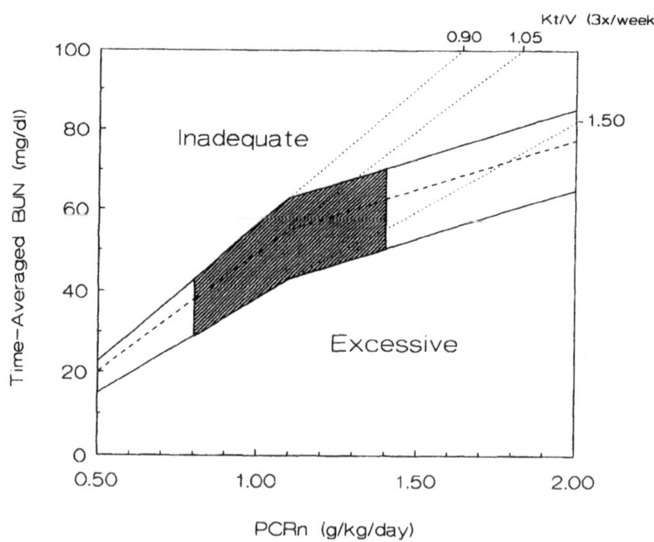

Figure 8.13. The shaded treatment outcome zone, derived from mechanistic analysis of the NCDS data, assures a favorable response to dialysis. See text for a definition of the boundaries.

data agreed that this study could offer no conclusions regarding the adequacy of dialysis when PCRn falls outside the range 0.8 to 1.4 g/kg/day (3,11). When PCRn is consistently below 0.8 g/kg/day, efforts should be made to uncover other health risk factors that might contribute to morbidity and mortality. When PCRn falls above 1.4 g/kg/day, dietary counseling may be necessary to temper excessive protein intake that can intensify the need for dialysis.

When we examined more closely those patients with PCRn values less than 0.8 g/kg/day in our patient population, we found no difference in age, sex ratios, or outcome compared to the whole group. These patients did not appear to be more ill than the others in our population. At other times and in other dialysis populations, this group may consist of more uremic patients or of those with hepatic disease and other illnesses that cause reduction in protein intake, as discussed below.

When PCRn is below 0.8 g/kg, two causes may be found. The patient may be voluntarily restricting protein intake, possibly through a diet of high-biological-value protein (increased proportion of essential amino acids), and is relatively healthy. On the other hand, the patient may be ill, possibly uremic, and has reduced protein (and likely reduced calorie) intake because of anorexia associated with the illness. In the latter case, to account for the low BUN, the illness must be one (like uremia) that is not associated with hypercatabolism. Catabolic illnesses may cause the PCRn to appear in the normal or above normal range in which urea kinetic analysis appears to reflect an appropriate dialysis prescription. All models of urea mass-balance (figures 4.1 and 5.1) assume that G, the rate of urea appearance, is constant, but none of the models makes any assumption about an endogenous or exogenous source of urea and none demands a zero net nitrogen balance. But to equate PCRn to protein intake and to allow application of equations 3.2 and 3.3, near steady-state zero nitrogen balance must be in effect. Demonstration of ideal urea kinetics and a "normal" PCRn is not sufficient evidence that the patient is healthy or in ideal nitrogen balance. When PCRn is low, it is important to distinguish the two possible causes listed above and, if the second cause is suspected, to attempt to identify and treat the underlying illness.

For patients with high PCRn, overindulgence in meat, eggs, milk products, or other high-protein foods is an obvious possibility; a catabolic illness (e.g., corticosteroid therapy or a febrile illness) is another. Similar to the patient with low PCRn, it is important to attempt to distinguish these two possibilities in each patient and to correct the cause of hypercatabolism if it exists.

Only after examining the patient and the dietary history will a course of action become apparent. Urea kinetic analysis is a valuable tool that helps the clinician understand the patient's need for and response to dialysis and includes a measure of PCR. The reasons for an abnormal PCR must be sought through other means.

Comparing dialysis outcome with the prescription

Target prescriptions are insensitive to errors in Kd

Misjudging dialyzer clearance (K_d) is a common source of error in urea modeling. The sources of this error are inaccurate measurement of blood or dialysate flow, incorrect values for *KA*, clotting or malfunctioning of the dialyzer, and/or recirculation in the access device. Overestimation of clearance causes a proportionate error in V, i.e., if the patient's dialyzer clears urea less efficiently than expected, V will be overestimated. The model sees a relatively small change in urea concentration induced by a device that has a (falsely) high clearance, so it assumes the patient (urea volume) must be larger than real. Despite the proportionate error in V, there will be little effect on the estimate of ideal dialysis duration. In the absence of residual function or fluid losses, variations in K_d will have essentially no effect on target dialysis time if the single-compartment model is valid (see chapter 9, case 2). Even when significant residual function remains or fluid is lost during dialysis, the changes in ideal dialysis time are minor compared with the error in K_d. The reason for this insensitivity to errors in K_d is discovered by examining the determinants of time-averaged BUN. The major determinants are urea generation rate and the ratio of dialyzer clearance to patient size. The model determines V primarily from the intradialysis change in BUN and the estimate of K_d (see equation 5.9a). Urea nitrogen generation rate is determined from the interdialysis change in BUN and V (see equation 5.10). Overestimation of K_d causes a proportionate inflation of both V and G. If K_d/V and G/V do not change, the patient's BUN/time profile is not changed and the area under it (TAC) remains constant. These effects are readily demonstrated with most modeling software by changing K_d and observing little change in target dialysis duration.

In clinical practice, this insensitivity of target dialysis duration to K_d means that if the only goal of urea modeling is to establish the ideal dialysis time, there is little need to worry about accurate measurements of K_d. The program automatically compensates for errors in K_d and gives an accurate estimate of target dialysis time in nearly all cases. On the other hand, if urea modeling is to be used for assessment of equipment or evaluation of blood access function, an accurate estimate of K_d is essential.

Discrepancies between estimated clearance and modeled clearance

Regardless of the model employed, a value for effective dialyzer clearance can be generated if the patient's urea volume is known. The latter can be obtained from repeated modeling in the same patient or from anthropometric measurements. The dialyzer clearance is usually estimated from specifications provided by the manufacturer or calculated from the measured mass transfer coefficient for the dialyzer model. If a discrepancy exists between the estimated dialyzer clearance and the modeled clearance, it will be mirrored by a similar discrepancy between the estimated urea volume and the modeled volume. Although no concern exists for the accuracy of target predictions as discussed above, these inconsistencies help to pinpoint problems with equipment. For example, recirculation in the blood access device will reduce modeled clearance. The recirculation can be quantified by translating the difference between modeled and estimated clearance translated into a fractional reflow using equation 7.21.

Forecasting dialysis outcome from urea modeling

Modeling can be used to forecast changes in the dialysis prescription when alterations are made in single or multiple dialysis parameters, eliminating the need for a trial-and-error approach with the patient. For example, how much more time on dialysis would be required if blood flow rate is reduced from 400 to 300 ml/min or how much less time would be required if dialyzer clearance is increased from 200 to 250 ml/min? A change in schedule also can be introduced (e.g., from Monday-Wednesday-Friday to Tuesday-Friday) to see the effect on required dialysis time and dialyzer clearances. Urea modeling software will provide answers to these and other prescription modification questions within seconds.

References

1. Barth RH: Dialysis by the numbers: the false promise of Kt/V. Semin Dial 2:207-212, 1989.
2. Jindal KK, Goldstein MB: Urea kinetic modeling in chronic hemodialysis: benefits, problems and practical solutions. Semin Dial 1:82-85, 1988.
3. Laird NM, Berkey CS, Lowrie EG: Modeling success or failure of dialysis therapy: the National Cooperative Dialysis Study. Kidney Int 23 (Suppl 13):S101-106, 1983.
4. Bennett WM, Plamp CE, Gilbert DN, Parker RA, Porter GA: The influence of dosage regimen on experimental gentamicin nephrotoxicity: dissociation of peak serum levels from renal failure. J Infect Dis 140:576-580, 1979.
5. Dedrick RL: Pharmacokinetic and pharmacodynamic considerations for chronic hemodialysis. Kidney Int 7:S2-S15, 1975.
6. Gotch FA, Keen ML: Care of the patient on hemodialysis, in *Introduction to Dialysis*, Cogan MG, Garovoy MR (eds), New York, Churchill Livingstone, pp 73-143, 1985.
7. Gotch FA: Kinetic modeling in hemodialysis, in *Clinical Dialysis* (2ed), Nis sensen AR, Gentile DE, Fine RN (eds), Norwalk CT, Appleton and Lange, pp 118-146, 1989.
8. Gotch FA, Krueger KK (eds): Proceedings of a conference on adequacy of dialysis, Kidney Int 7 (Suppl 2), pp S1-S266, 1975.
9. Scribner BH, Farrell PC, Milutinovic J, Babb AL: Evolution of the middle molecule hypothesis, in *Proceedings of the Fifth International Congress of Nephrology*, Villarreal H (ed), Basel, Karger, pp 190-199, 1974.
10. Lowrie EG, Teehan BP: Principles of prescribing dialysis therapy: Implementing recommendations from the National Cooperative Dialysis Study. Kidney Int 23 (Suppl 13):S113-122, 1983.

11. Gotch FA, Sargent JA: A mechanistic analysis of the National Cooperative Dialysis Study (NCDS). Kidney Int 28:526-534, 1985.
12. Keshaviah P, Collins A: A re-appraisal of the National Cooperative Dialysis Study (abstract). Kidney Int 33:227, 1988.
13. Henderson LW, Leypoldt JK: Quantitation and prescription of therapy, in *Hemofiltration*, Henderson LW, Quellhorst EA, Baldamus CA, Lysaght MJ (eds), Berlin, Springer-Verlag, pp 129-145, 1986.
14. Giordano C, DePascale C, DeCristofaro D, Capodiscasa G, Balestrieri C, Baczyk K: Protein malnutrition in the treatment of chronic uremia, in *Nutrition in Renal Disease*, Berlyne GM (ed), Baltimore, The Williams & Wilkins Co., pp 23-37, 1968.
15. Giovannetti S, Maggiore Q: A low nitrogen diet with proteins of high biological value for severe chronic uremia. Lancet 1:1000, 1964.
16. Lowrie EG, Sargent JA: Clinical example of pharmacokinetic and metabolic modeling: Quantitative and individualized prescription of dialysis therapy. Kidney Int 18 (Suppl 10):S11-S16 1980.
17. Streicher E, Schneider H: Polysulfone membrane mimicking human glom erular basement membrane. Lancet 2:1136, 1983
18. Henderson LW: Hemofiltration. The Kidney 20:25-30, 1987.
19. Editorial: Haemofiltration. Lancet 1:1196-1197, 1983.
20. Francisco LL: High efficiency dialysis. The Kidney 21:7-11, 1988.
21. Hakim RM, Lowrie EG: Hemodialysis-associated neutropenia and hypoxemia: the effect of dialyzer membrane materials. Nephron 32:32, 1982.
22. Hakim RM, Fearon DT, Lazarus MJ: Biocompatibility of dialysis membranes: effects of chronic complement activation. Kidney Int 26:194-200, 1984.
23. Lysaght M, Pollock C, Schindhelm K, Ibels L, Farrell P: The relevance of urea kinetic modeling (UKM) to CAPD. Trans Am Soc Artif Intern Organs, Abstracts, p 84, 1988.

Chapter 9

EXAMPLES OF UREA MODELING

The following examples illustrate typical applications of urea modeling. Clinical "cases" represented by all data pertinent to urea modeling, such as dialyzer clearance and BUN, are presented with variables calculated from different models. Sometimes the technical conditions are exaggerated to emphasize a point, such as the effect of fluid gain. The cases illustrate practical clinical problems and how urea modeling is used to solve them. An "ideal" patient is presented first, followed by selected modifications to demonstrate the effect of residual clearance, schedule variations, large weight changes, dietary noncompliance, high-flux dialysis, errors in measurement of dialyzer clearance, and variations in patient size. All data are displayed in tables, so no calculator or computer is required to examine the cases in this chapter. In the tables, significant changes due to manipulation of patient or dialysis variables are underlined. Explanations for the underlined changes are found in the text.

CASE 1: EXPECTED RESULTS IN AN AVERAGE ADULT PATIENT

Table 9.1 lists data from a "normal" 70 kg adult with end-stage renal disease who lacks significant residual renal function and is dialyzed with conventional equipment three times/week. This patient eats 70 to 80 grams of protein each day, gains 1.0 kilograms of body fluid each day, and requires four hours of conventional hemodialysis thrice weekly. The conventional dialyzer has a surface area of 1.3 m^2 and a urea clearance of 178 ml/min when blood flow is 300 ml/min and dialysate flow is 500 ml/min. The dialyzer mass transfer coefficient is 360 ml/min. Typical midweek predialysis and postdialysis BUN values are 75 and 31 mg/dl. Note that increasing dialysate flow from 500 to 800 ml/min causes a small increase in clearance (178 to 188 ml/min). Similarly, increasing blood flow by 33% to 400 ml/min increments dialyzer urea clearance by only 11% (178 to 198 ml/min), as shown in table 9.1, data column 2. The nonlinear increase in clearance as blood or dialysate flow increases is illustrated in figure 7.3. Case 7 shows a more significant improvement in clearance for the same flow rate change in a patient receiving high-flux dialysis.

Table 9.1 Case 1: a standard idealized patient

	Q_b=300 ml/min	Q_b=400 ml/min	Q_d=800 ml/min	Units
Patient:				
Weight	70	70	70	kg
Weight gain/day	1.0	1.0	1.0	kg
Volume of urea distribution (*V*)				
absolute	41	41	41	liters
% body weight	58%	58%	58%	
Residual clearance (K_r)	0	0	0	ml/min
Urea nitrogen generation rate (*G*)	7.0	7.0	7.0	mg/min
Protein catabolic rate (PCRn)	1.1	1.1	1.1	g/kg/day
Dialyzer:				
Blood flow (Q_b)	300	<u>400</u>	300	ml/min
Dialysate flow (Q_d)	500	500	<u>800</u>	"
Mass transfer coefficient (*KA*)	360	360	360	"
Urea clearance (K_d)	178	<u>198</u>	<u>187</u>	"
Dialysis:				
Schedule	MWF	MWF	MWF	
Duration	4.0	<u>3.6</u>	<u>3.8</u>	hours
BUN:				
Midweek predialysis	75	75	75	mg/dl
Midweek postdialysis	31	31	31	"
Average predialysis	77	77	77	"
Time-averaged	56	56	56	"
Kt/V	1.05	1.05	1.05	/dialysis
Target values:				
Time-averaged BUN	56	56	56	mg/dl
Average predialysis BUN	77	77	77	"
Kt/V	1.05	1.05	1.05	/dialysis
Dialysis duration	4.0	<u>3.6</u>	<u>3.8</u>	hours

CASE 2: A CASE WITH NO RESIDUAL FUNCTION AND NO WEIGHT GAIN BETWEEN DIALYSES; THE EFFECT OF CHANGING K_D

This case, shown in table 9.2, illustrates a point raised in chapter 3 regarding the significance of the ratio K_d/V. If the only goal of modeling is to decide how well the patient is dialyzed, then it is not necessary to resolve K_d or V but only to determine their ratio. Since K_r is zero, the K term in Kt/V is the dialyzer clearance (K_d). Data columns 2 and 3 of table 9.2 show that there is no change in expected dialysis duration or Kt/V when K_d is set at widely divergent values. This is not surprising, since Kt/V is the major determinant of expected time-averaged BUN and its value is largely determined by the ratio of predialysis BUN to postdialysis BUN, as shown in equation 8.2. In table 9.2 the ratio of K_d/V remains constant at 4.3 ml/l/min while entered values for estimated K_d vary from 100 to 300 ml/min.

Note also that the error in estimated urea generation rate (G) that results from incorrect assessment of dialyzer urea clearance does not affect PCRn. The reason for this is that PCRn is a function of G factored for V. The two errors (in G and V) cancel each other. This canceling of errors is also consistent with our understanding that PCRn is largely determined by predialysis BUN. This is an important concept, because PCRn is the major determinant of ideal dialysis time (see equations 8.7 and 8.8). Like K_d/V, PCRn is protected from this type of error in dialyzer clearance. Both ideal dialysis time and K_d/V are insensitive to errors in K_d, so the dialysis prescription is usually valid, despite errors in dialyzer clearance estimates. This means that clotting in the dialyzer, poor data from the manufacturer regarding expected clearance, poor blood pump calibration, and other causes of dialyzer variance will not significantly affect the modeled estimate of dialysis adequacy or recommended time on dialysis in this setting.

The clinical setting depicted here, where the patient gains no weight and lacks residual function, is rarely seen. In patients who can avoid weight gain between dialyses (a rare minority in our experience), significant native kidney function likely exists. These additional variables usually have a small effect, but they can be significant, as shown in case 8. When absolute values for K_d or V are required, weight gain and residual clearance become more critical elements of urea modeling.

Measuring ideal dialysis time is not the only goal of urea modeling. Resolution of V by determining an expected dialyzer urea clearance from blood and dialysate flows adds much more to the benefit of these studies; therefore, accurate estimates of dialyzer urea clearance should be part of the quality assurance program.

Table 9.2 Case 2: no residual function and no interdialysis weight gain; estimates of K_d have no effect on ideal dialysis time

	K_d=178 ml/min	K_d=100 ml/min	K_d=300 ml/min	Units
Patient:				
Weight	70	70	70	kg
Weight gain/day	0	0	0	kg
Volume of urea distribution (V)				
absolute	41	23	58	liters
% body weight	58%	33%	82%	
Residual clearance (K_r)	0	0	0	ml/min
Urea nitrogen generation rate (G)	7.0	4.0	10.0	mg/min
Protein catabolic rate (PCRn)	1.1	1.1	1.1	g/kg/day
K_d/V	4.3	4.3	4.3	ml/l/min
Dialyzer:				
Blood flow (Q_b)	300	118	830	ml/min
Dialysate flow (Q_d)	500	500	900	"
Mass transfer coefficient (KA)	360	360	360	"
Urea clearance (K_d)	175	100	250	"
Dialysis:				
Schedule	MWF	MWF	MWF	
Duration	4.0	4.0	4.0	hours
BUN:				
Midweek predialysis	78	78	78	mg/dl
Midweek postdialysis	30	30	30	"
Average predialysis	81	81	81	"
Time-averaged	56	56	56	"
Kt/V	1.05	1.05	1.05	/dialysis
Target values:				
Time-averaged BUN	56	56	56	mg/dl
Average predialysis BUN	81	81	81	"
Kt/V	1.05	1.05	1.05	/dialysis
Dialysis duration	4.0	4.0	4.0	hours

CASE 3: THE PATIENT WITH SIGNIFICANT RESIDUAL RENAL FUNCTION (K_R)

Effects of K_r on time-averaged BUN, ideal dialysis time

If the above patient has native kidneys that are partially functioning but at a level insufficient to sustain life, we can expect that predialysis and postdialysis BUN levels will be lower, as shown in table 9.3, data column 2. If residual urea clearance is 3.0 ml/min, dialysis duration can be reduced to 2.9 hours. Although a residual clearance of 3.0 ml/min is a small fraction of dialyzer clearance, it has a marked effect on all measured BUN values because it continues to play a role during the long interdialysis interval.

Note that time-averaged BUN changed by adding residual function but that Kt/V did not. Also, ideal Kt/V fell to 0.77/dialysis, whereas target time-averaged BUN is unchanged. This serves to illustrate the advantage time-averaged BUN has over Kt/V as an indicator of dialysis outcome. The commonly stated rule of thumb that

Kt/V should be kept above or about 1.0/dialysis applies only when there is no residual function, dialysis is scheduled 3 times/week, and PCRn is less than 1.1 g/kg/day. In contrast, target values for time-averaged BUN (ideal TAC) depend only on PCRn, so at a constant level of PCRn the ideal TAC applies to any patient regardless of residual function or dialysis schedule.

Instead of reducing dialysis time, another option is to change the dialysis schedule to twice/week as shown in data column 3 of table 9.3. If this is done, dialysis time must increase to 4.7 hours to maintain time-averaged BUN at 56 mg/dl. Here again, target *Kt/V* has increased to 1.22, this time due to a change in schedule. Target time-averaged BUN remains constant at 56 mg/dl. The meaning of midweek predialysis BUN also disappears, since there is no corresponding blood sample. Predialysis and postdialysis values are given for both dialyses (Monday and Thursday) in data column 3.

Table 9.3 Case 3: the effect of residual native kidney function

	$K_r = 0$ ml/min	$K_r = 3$ ml/min	2x/wk	Disregard K_r	Units
Patient:					
Weight	70	70	70	70	kg
Weight gain/day	1.0	1.0	1.0	1.0	kg
Volume of urea distribution (V)					
absolute	41	41	41	41	liters
% body weight	58%	58%	58%	58%	
Residual clearance (K_r)	0	3.0	3.0	0	ml/min
Urea nitrogen generation rate (G)	7.0	7.0	7.0	5.7	mg/min
Protein catabolic rate (PCRn)	1.1	1.1	1.1	0.91	g/kg/day
Dialyzer:					
Blood flow (Q_b)	300	300	300	300	ml/min
Dialysate flow (Q_d)	500	500	300	500	"
Mass transfer coefficient (KA)	360	360	360	360	"
Urea clearance (K_d)	178	178	178	178	"
Dialysis:					
Schedule	MWF	MWF	MTh	MWF	
Duration	4.0	4.0	4.7	4.0	hours
BUN:					
Midweek predialysis	75	60	75.85	60	mg/dl
Midweek postdialysis	31	25	26.30	25	"
Average predialysis	77	63	80	63	"
Time-averaged	56	45	56	45	"
Kt/V	1.05	1.05	1.22	1.05	/dialysis
Target values:					
Time-averaged BUN	56	56	56	45	mg/dl
Average predialysis BUN	77	73	80	63	"
Kt/V	1.05	0.77	1.22	1.05	/dialysis
Dialysis duration	4.0	2.9	4.7	4.0	hours

Conversely, omitting measurement of residual clearance in this patient and assuming a value of zero for modeling causes a substantial underestimation of PCR, as shown in table 9.3, data column 4. This is the expected result from ignoring a significant route of nitrogen elimination. The program can only conclude that the patient is taking in less protein to account for low predialysis BUN levels. This exercise emphasizes the importance of measuring residual function in patients with significant urine output.

Is once-a-week dialysis ever justified?

Sometimes we are tempted to start patients on once-a-week dialysis early in the course of end-stage renal disease to ease the psychological pain of dialysis dependency and to take advantage of residual renal function. Urea modeling provides some interesting insights to this approach. For a 70 kg patient with 50% body water, PCRn of 1.1 g/kg/day, and a dialyzer clearance of 300 ml/min, dialysis once-a-week for four hours is feasible only if K_r exceeds 6 ml/min. To achieve a dialyzer clearance of 300 ml/min, blood flow would have to be pushed to a maximum even with the best of modern high-flux dialyzers. Both high clearance and extended time are contraindicated when dialysis is initiated because of the risks from dialysis disequilibrium. It seems unreasonable, therefore, to begin with once-a-week dialysis. On the contrary, dialysis three times/week or daily dialysis is more appropriate to quickly control the uremic state, falling back later to the less frequent schedule. Since residual function continues to decline, perhaps in an accelerated fashion after commencing hemodialysis, prescription of infrequent dialysis will be short-lived in most patients. In smaller patients, a residual clearance of 3 or 4 ml/min may permit once-a-week dialysis, but this degree of retained renal function is a greater fraction of normal native kidney function in smaller patients, some of whom may not require dialysis. It is thus a rare patient who comes to dialysis and can tolerate once-a-week treatments. If such a regimen is used, it is important to measure residual function frequently to avoid underdialysis when native kidney function inevitably deteriorates.

Case 4: The patient with habitually large weight gains between dialyses

Large weight gains between dialyses and their requisite high ultrafiltration rates during dialysis can cause problems with patient tolerance of dialysis and create difficulty with fluid management. Patients cycle between the verge of pulmonary edema predialysis (especially after the long cycle) and volume depletion with cramps and hypotension at the end of dialysis. With respect to urea exposure, however, the cyclical weight gain and loss have an effect much like residual clearance, as shown in table 9.4. Predialysis BUN values tend to be lower due to dilution, while postdialysis BUN values are unchanged.

One can quickly show that weight gain from fluid ingestion between dialyses, requiring fluid losses during dialysis, causes a significant reduction in target dialysis

Table 9.4 Case 4: The effect of large interdialytic weight gains

	Gain = 1.0 kg/day	Gain = 4.0 kg/day	Units
Patient:			
Weight	70	70	kg
Weight gain/day	1.0	4.0	kg
Volume of urea distribution (V)			
absolute	41	41	liters
% body weight	58%	58%	
Residual clearance (K_r)	0	0	ml/min
Urea nitrogen generation rate (G)	7.0	7.0	mg/min
Protein catabolic rate (PCRn)	1.1	1.1	g/kg/day
Dialyzer:			
Blood flow (Q_b)	300	300	ml/min
Dialysate flow (Q_d)	500	500	"
Mass transfer coefficient (KA)	360	360	"
Urea clearance (K_d)	178	189	"
Dialysis:			
Schedule	MWF	MWF	
Duration	4.0	4.0	hours
BUN:			
Midweek predialysis	75	66	mg/dl
Midweek postdialysis	31	31	"
Average predialysis	77	66	"
Time-averaged	56	51	"
Kt/V	1.05	1.14	/dialysis
Target values:			
Time-averaged BUN	56	56	mg/dl
Average predialysis BUN	77	72	"
Kt/V	1.05	1.05	/dialysis
Dialysis duration	4.0	3.5	hours

duration. This change results from two factors: 1) the additional clearance gained from required ultrafiltration during dialysis, and 2) the expansion and contraction of the urea distribution space caused by fluid gain and loss. Higher clearances result from addition of ultrafiltration to dialyzer clearance. This addition, however, is not a simple arithmetic one; added clearance from ultrafiltration is often less than one might expect (see chapter 7 for a more detailed discussion of the effect of ultrafiltration on urea clearance). The expansion of body water that results from fluid intake between dialyses benefits the patient by lowering the predialysis BUN (dilution effect). Less obvious is the beneficial effect of fluid loss during dialysis. Contraction of the urea volume during dialysis serves to concentrate urea, the converse of the dilution effect between dialyses. This concentrating effect maintains high dialyzer inlet BUN levels and improves urea removal during the later phases of dialysis, even when dialyzer clearance is only minimally improved by ultrafiltration. These effects of expansion and contraction of the urea space are included in the variable-volume model and serve to emphasize the importance of this more complex model when ultrafiltration rates are high.

Case 5: The patient with high protein intake

The patient who relishes high-protein foods requires more dialysis. The relationship between protein intake (equated to net PCRn rate in the stable patient) and dialysis intensity is shown as the target therapy line in figures 8.7 and 8.8. As PCRn increases above 1.1 g/kg/day, Kt/V must increase to maintain time-averaged values within the "safe" range of the NCDS patients with favorable outcomes. Since V cannot be changed and K_d is most likely at its maximum already, the only change that can accommodate the high protein intake is an increase in dialysis duration (t). The price that both the patient and the dialysis center pay for the higher protein intake is illustrated in table 9.5. The increase in dialysis duration is linearly related to the increase in PCRn, provided that dialyzer clearance remains constant.

Examination of the target values in table 9.5 reveals that a higher time-averaged BUN is permitted in patients with high protein intakes. This allowance results from the NCDS findings that patients with high PCRn values had lower morbidity than those with the same time-averaged BUN but with low PCRn values. Although controversial, this negative correlation between BUN and morbidity may relate to our understanding that urea is not very toxic and that high urea levels alone are not responsible for uremic symptomatology. Urea is merely a marker for the uremic state with limited application. Still, it is the best measure of uremia in dialyzed patients that we have at present.

At PCRn values above 1.8 g/kd/day, the model used to generate this data establishes a ceiling on time-averaged BUN (TAC) at 70 mg/dl. Although a PCRn of 1.8 is well above the "normal" range within which the NCDS data can be applied, we occasionally encounter patients with protein intakes at this level, especially in the younger age groups. Because some of these patients do not accept dietary protein restriction, and because we have no data to tell us about risks of morbidity and mortality at this level of protein intake, a safe approach was taken by limiting the ceiling on TAC to 70 mg/dl. This means that for all PCRn levels above 1.8 g/kg/day, TAC will be fixed at 70 mg/dl and target dialysis time will increase exponentially with PCRn. The reason for the exponential increase, rather than a linear increase in dialysis time, is related to diminishing urea removal rates as the BUN falls toward the end of dialysis. The dialyzer becomes less efficient in removing urea as the BUN (driving force) falls.

Table 9.5 Case 5: The effect of high protein intake

	PCRn = 1.1 g/kg/day	PCRn = 1.4 g/kg/day	PCRn = 1.7 g/kg/day	Units
Patient:				
Weight	70	70	70	kg
Weight gain/day	1.0	1.0	1.0	kg
Volume of urea distribution (*V*)				
absolute	41	41	41	liters
% body weight	58%	58%	58%	
Residual clearance (K_r)	0	0	0	ml/min
Urea nitrogen generation rate (*G*)	7.0	9.3	11.6	mg/min
Protein catabolic rate (PCRn)	1.1	1.4	1.7	g/kg/day
Dialyzer:				
Blood flow (Q_b)	300	300	300	ml/min
Dialysate flow (Q_d)	500	500	500	"
Mass transfer coefficient (*KA*)	360	360	360	"
Urea clearance (K_d)	178	178	178	"
Dialysis:				
Schedule	MWF	MWF	MWF	
Duration	4.0	4.0	4.0	hours
BUN:				
Midweek predialysis	75	99	123	mg/dl
Midweek postdialysis	31	41	51	"
Average predialysis	77	102	127	"
Time-averaged	56	73	91	"
Kt/V	1.05	1.05	1.05	/dialysis
Target values:				
Time-averaged BUN	56	62	69	mg/dl
Average predialysis BUN	77	102	127	"
Kt/V	1.05	1.27	1.48	/dialysis
Dialysis duration	4.0	4.9	5.7	hours

Case 6: The patient with low protein intake

Patient with low protein intake have been the subject of considerable controversy. The National Cooperative Dialysis Study (NCDS) found that this group had a higher morbidity, presumed to be related to underdialysis. Another obvious cause of morbidity might be malnutrition, but malnutrition can result from underdialysis. A downward spiral of malnutrition leading to underdialysis and vice versa can result if the time-averaged or predialysis BUN is kept constant as it was in this study.

Interpretation of the results of the NCDS, using a mechanistic analysis, led to the recommendation that *Kt/V* be kept above 1.0 and that PCRn be kept above 0.8 g/kg/day (1). For patients with PCRn below 0.8 g/kg/day, no recommendation can be made, since the NCDS did not include significant numbers of patients in this range. Because not all patients will fall in the ideal range for PCRn, in practical applications we find ourselves extrapolating the *Kt/V* line below the 0.8 g/kg/day limit and prescribing dialysis on this somewhat unproven basis. The result for patients with low PCRn is a surprisingly intense dialysis that keeps the predialysis BUN at very low levels. This is probably a safe albeit unproven approach, since there is little evidence for deleterious effects of overdialysis.

Table 9.6 shows the target values for our idealized patient whose steady-state PCRn falls to 0.6 g/kg/day. Target predialysis BUN falls to 35 mg/dl and time-averaged BUN falls to 26 mg/dl. Note that no relaxation in dialysis intensity, shown as target dialysis duration, is allowed for low protein intake. This simply reflects the constant value of *Kt/V*, established as slightly over 1.0 for all patients dialyzed three times/week, when PCRn is less than 1.1 g/kg/day (refer to figure 8.8). This means that patients with low protein intake require the same intensity of dialysis as those patients with adequate but not excessive protein intakes. Patients with high protein intake require even more. The mechanistic approach to dialysis prescription dictates that lowering protein intake improves the fiscal economy of dialysis only if protein intake is high before lowering it. If dialyzer clearance cannot be improved, the only way to economize with fixed per capita reimbursement (by reducing dialysis time) is to reduce protein intake in patients with high PCRn (>1.1 g/kg/day) or to accept only smaller patients. The former technique lowers *t* by lowering *Kt/V* and the latter lowers *t* by lowering *V* at a fixed *Kt/V*. Both of these techniques are either unreliable or impractical; the better method is to improve dialyzer urea clearance (K_d).

Case 7: The patient treated with high-flux dialysis

High-flux dialysis induces a more rapid rate of urea removal from the patient and generally permits shortening of dialysis duration. The rapid flux of urea causes an accentuation of intercompartment urea disequilibrium, leading to plasma levels significantly below intracellular levels. In such states, the two-compartment model is sometimes more appropriate than the one-compartment model to generate urea/time profiles that correlate closely with measured data and to provide an accurate measure of urea generation.

Table 9.6 Case 6: Effect of low protein intake

	PCRn=1.1 g/kg/day	PCRn=0.6 g/kg/day	Units
Patient:			
Weight	70	70	kg
Weight gain/day	1.0	1.0	kg
Volume of urea distribution (V)			
absolute	41	41	liters
% body weight	58%	58%	
Residual clearance (K_r)	0	0	ml/min
Urea nitrogen generation rate (G)	7.0	3.2	mg/min
Protein catabolic rate (PCRn)	1.1	0.6	g/kg/day
Dialyzer:			
Blood flow (Q_b)	300	300	ml/min
Dialysate flow (Q_d)	500	500	"
Mass transfer coefficient (KA)	360	360	"
Urea clearance (K_d)	178	178	"
Dialysis:			
Schedule	MWF	MWF	
Duration	4.0	4.0	hours
BUN:			
Midweek predialysis	75	34	mg/dl
Midweek postdialysis	31	14	"
Average predialysis	77	35	"
Time-averaged	56	26	"
Kt/V	1.05	1.05	/dialysis
Target values:			
Time-averaged BUN	56	26	mg/dl
Average predialysis BUN	77	35	"
Kt/V	1.05	1.05	/dialysis
Dialysis duration	4.0	4.0	hours

Table 9.7 shows the results of modeling with both single- and double-compartment models in our idealized 70 kg patient undergoing high-flux dialysis. In each case the dialysis is adjusted to be adequate according to the dictates of the model (TAC = ideal TAC). The increase in dialyzer clearance from the conventional 178 ml/min to 306 ml/min with high-flux equipment allows shortening the dialysis to 2.3 hours when the single-compartment model is applied. Other parameters remain unchanged except for a slight lowering of postdialysis BUN, a requirement dictated

Table 9.7 The effect of high-flux dialysis

	Standard	high-flux 1-pool	high-flux 2-pool	Units
Patient:				
Weight	70	70	70	kg
Weight gain/day	1	1	1	kg
Volume of urea distribution (V)				
absolute	41	41	41	liters
% body weight	58%	58%	58%	
Residual clearance (K_r)	0	0	0	ml/min
Urea nitrogen generation rate (G)	7.0	7.0	7.0	mg/min
Protein catabolic rate (PCRn)	1.1	1.1	1.1	g/kg/day
Dialyzer:				
Blood flow (Q_b)	300	400	400	ml/min
Dialysate flow (Q_d)	500	800	800	"
Mass transfer coefficient (KA)	360	900	900	"
Urea clearance (K_d)	175	306	306	"
Dialysis:				
Schedule	MWF	MWF	MWF	
Duration	4.0	2.3	2.6	hours
BUN:				
Midweek predialysis	75	75	75	mg/dl
Midweek postdialysis	31	30	26	"
Average predialysis	77	77	77	"
Time-averaged	56	56	56	"
Kt/V	1.05	1.05	1.17	/dialysis
Target values:				
Time-averaged BUN	56	56	56	mg/dl
Average predialysis BUN	77	77	77	"
Kt/V	1.05	1.05	1.17	/dialysis
Dialysis duration	4.0	2.3	2.6	hours

by slight lengthening of the interdialysis interval. When the two-compartment model is applied, a somewhat longer dialysis of 2.6 hours is required to maintain time-averaged BUN the same at 56 mg/dl. The longer dialysis duration is necessary because the dialysis is less efficient. The loss of efficiency results from lower BUN values throughout dialysis, a consequence of delayed diffusion from the second (e.g., intracellular) compartment. The lower extracellular levels reduce the driving force for diffusion of urea, which reduces the urea removal rate. More dialysis time is required to remove the same amount of urea. The two-compartment model generates a significantly lower postdialysis BUN. The discussion of postdialysis rebound in chapter 5 helps to explain these differences between the two models. Predialysis BUN is unchanged because its value is largely determined by the patient's PCR, which is unchanged. The intercompartment mass transfer coefficient was set at 800 ml/min for these comparisons.

If we increase blood flow from 300 to 400 ml/min, a substantial increase in urea clearance is observed for the high-flux dialyzer. The increase in K_d from 251 to 306 ml/min should be contrasted with a similar increase in blood flow shown in case 1, a patient receiving conventional dialysis. The 33% increase in blood flow nets a 22% increase in clearance for the high-flux dialyzer compared to 11% for the conventional dialyzer. At higher blood flows, the high-flux dialyzer's clearance remains dependent on blood flow because of its higher urea mass transfer coefficient. Clearance for the conventional dialyzer is less improved at higher blood flow rates because urea flux is membrane limited.

Case 8: The patient whose dialyzer clearance (K_d) varies from the expected clearance

Dialyzer clearance is more often overestimated than underestimated. Overestimation of clearance means that actual clearance is significantly less than either the manufacturer's claims or previous measured clearances for the specific model of dialyzer. Several causes of this type of error are listed in chapter 3, table 3.2. While dialysate flow is usually fairly accurate, blood flow is often less than expected when the blood pump is turning at higher speeds. When blood pump flow is turned up, prepump pressure tends to fall, creating a sucking effect that progressively collapses the pump tubing segment (see figure 3.4) (2). As the pump segment collapses, the blood pump RPM meter may not accurately reflect actual flow in the blood tubing. Clearance may also be diminished by clotting in new or reused dialyzers, but the effect is not linearly related to loss of dialyzer volume. For a 20% loss of fiber volume, the hollow fiber kidney loses approximately 10% of its original urea clearance (3). Another cause of reduced clearance is recirculation in the blood access device discussed in chapter 7.

Data columns 2 and 3 of table 9.8 show the effects of overestimation and underestimation of dialyzer urea clearance when the patient has significant residual renal function and weight gain between dialyses. This case should be compared to case 2, where no residual function or weight gain was included. Although the major effect of these clearance errors is a corresponding error in the urea volume (V), significant errors in PCRn, Kt/V, and target dialysis time also appear. In contrast to case 2, an error occurs in PCRn due to the effect of residual clearance and weight gain. An error in target dialysis time results from the error in PCRn. Although not shown here, the errors are magnified when the two-compartment model is used. These observations mean that accurate or near accurate estimates of K_d are necessary to determine the effectiveness of dialysis and required dialysis intensity when significant residual function or weight changes occur.

Deviation from what should be a constant urea volume in a single patient serves as a flag for errors in dialyzer clearance. These aberrancies in V will become more certain after repeated urea kinetic analyses establish the patient's mean V. When found, errors in V should stimulate further investigation into possible causes of dialyzer variance. A typical example of an error in V that required additional studies to resolve is shown in figure 9.1. This patient was studied because she seemed to require more intensive dialysis than usual to maintain acceptable BUN levels. During a four-hour dialysis with a large conventional dialyzer (K_d = 195 ml/min), the BUN dropped from 84 to 31 mg/dl, a reasonable accomplishment. However, the patient weighed only 50 kg. Urea modeling showed that the BUN should have fallen more and that the urea volume would be 86% of body weight if the dialyzer clearance was correct. If V is a more reasonable 60% of body weight, then the modeled whole-body dialyzer clearance is 137 ml/min. Thirty minutes after starting dialysis, inlet, outlet and peripheral blood samples were drawn to examine dialyzer

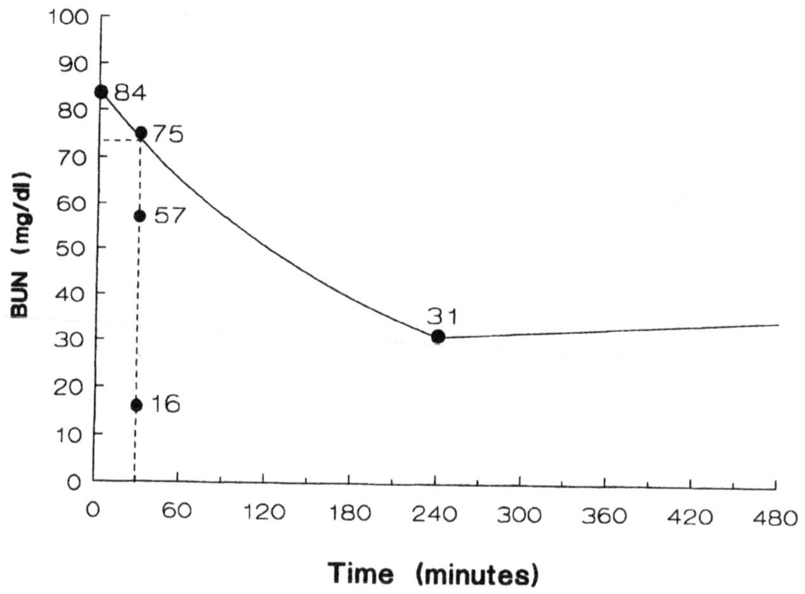

Figure 9.1. Venous recirculation exaggerates the decrease in dialyzer arterial BUN. BUN values are printed alongside solid circles representing blood samples. Thirty minutes after starting dialysis in this patient, peripheral BUN is 75 mg/dl, nearly identical to the modeled estimate (73 mg/dl). Dialyzer inlet BUN is lower (57 mg/dl) due to recirculation of venous blood with BUN 16 mg/dl. Postdialysis peripheral BUN is 31 mg/dl.

Table 9.8 Case 8: The effect of errors in dialyzer clearance estimates

	Baseline	Overestimate	Underestimate	Units
Patient:				
Weight	70	70	70	kg
Weight gain/day	1.0	1.0	1.0	kg
Volume of urea distribution (V)				
absolute	41	59	21	liters
% body weight	58%	84%	31%	
Residual clearance (K_r)	3.0	3.0	3.0	ml/min
Urea nitrogen generation rate (G)	8.9	11.7	5.8	mg/min
Protein catabolic rate (PCRn)	1.3	1.2	1.6	g/kg/day
Dialyzer:				
Urea clearance (K_d)	178	250	100	ml/min
Dialysis:				
Schedule	MWF	MWF	MWF	
Duration	4.0	4.0	4.0	hours
BUN:				
Midweek predialysis	75	75	75	mg/dl
Midweek postdialysis	31	31	31	"
Average predialysis	77	77	77	"
Time-averaged	56	56	56	"
Kt/V	1.05	1.02	1.13	/dialysis
Target values:				
Time-averaged BUN	61	59	68	mg/dl
Average predialysis BUN	82	81	85	"
Kt/V	0.93	0.95	0.84	/dialysis
Dialysis duration	3.5	3.7	3.0	hours

performance and recirculation (see chapter 7). The inlet BUN was 57 mg/dl, considerably lower than the predicted value of 73 mg/dl read from the modeled curve in figure 9.1. The peripheral BUN obtained from the other arm was 75 mg/dl, much closer to the predicted value, and the dialyzer venous BUN was 16 mg/dl. Dialyzer clearance from mass balance across the dialyzer (equation 7.6) is 194 ml/min, in agreement with the expected clearance generated at these flow rates in this model of dialyzer. Venous recirculation, calculated from equation 7.17 is 30%, a value that agrees with the model's prediction of 35% recirculation. This analysis pinpointed a correctable problem with the patient's dialysis access site.

Less frequently encountered *under* estimation of dialyzer clearance (data column 3, table 9.8) often results from drawing the postdialysis BUN from the venous rather than the arterial blood line. For staff members responsible for drawing blood samples, a clear distinction should be made between methods for measuring urea clearance and methods for modeling urea kinetics. Both require two blood samples. For clearance measurements, blood is drawn from the arterial (inlet) and venous (outlet) ports. For modeling, blood is always drawn from the *arterial port* before and after dialysis.

CASE 9: SMALL PATIENTS AND THE PEDIATRIC PATIENT

Patient size but not age is a significant variable in most urea models. Younger patients generally have a higher PCRn, but there is no allowance for this in the equations or programs. During growth spurts, protein intake will exceed the protein catabolic rate, so no adjustment in PCRn seems to be necessary for young growing patients. Urea modeling simply looks at the rate of urea appearance as an index of protein catabolism. Anabolic processes are ignored or simply not discoverable with urea modeling.

Patient size is a major determinant of Kt/V. At a fixed K_d, smaller patients require less time on dialysis if we can extrapolate the mechanistic interpretation of the NCDS to pediatric patients and maintain Kt/V at approximately 1.0/dialysis. In practice, however, K_d is usually reduced because of limitations to the maximum blood flow attainable through pediatric shunts and fistulas. Pediatric patients are generally considered to be more difficult to maintain with hemodialysis for a variety of reasons, so most are maintained with peritoneal dialysis or receive renal transplants. Hemodialysis urea modeling predicts that shortening or at least no prolongation of dialysis time should be expected in these smaller patients.

Table 9.9 shows a comparison of our idealized 70 kg patient to a smaller 30 kg patient. Smaller adult patients with adequate blood access devices should be more economical to dialyze. Earlier in this chapter we discussed the lack of economical advantage to lowering protein intake in patients with PCRn values of 1.1 g/kg/day or less. Smaller size, however, has a big advantage, as shown in table 9.9. In practice, under per capita reimbursement, dialysis centers benefit economically from smaller patients and tend to lose with larger patients. Fortunately, the mix of patients is relatively constant, so no deliberate discrimination against larger patients is seen and the extra effort required to dialyze larger patients is offset by time saving during dialysis of smaller patients. However, if some form of urea modeling is not part of the dialysis prescription and the dialysis prescription or duration is kept constant for all patients, a significant discrimination against larger patients is guaranteed.

Table 9.9 Case 9: The impact of patient size

	Wt = 70 kg	Wt = 35 kg	Units
Patient:			
Weight	70	35	kg
Weight gain/day	1.0	0.5	kg
Volume of urea distribution (*V*)			
absolute	41	20	liters
% body weight	58%	58%	
Residual clearance (K_r)	0	0	ml/min
Urea nitrogen generation rate (*G*)	7.0	3.5	mg/min
Protein catabolic rate (PCRn)	1.1	1.1	/kg/day
Dialyzer:			
Blood flow (Q_b)	300	300	ml/min
Dialysate flow (Q_d)	500	500	"
Mass transfer coefficient (*KA*)	360	360	"
Urea clearance (K_d)	178	178	"
Dialysis:			
Schedule	MWF	MWF	
Duration	4.0	2.0	hours
BUN:			
Midweek predialysis	75	77	mg/dl
Midweek postdialysis	31	31	"
Average predialysis	77	79	"
Time-averaged	56	56	"
Kt/V	1.05	1.05	/dialysis
Target values:			
Time-averaged BUN	56	56	mg/dl
Average predialysis BUN	77	79	"
Kt/V	1.05	1.05	/dialysis
Dialysis duration	4.0	2.0	hours

Standards for urea modeling

Simulated clinical data also serves as a basis for comparison of results obtained from one program with results from another program. There is need for an industry standard against which software developers can compare their programs to uncover flaws in algorithms and written code. While these examples are not proposed as gold standards, it is hoped that they might serve as a springboard for future establishment of standard expected results.

References

1. Gotch FA, Sargent JA: A mechanistic analysis of the National Cooperative Dialysis Study (NCDS). Kidney Int 28:526-534, 1985.
2. Depner TA, Rizwan S, Stasi T: Prepump vacuum effect of hemodialysis blood flow: in vitro and in vivo studies. Trans Am Soc Artif Intern Organs, in press, 1990.
3. Gotch F: Mass transport in reused dialyzers. Proc Clin Dial Transplant Forum 10:81-85, 1980.

Chapter 10

THE FUTURE

DIALYSIS VERSUS OTHER TREATMENTS FOR END-STAGE RENAL FAILURE

It is hazardous to predict the direction that management of end-stage renal disease will take in the future, in full view of the unexpected turns taken in the past. Nephrologists once expected that home dialysis would dominate as the preferred treatment modality. At other times it appeared that transplant or peritoneal dialysis would replace other treatments. The anticipated breakthrough in the field of tissue compatibility promised to enhance the outcome of renal allografts and deemphasize artificial organs except as a temporizing measure before or between renal transplants. Unfortunately, this highly desired advancement has not occurred and is not yet on the horizon. When and if the transplant rejection problem is solved, dialysis replacement therapy will continue to be necessary as a life-saving treatment in emergencies, both for patients with acute renal failure and for those who cannot be transplanted. In this hypothetical era of future transplant popularity, those patients who cannot be transplanted will appreciate the efforts expended during the current era that have led to major improvements in the performance and biocompatibility of extracorporeal equipment. The time is ripe for refining replacement therapy, since most patients around the world with renal function insufficient to sustain life continue to depend on dialysis in one form or another.

A breakthrough in the field of hemofiltration therapy would be welcome. This method of renal replacement therapy employs the same membranes currently used for high-flux dialysis, but instead of dialysis uses pure convection to remove toxins, much like the normal glomerulus. When combined with adsorbents and recirculating dialysate, hemofiltration promises to miniaturize and render "wearable" the artificial kidney (1). This type of system might take advantage of cultured cells or membrane technology that generates active, instead of passive, transport of solutes selectively, much like the native kidney (2,3). If blood binding of significant toxins proves to be a problem, this type of bioartificial system may provide better clearance of toxins. Other techniques for unbinding toxins, such as treatment of plasma at a reduced pH or at high osmolarity, also might be worth pursuing.

Until these elusive goals are in sight, emphasis should be placed on improving the equipment and techniques that are now in place and have proven effective over the past 30 years. These methods should include monitoring of dialysis outcome, a goal met only by modeling urea kinetics at present. Following is a discussion of improvements in kinetic modeling that may be of use in the future.

Outcome parameters for high-flux dialysis

As mentioned in chapter 8, use of criteria for the adequacy of dialysis established by the National Cooperative Dialysis Study (NCDS) may not be appropriate for high-flux dialysis. All NCDS centers used cellulosic membranes, not the more biocompatible and more porous membranes used for high-flux dialysis. Increased permeablility of high-flux dialyzers improves middle molecule clearances and could reduce the need for additional time on dialysis independent of urea clearance. Patients whose uremic state is controlled with these dialyzers may behave more like those treated with peritoneal dialysis, who fare as well or better than hemodialysis patients despite higher BUN levels. On the other hand, the NCDS found that time on dialysis had a positive but weak independent effect on the probability of failure (4,5). Focus on this aspect of the data might lead us to conclude that shortening dialysis time without changing time-averaged BUN would worsen outcome. A study similar to the NCDS but more limited in scope would help to detect differences in outcome between abbreviated and more prolonged dialysis time in patients treated with high-flux dialyzers. These theories need to be tested not just for academic purposes but to resolve whether or not the psychological and economic gains of shortened dialysis are eclipsed by medical complications, both short-term and long-term.

More complex models

Two-compartment variable-volume model

Better computer simulation of hemodialysis may require more complex models that account for changes in plasma volume and intercomparmental solute movement during dialysis. Patients may differ in their response to dialysis depending on the efficiency of peripheral circulation. Fluid removal by ultrafiltration during dialysis could decrease the efficiency of solute removal by causing vasoconstriction that reduces the effective intercompartment mass transfer coefficient. The patient's hematocrit and binding capacity of tissue and plasma may influence the removal of toxins yet to be identified. These factors can be added to existing models to simulate the target solute's kinetics.

Two-compartment models are available but are not commonly used in routine clinical dialysis (6,7,8,9). This type of model may be more appropriate for high-flux dialysis, where high clearances exaggerate urea disequilibrium between compartments.

Two compartments with variable intracellular volume

A two-compartment model that allows the second compartment as well as the first to expand and contract is described briefly in chapter 5 and appendix E. It has been successfully developed but has not been routinely applied because of the expensive hardware required and slowness of the software (9). Setting up the model also requires a more advanced understanding of modeling in general and of higher mathematics. But once the model is operative, the only extra effort required is entry of the additional parameters (e.g., serum and dialysate sodium concentrations). Processing time for modern desktop computers has decreased tenfold in the past ten years, so the time required to process complex models such as this may not be limiting in the future.

Refinement of the intercompartment mass transfer coefficient for urea

There is need for a better understanding of intercompartment fluid shifts during dialysis. Newer multicompartment models may be required that use different mass transfer coefficients for different compartments, e.g., ischemic limbs, transcellular compartments, or relatively stagnant compartments such as peritoneal or pleural fluid. Intracellular urea pools might be examined during dialysis using nuclear magnetic resonance. Intracellular and extracellular volumes can be measured with whole-body impedance techniques (10). There may be opportunities to improve equilibration between intravascular and the slowly equilibrating urea pools using drugs, exercise, or manipulation of the environment.

Modeling other solutes

The kinetics of heparin, bicarbonate, sodium, water, calcium, urea, creatinine, inulin, potassium, polypeptides, beta-2 microglobulin, and heat transfer during and between dialyses have been successfully modeled (6,7,9,11,12,13,14,15,16,17). Application of kinetic modeling to other solutes may help to examine responses to exercise and dietary manipulation, and the effects of different dialysis membranes. Further studies of the kinetics of phosphorus, potassium, and aluminum may provide insights that will eventually help control the extracellular concentrations of these life-threatening elements. As each new potential uremic toxin is identified in the future, models can be developed to describe its dialyzability and kinetics during and between dialysis. Better understanding of its kinetic behavior will help design techniques to control its concentration and optimize removal from the body.

Better markers for uremia

Table 10.1 shows a list of plasma markers currently in use for clinical assessment of patients with end-stage renal failure. They reflect complications attributed to renal failure and its treatment with dialysis. Only the first two, serum urea and creatinine, are measures of bulk solute removal, i.e., the dialyzer membrane's capacity to remove small-molecular-weight compounds. Other markers have little to do with the efficiency of dialysis, e.g., ferritin levels measure accumulation of iron, a consequence of blood replacement therapy. Serum aluminum concentrations reflect water contamination, a risk of dialysis that has no relationship to its effectiveness. Beta-2 microglobulin is the first recognized high-molecular-weight toxin that accumulates as a direct consequence of loss of renal elimination. Other high-molecular-weight proteins accumulate from lack of renal elimination, (e.g., amylase, insulin) but are not known to be toxic. Although not part of the classically described uremic syndrome, beta-2 microglobulin is an important toxin to address in patients dialyzed for more than one year. Its removal will almost certainly shape treatment of end-stage renal disease, dialytic or otherwise, in the future.

Table 10.1 Clinical markers for end-stage renal failure

Stage	Marker	Measure of
Classical (acute)	BUN	Uremia
	Creatinine	Uremia
	Potassium	Potassium intoxication
	Bicarbonate	Acidosis
	Sodium	Water balance
	Magnesium	Magnesium intoxication
Intermediate	Calcium/phosphorus/PTH	Osteodystrophy
	Hematocrit	Anemia
Extended	Ferritin	Hemosiderosis
	Aluminum	Aluminum intoxication
	Beta-2 microglobulin	Dialysis amyloidosis

Even if a more specific toxin is identified in the future, it is likely that urea kinetic modeling will continue to be the preferred technique for quantifying overall dialysis effectiveness because urea is more abundant, easily measured, and a good measure of the dialyzer's capacity for mass transport of small molecular weight solute.

Improved Blood Flow Monitoring

Roller pumps have a good reputation for accuracy, but they fail when subjected to pressure effects. A good roller pump records actual revolutions per minute (RPM) rather than torque or voltage as a measure of blood flow. When properly calibrated and adjusted for occlusion of the pump segment, these recordings are very close to the actual flow, provided upstream pressures are not too low and downstream pressures are not too high. The upstream (prepump) pressure is more likely to cause inconsistencies between recorded and actual flow.

Prepump monitoring of pressure is important because very low pressures (below -300 mm) can cause pump segment collapse and low flow despite no change in RPM (see figure 3.4). Most pump segments can tolerate pressures as low as -500 mm pressure before the tubing collapses completely. Low pressure and flow will not be detected except perhaps as an increase in vibration of the tubing, unless a pressure monitor is in place. The vibration is due to a water hammer effect as the partially collapsed pump segment under low pressure suddenly fills from the relatively high pressure on the upstream side. Depending on the quality of the pump segment, significant deviation from the true flow can occur when prepump pressure is low. The effect of low prepump pressure is discussed in chapter 3. Downstream pressures

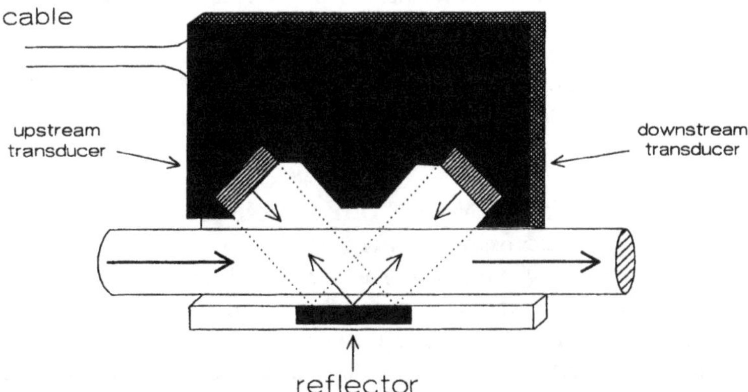

Figure 10.1. Transit-time flowmeter (Transonic Systems Inc., Ithaca, N.Y.). Two transducers (upstream and downstream) act alternately as transmitter and receiver of a wide beam of ultrasound that is reflected off a metal plate on the other side of the blood tubing. Velocity of the sound beam is retarded in the upstream direction and accelerated in the downstream direction. The difference in velocity, detected by sensitive recording equipment, correlates linearly with flow. Because it emits a wide beam of sound that encompasses the entire tubing diameter, the device measures flow, not velocity.

are usually high, but they do not have an adverse effect unless pump occlusion is incomplete. When a pump roller fails to properly occlude the pump segment, blood continually leaks back into the low pressure prepump segment behind the roller.

To avoid errors frequently encountered in estimating urea clearance, an on-line blood flowmeter would be helpful. Work in the past with the classic bubble tract technique has proven this method applicable only in the research setting; it is almost impossible to set up in the busy outpatient clinic. In addition, the high blood flow common to high-flux dialysis requires very long tracts to give timing intervals under five seconds between the marks. When blood flow is 400 ml/min, the velocity of blood through 3/16 inch ID tubing is 1.2 feet/sec. A variety of ultrasonic flowmeters are appearing that provide instantaneous reading of flow. Most use the Doppler effect to measure flow velocity. One that we have found useful is based on ultrasonic transit-time shown in figure 10.1 (18,19). If this type of flowmeter can be combined with an integrator to provide average flow during the entire dialysis (instead of an instantaneous RPM meter reading), errors in clearance estimates might be eliminated.

Dialysate modeling

When urea modeling is used for quality assurance monitoring of equipment, the most troublesome variable is the dialyzer clearance. When expected clearance does not agree with modeled clearance, several reasons can be proposed, as outlined in appendix F. Even when blood flow is accurate, dialyzer clearance can be overestimated because of clotting or recirculation of venous blood. Several investigators have suggested the *direct dialysate method* to avoid these problems (20,21). This procedure uses collection of the entire dialysate volume during dialysis to get a true whole-body clearance, averaged for the duration of the studied dialysis (22). The methodology is too cumbersome for routine application in the outpatient clinic, but sampling dialysate instead of blood may prove fruitful if the samples are taken at opportune times (23).

Real-time monitoring of urea kinetics

Studies of urea kinetics using dialysate analysis have uniformly shown discrepancies when compared with conventional kinetic modeling using blood-compartment measurements of urea nitrogen. Disequilibrium between the intracellular fluid (ICF) and extracellular fluid (ECF) urea concentration is often invoked to explain these differences, often as much as 30% in the urea distribution volume (20,21,22,24,25,26). When the two-compartment model described in chapter 5 is applied, only small differences in urea volume are seen. To account for the larger differences, values of K_c, the ECF/ICF mass transfer coefficient for urea must be lowered, sometimes to less than 200 ml/min, inconsistent with measured values. When such low transfer coefficients are employed, the rebound in urea concentration postdialysis is exaggerated beyond values measured in patients postdialysis (27). This suggests that factors other than disequilibrium between ECF and ICF compartments account for the differences between direct dialysate modeling and conventional modeling using BUN.

A possible cause of these discrepancies is a decline in either dialyzer or whole-body urea clearance during dialysis. The potential loss of surface area for dialysis toward the end of the dialysis procedure has not been adequately addressed. Hypercoagualability or insufficient heparin may cause clotting of dialyzer fibers sufficient to reduce clearance but not enough to impair blood flow. Studies of this phenomenon have been hampered by inaccuracies inherent in urea clearance measurements toward the end of dialysis. Greater errors in clearance measurements can be expected because of the lower blood and dialysate urea concentrations as dialysis nears completion.

Recirculation in the blood access device, as discussed in chapter 7, will decrease whole-body urea clearance, causing a discrepancy between measured and modeled clearance. Even when recirculation is measured at the beginning of dialysis, an additional error may appear as the recirculation fraction increases toward the end of dialysis.

Real-time monitoring of solute removal may be the best answer to the problem of clearance variability. If urea clearances could be continuously monitored, problems such as clotting in the dialyzer or reflow from poor access function could be spotted instantly. Dialysis staff could make adjustments in heparin dosage and dialysis time when the problem begins to arise or during the next dialysis instead of several weeks or months later as is the current practice. Continuous monitoring of dialysate urea concentration is feasible. The possibility is foreseeable of evaluating urea kinetics continuously in every patient during each dialysis using fully integrated software and hardware that interfaces with urea-specific electrodes like those under development in several laboratories (28,29,30). Disposable electrodes might be incorporated into the dialysate tubing, eliminating the problem of electrode deterioration with time. Dialysate monitoring offers the advantage of directly measuring removal rates, instead of estimating it from blood-side measurements. It also allows measurement of urea concentrations in a pure solution without protein or red cells that can interfere with electrode function. This type of equipment has the potential to reduce the effort and cost of modeling while providing continous on-line data upon which to base decisions regarding blood and dialysate flow and dialysis duration. Such decisions could be made in real time, resulting in optimal conditions for each dialysis instead of an average for the month or quarter. Future dialysis goals might be to remove 25 grams of urea instead of stopping at 2 1/2 hours. This would make allowances for recirculation and dialyzer clotting, both of which may change significantly during a single dialysis treatment.

Infrequent (monthly or quarterly) analyses of urea kinetics can be hazardous, because they assume that the single measured predialysis BUN typifies the entire interval. Patients may vary protein intake considerably from day to day depending on individual cravings, availability of food, and the provider of meals. This will be reflected in the predialysis BUN, altering the modeled value for PCRn. A benefit of frequent or on-line monitoring of urea kinetics is better averaging of urea generation rate and hence of protein intake. This will provide appropriate compensation for those short bursts of dietary protein overindulgence and help to avoid prolonged periods of underdialysis.

UREA MODELING FROM THE TOTAL-CARE PERSPECTIVE

In the United States, dialysis centers receive payment from the federal health care system for treatment of patients with end-stage renal disease. Implicit, although not stated in the provision of funds for dialysis is the understanding that the treatment will be adequate. The patient, though he or she is the recipient of these benefits, often favors inadequate treatment, i.e., shortened dialysis time. This is not a unique occurrence in medical clinics, where it is the physician who usually assumes the role of taskmaster, encouraging and admonishing the patient to accept the treatment. For the treatment of end-stage renal disease, the physician also may favor inadequate treatment for economic reasons. Justifications often invoked are the failure of the system to define a threshold for adequate dialysis or even to provide a yardstick to measure dialysis.

The consequences of inadequate dialysis are subtle and may go unnoticed by the patient and the dialysis staff. Indeed, if all patients in a single dialysis center receive borderline or inadequate treatment, the complications experienced by each individual patient will compare favorably with the norm. Only by applying a uniform measure of dialysis outcome can results of newer equipment and techniques be meaningfully evaluated and results compared from center to center.

Lastly, it is important to have a proper perspective when including urea modeling in the total evaluation of patients with end-stage renal disease. One theme that seems to have blossomed over the past 30 years is that chronic renal failure is not one disease, or one condition, nor the consequence of one toxin or one hormone. There appear to be many disease states that combine in a single patient to create what we know as uremia. While we can easily see that lack of renal function can be quickly fatal, selective removal of the toxins that cause early fatality is not enough to manage these patients completely. When the goal is simply to prevent death from uremia, the patient may trade rapid death for a slower demise. Inadequate attention to all aspects of the uremic state can lead to a long course fraught with such complications as severe anemia, immunosuppression, bone disease, derangements in multiple hormonal systems, accumulation of calcium, iron, and amyloid deposits, and severe psychological disability. Urea modeling helps to control removal of small easily dialyzed solutes. Within the span of a few months, control of these solutes seems to correlate with a favorable outcome. It would be naive to think that, after such control is achieved, we can walk away from the patient.

REFERENCES

1. Kolff WJ, Jacobsen S, Stephen RL, Rose D: Towards a wearable artificial kidney. Kidney Int 10 (Suppl 7):S300-304, 1976.
2. Aebischer P, Ip TK, Miracoli L, Gallenti PM: Renal epithelial cells grown on semi-permeable hollow fibers as a potential ultrafiltrate processor. Trans Am Soc Artif Intern Organs 33:96-102, 1987.
3. Ip TK, Aebischer P, Galleti PM: Cellular control of membrane permeability: Implications for a bioartificial tubule. Trans Am Soc Artif Intern Organs 34:351, 1988.
4. Laird NM, Berkey CS, Lowrie EG: Modeling success or failure of dialysis therapy: the National Cooperative Dialysis Study. Kidney Int 23 (Suppl 13):S101-106, 1983.
5. Henderson LW: Of time, TACurea, and treatment schedules. Kidney Int 33 (Suppl 24):S105-106, 1988.
6. Frost TH, Kerr DNS: Kinetics of hemodialysis: A theoretical study of the removal of solutes in chronic renal failure compared to normal health. Kidney Int 12:41-50, 1977.
7. Dombeck DH, Klein E, Wendt RP: Evaluation of two pool model for predicting serum creatinine levels during intra- and inter-dialytic periods. Trans Am Soc Artif Intern Organs 21:117-124, 1975.
8. Guthke R, Gunther K, Stein G, Knorre WA: Two-pool model analysis of data in hemodialysis by means of programmable pocket calculator TI 59. Comput Prog Biomed 19:189-195, 1985.
9. Heineken FG, Evans MC, Keen ML, Gotch FA: Intercompartmental fluid shifts in hemodialysis patients. Biotechnol Progr 3:69-73, 1987.
10. Kushner RF, Schoeller DA: Estimation of total body water by bioelectrical impedance analysis. Am J Clin Nutr 44:417-424, 1986.
11. Gotch FA: Kinetic modeling in hemodialysis, in *Clinical Dialysis* (2ed), Nissensen AR, Gentile DE, Fine RN (eds), Norwalk CT, Appleton and Lange, pp 118-146, 1989.
12. Kimura G, Satani M, Kojima S, Ito K: Optimal dialysate concentrations based on kinetic modeling. Trans Am Soc Artif Intern Organs 30:236, 1984.
13. Kimura G, Van Stone JC, Bauer JH: The amount of sodium removed by hemodialysis. Am J Kidney Dis 8:253-256, 1986.
14. Murisasco A, Leblond G, Elsen R, Stroumza P, Durand C, Jeanningros E, Crevat A, Reynier JP: Equilibration of body water distribution and Na+ balance during hemodialysis with an ion specific electrode feedback system and integrated computer. Trans Am Soc Artif Intern Organs 30:254-258, 1984.
15. Gotch FA, Heineken F, Keen M, Evans M: Measurement of whole body ultrafiltration (Kf) and urea mass transfer (KU) coefficients: implications for Na & U Modeling in dialysis. Kidney Int 27:162, 1985.
16. Bowsher DJ, Krejcie TC, Avram MJ, Chow MJ, del Greco F, Atkinson AJ:

Reduction in slow intercompartmental clearance of urea during dialysis. J Lab Clin Med 105:489-497, 1985.

17. Gotch FA, Keen ML, Yarian SR: An analysis of thermal regulation in hemodialysis with one and three compartment models. Trans Am Soc Artif Intern Organs 35:622-624, 1989.
18. Drost CJ, Thomas GG, Sellers AF: In vivo validation of the transit-time ultrasonic volume flow meter. Proc 7th N Engl Bioeng Conf, Elmsford, NY, Pergamon Press, 1981, pp 387-390.
19. Burton RG, Gorewit RC: Ultrasonic flowmeter uses wide-beam transit-time technique. Med Electron 15:68-73, 1984.
20. Ellis PW, Malchesky PS, Magnusson MO, Goormastic M, Nakamoto S: Comparison of two methods of kinetic modeling. Trans Am Soc Artif Intern Organs 30:60-64, 1984.
21. Lankhorst BJ, Ellis P, Nosse C, Malchesky P, Magnusson MO: A practical guide to kinetic modeling using the technique of direct dialysis quantification. Dial Transplant 12:694, 1983.
22. Malchesky PS, Ellis P, Nosse C, Magnusson M, Lankhorst B, Nakamoto S: Direct quantification of dialysis. Dial Transplant 11:42-44, 1982.
23. Depner TA, Singh J: Dialysate urea modeling: a more precise method (abstract). Int Soc Nephrol Abstracts, 1990.
24. Asad SN, Olmer J: Comparison of two mathematical models for urea kinetics in chronic hemodialysis patients (abstract). Clin Res 35:808A, 1987.
25. Aebischer P, Schorderet D, Juillerat A, Wauters JP, Fellay G: Comparison of urea kinetics and direct dialysis quantification in hemodialysis patients. Trans Am Soc Artif Intern Organs 31:338-341, 1985.
26. Ilstrup K, Hanson G, Shapiro W, Keshaviah P: Examining the foundations of urea kinetics. Trans Am Soc Artif Intern Organs 31:164-168, 1985.
27. Pedrini LA, Zereik S, Rasmy S: Causes, kinetics and clinical implications of post-hemodialysis urea rebound. Kidney Int 34:817-824, 1988.
28. Karbue I, Tamiya E, Dicks JM: A microsensor for urea based on an lon-selective field effect transistor. Anal Chim Acta 185:195-200, 1986.
29. Ihn GS, Woo ST, Sohn MJ, Buck RP: Preparation of the Proteus mirabilis bacterial electrode for the determination of urea and its clinical applications. Anal Lett 22:1-13, 1989.
30. Klein E, Montalvo JG, Wawro R, Holland FF, Lefeouf A: Continuous monitoring of urea levels during hemodialysis. Int J Artif Organs 1:116-122, 1978.

APPENDICES

APPENDIX A

SOURCE CODE FOR A SINGLE-COMPARTMENT, VARIABLE-VOLUME MODEL: THREE-BUN METHOD

```
'BASIC code for calculating urea volume (V) and urea nitrogen generation rate (G)

'C1 = predialysis BUN (mg/ml)              W1 = weight predialysis (kg)
'C2 = postdialysis BUN (mg/ml)             W2 = weight postdialysis (kg)
'C3 = pre-2nd dialysis BUN (mg/ml)         W3 = weight pre-2nd dialysis (kg)
'Kd = dialyzer urea clearance (ml/min); must be true blood water clearance plus
    ultrafiltration component
'Kr = patient's residual urea clearance (ml/min)
'Td = duration of dialysis (minutes)
'Ti = time interval between dialyses (minutes)
'G = urea nitrogen generation rate (mg/min)
'V = volume of urea distribution (ml)

G=6                              'set initial generation rate to 6 mg/min
WtLoss = (W1 - W2) * 1000        'weight loss during dialysis (ml); cannot be zero
WtGain = (W3 - W2) * 1000        'weight gain between dialyses (ml); cannot be zero
dW1 = -WtLoss / Td               'rate of weight change during study dialysis (ml/min)
dV = WtGain / Ti                 'rate of weight change between dialyses (ml/min)
CA = C2 * (Kr + Kd + dW1)        'CA, CB, CC, CD are internal variables (see equation
                                   4.28)
CB = C1 * (Kr + Kd + dW1)
CC = -(Kr + dV) / dV
CD = -dW1/(Kr + Kd + dW1)
FOR J = 1 TO 10
   Z = ((CA - G)/(CB - G))^CD
   V = dW1 * Td / (Z - 1) + dW1 * Td    'V from intradialysis data (see equation 4.29)
   GOSUB GETG                           'G from interdialysis data & V (see equation
                                          4.30)
NEXT J
END

GETG:
   Z1 = ((V + dV * Ti) / V)^CC
   G = (Kr + dV) * ((C3 - C2 * Z1) / (1 - Z1))
RETURN
```

APPENDIX B

SOURCE CODE FOR A SINGLE-COMPARTMENT, VARIABLE-VOLUME MODEL: TWO-BUN METHOD

```
'BASIC code for calculating urea volume (V) and urea nitrogen generation rate (G)

'see Appendix A for explanation of symbols

dW1 = -WtLoss / Td              'calculation of V is identical to the fixed-volume model
CA = C2 * (Kr + Kd + dW1)
CB = C1 * (Kr + Kd + dW1)
CD = -dW1 / (Kr + Kd + dW1)

DO
   GOSUB GETV
   GOSUB Calc
   FractionalErr = ABS((C - C1) / C1)
   IF C > C1 THEN
      G = G - FractionalErr * G
   ELSE
      G = G + FractionalErr * G
   END IF
LOOP WHILE FractionalErr > .01

PRINT "G is established at "; G
PRINT "V is established at "; Vpost
END

Calc:
   'Calculate predialysis and postdialysis BUN values during the week
   '   returning a new predialysis BUN (C), one week later.

   'input variables:
   '   C1          measured predialysis BUN (mg/ml)
   '   Ndays       number of dialyses/week
   '   Interval()  array of days between dialyses
   '   Td          time on dialysis (min)
   '   WtLoss      filtration loss during study dialysis (ml); this cannot be zero
   '   Qb          dialyzer blood water flow (ml/min)
   '   Kd          dialyzer urea clearance with no ultrafiltration (ml/min)
   '   Kr          patient's residual urea clearance (ml/min)
   '   Vpost       estimated urea volume (ml)
   '   G           estimated urea nitrogen generation rate (mg/min)

   'output variable:
   '   C           new predialysis BUN (mg/ml)
```

```
'internal variables:
'  C          predialysis or postdialysis BUN
'  T          dialysis duration or interdialysis interval (min)
'  V0         V at start of interval (ml)
'  dW         rate of fluid gain (ml/min); negative during dialysis, varies with interval
'  dV         rate of fluid gain between dialyses (ml/min); this is constant
'  K          total clearance (ml/min); Kr + Kd + ultrafiltration component for the
                 interval

'Kd must be blood water clearance without Qf
'Qb must be blood water flow
'all urea nitrogen concentrations must be corrected to serum water
'   concentrations (BUN/0.93)

C = C1
dV = WtLoss / (Interval(Ndays) * 1440 - Td)
FOR I = 1 TO Ndays
IF I = 1 THEN
   dW = -WtLoss / Td
   V0 = Vpost + WtLoss                          'V0 is predialysis V
ELSE
dW = -dV * (Interval(I - 1) * 1440 - Td) / Td  'rate of weight loss during dialysis
                                               ' depends on weight gain during
                                               ' previous interdialysis interval
V0 = Vpost - dW * Td                           'V0 is predialysis V
END IF
T = Td                                         'T is time on dialysis
K = Kr + Kd - dW * (1 - Kd / Qb)               'calculates total K from Kd, dW and Qb
GOSUB GetC                                     'get postdialysis BUN

dW = dV                                        'rate of weight gain between dialyses
T = Interval(I) * 1440 - T                     'T is interdialysis time
K = Kr
V0 = Vpost                                     'V0 is postdialysis V
GOSUB GetC                                     'get predialysis BUN
NEXT I
RETURN

GetC:
'calculate BUN at time T
'a multicompartment calculation may be substituted here

'input variables:
'  V0         urea volume at start of interval (ml)
'  G          urea nitrogen generation rate (mg/min)
'  dW         rate of weight gain (ml/min); negative during dialysis
'  T          duration of interval (min)
'  K          total urea clearance during interval (ml/min)
'  C          BUN at start of interval (mg/ml)
'output variable:
```

```
'output variable:
'  C          BUN at end of interval (mg/ml); uses same variable as starting BUN

'internal variables:
'  Vf         fractional increment (or decrement) in V
'  Expn       exponent
Vf = (V0 + dW * T) / V0                                 'see equation 4.28
Expn = -(K + dW) / dW                                   'see equation 4.28
C = C * Vf ^ Expn + (G / (K + dW)) * (1 - Vf ^ Expn)    'see equation 4.28
RETURN

GETV:
    'calculate urea volume using an explicit formula (single pool)
    'V is calculated primarily from intradialysis data
    'this is the same routine as that used for the three-BUN model
    Z = ((CA - G) / (CB - G)) ^ CD
    Vpost = dW1 * Td / (Z - 1) + dW1 * Td
    RETURN
```

APPENDIX C

SOURCE CODE FOR A TWO-COMPARTMENT, FIXED-VOLUME MODEL

```
'Basic code for calculation of Ce and Ci: two-pool model
' see equations 5.3 to 5.8 (1)
'Ve = extracellular compartment volume (ml)
'Vi = intracellular compartment volume (ml)
'V = Vi + Ve
'Vi/Ve is fixed at 2.0
'Ce = extracellular urea nitrogen concentration (mg/ml)
'Ci = intracellular urea nitrogen concentration (mg/ml)
'G = urea nitrogen generation rate (mg/min)
'V = initial total volume of urea distribution
'Kc = intercompartment mass transfer coefficient (ml/min)
'Kd = dialyzer urea clearance (ml/min)
'Kr = patient's residual urea clearance (ml/min)
'K = total clearance (Kd + Kr)
't = time (min) during or between dialyses
'A, B, S, Z1, Z2, Z3, Z4 are internal variables

    Vi = 2 * V / 3: Ve = Vi / 2
    K = Kr + Kd
    S = SQR((Kc / Vi + (Kc + K) / Ve)^2 - 4 * Kc * K / (Ve * Vi)) / 2
    Z1 = (Kc / Vi + (Kc + K) / Ve) / 2
    A = Z1 - S
    B = Z1 + S
    Z2 = Ci0 * Kc / Ve + Ce0 * Kc / Vi + G / Ve
    Z3 = G * Kc / (Ve * Vi)
    Ce =(Z2 - Z3 / B - Ce0 * B) * EXP(-B * t) / (A - B) -
          (Z2 - Z3 / A - Ce0 * A) * EXP(-A * t) / (A - B) + Z3 / (A * B)
    Z4 = Ce0 * Kc / Vi + Ci0 * (Kc + K) / Ve
    Ci = (Z4 - Z3 / B - Ci0 * B) * EXP(-B * t) / (A - B) -
          (Z4 - Z3 / A - Ci0 * A) * EXP(-A * t) / (A - B) + Z3 / (A * B)
```

APPENDIX D

NUMERICAL SOLUTION FOR A TWO-COMPARTMENT, VARIABLE-ECV MODEL

Variables:

V_e = extracellular compartment volume (ml)
V_i = intracellular compartment volume (ml)
C_e = extracellular urea nitrogen concentration (mg/ml)
C_i = intracellular urea nitrogen concentration (mg/ml)
G = urea nitrogen generation rate (mg/min)
KC = intercompartment mass transfer coefficient (ml/min)
K_d = dialyzer urea clearance (ml/min)
K_r = patient's residual urea clearance (ml/min)
t = time (min)
Q_f = rate of weight gain during or between dialyses (ml/min)
dt = small interval of time (min)

Initially, when $t = 0$:

$C_{e0} = C_e(t)$
$C_{i0} = C_i(t)$
$V_{e0} = V_e(t)$

Expanding equation 4.12 to two compartments:

$$\frac{d(C_e \cdot V_e)}{dt} = V_e\frac{dC_e}{dt} + C_e\frac{dV_e}{dt} = G - KC\,(C_e - C_i) - C_e(K_d + K_r) \qquad \text{D.1}$$

$$\frac{dV_e}{dt} = Q_f \qquad \text{(Q is negative during dialysis)} \qquad \text{D.2}$$

Integrate:

$$V_e = V_{e0} + Q_f \cdot t \qquad \text{D.3}$$

equation 5.2:

$$V_i\frac{dC_i}{dt} = KC(C_e - C_i) \qquad \text{D.4}$$

Discrete solution of equation D.1:

$$C_e(t + dt) = C_e(t) + \frac{dt}{Q_f \cdot t + V_{e0}}\left[G - KC\left[C_e(t) - C_i(t)\right] - C_e(t)\cdot(K_d + K_r) - C_e(t)\cdot Q_f\right] \qquad \text{D.5}$$

Discrete solution of equation D.4:

$$C_i(t + dt) = C_i(t) + \frac{dt}{V_i}\left[KC\,(C_e(t) - C_i(t))\right] \qquad \text{D.6}$$

Following is a description of the algorithm for calculating V and G described in figure 5.6:

To solve for C_e and C_i at the end of dialysis:

Divide total dialysis time into a selected number of time intervals (n)
Divide total dialysis time by (n) to obtain a discrete value for dt
Solve equation D.5 for $t = 0$ conditions
Solve equation D.6 for $t = 0$ conditions
Replace $C_e(t)$ and $C_i(t)$ with new values for C_e and C_i respectively
Loop through solutions of equations D.5 and D.6 (n) times

C_e should match the measured postdialysis BUN (corrected for plasma water).
If C_e does not match, adjust V_e and V_i and begin again.
This establishes V.

To solve for C_e and C_i at the end of a week of dialysis:

Select the next interdialysis interval
Divide this total time into discrete small (n) intervals as above
Change Q_f to reflect rate of weight gain between dialyses
Use previous solutions for C_e and C_i above for C_{e0}, C_{i0}, and V_{e0}
Loop through solutions for equations D.5 and D.6 (n) times
Repeat intradialysis and interdialysis solutions for C_e and C_i until a week of time has elapsed

C_e should match the measured predialysis BUN (corrected for plasma water).
If C_e does not match, adjust G and repeat from the beginning.
This establishes G.

APPENDIX E

DESCRIPTION OF A TWO-COMPARTMENT, OSMOTIC MODEL WITH VARIABLE ECV AND ICV

See figure 5.13 (2)

J_{du} = dialyzer urea nitrogen flux
$= K_d \bullet C_{eu}$
K_d is dialyzer urea clearance
C_{eu} is extracellular urea nitrogen concentration
K_d should include the filtration component (see chapter 7)
$K_d = K_d'(1 - Q_f/Q_b) + Q_f$
Q_f = dialyzer ultrafiltration rate
Q_b = dialyzer blood flow
K_d' = dialyzer urea clearance at zero Q_f

J_{d+} = dialyzer cation (sodium) flux
$= K_d \bullet C_{e+} - K_d \bullet C_{d+}$
C_{e+} is extracellular cation (sodium) concentration
C_{d+} is dialysate cation (sodium) concentration
dialyzer sodium clearance assumed equal to urea clearance

J_{dw} = dialyzer ultrafiltration rate also expressed as Q_f
J_{cu} = transcellular urea nitrogen flux (between compartments)
$= K_{cu}(C_{iu} - C_{eu})$, positive during dialysis
K_{cu} is the intercompartment mass transfer coefficient for urea
C_{iu} is intracellular urea nitrogen concentration

J_{ru} = flux of urea nitrogen due to residual renal function
$= K_r \bullet C_{eu}$
K_r is patient's residual urea clearance

J_{cw} = transcellular water flux (due to osmotic forces)
$= K_{cw}[(C_{eu} - C_{iu}) + (C_{e+} - C_{i+})]$
C_{i+} is intracellular cation concentration
K_{cw} is the cellular coefficient of fluid transfer
sodium and urea nitrogen concentrations must be expressed as mmol/L

J_{gu} = urea nitrogen generation rate
J_{g+} = dietary sodium intake
J_{cw} = dietary water intake + water produced by metabolism

During dialysis:

extracellular urea nitrogen flux:

$$\frac{d(V_e \bullet C_{eu})}{dt} = J_{cu} + J_{gu} - J_{du} \qquad (K_r \text{ negligible})$$

intracellular urea nitrogen flux:

$$\frac{d\,(V_i \cdot C_{iu})}{dt} = -J_{cu}$$

extracellular cation (sodium) flux:

$$\frac{d\,(V_e \cdot C_{e+})}{dt} = -J_{d+}$$

intracellular cation flux:

$$\frac{d\,(V_i \cdot C_{i+})}{dt} = 0 \quad \text{(assume no intracellular cation is removed)}$$

extracellular water flux:

$$\frac{d\,V_e}{dt} = J_{cw} - J_{dw} \quad \text{(intake negligible)}$$

intracellular water flux:

$$\frac{d\,V_i}{dt} = -J_{cw}$$

Between dialyses:

$$\frac{d\,(V_e \cdot C_{eu})}{dt} = J_{cu} + J_{gu} - J_{ru} \quad \text{(no dialysis)}$$

$$d\,(V_i \cdot C_{iu}) = -J_{cu}$$

$$\frac{d\,(V_e \cdot C_{e+})}{dt} = J_{g+}$$

$$\frac{d\,(V_i \cdot C_{i+})}{dt} = 0$$

$$\frac{d\,V_{ei}}{dt} = \frac{J_{cw}}{3} \quad \text{(assume uniform constant fluid gain)}$$

$$\frac{dV_i}{dt} = \frac{2 \cdot J_{cw}}{3}$$

APPENDIX F

INTERPRETING THE RESULTS OF UREA MODELING

True (nonerroneous) abnormalities

Causes of low PCRn, G:

- low dietary intake of protein due to:
  - anorexia from underdialysis
  - excessive dietary protein restriction
  - another illness
  - anabolic states

Causes of high PCRn, G:

- inadequate dietary protein restriction
- noncompliance with dietary protein restriction

Causes of low V:

- low body weight (low absolute V)
- excessive body fat (V = low fraction of weight)

Causes of high V:

- edematous states
- hypo-osmolar states

Modeling errors

Causes of low PCRn, G:

- false expansion of V (see below)
- underestimation of residual clearance
- falsely low predialysis BUN (e.g., drawing predialysis sample after starting blood pump)

Causes of high PCRn, G:

- false contraction of V (see below)
- overestimation of residual clearance
- low intercompartment mass transfer coefficient: when KC is low, a two-compartment model is more appropriate

Causes of low V:

- underestimation of dialyzer clearance
- false compression of intradialysis duration
- venous admixture during postdialysis blood sampling

drawing blood from the venous instead of the arterial port postdialysis
drawing blood from arterial port postdialysis while flushing blood compartment with saline
low intercompartment mass transfer coefficient: when *KC* is low, a two-compartment model is more appropriate

Causes of high *V*:
- venous reflow in blood access device
- overestimation of dialyzer clearance
  - dialyzer clotting
  - poorly functioning blood pump (e.g., nonoccluding rollers)
  - inaccurately calibrated blood pump
  - wrong pump segment tubing diameter
  - loss of surface area from reuse
  - channeling of blood and/or dialysate
  - dialyzer mass transfer coefficient too high
  - reversal of blood/dialysate directional flow
- false extension of intradialysis duration (wall clock syndrome)
- use of whole-blood instead of blood water urea nitrogen concentrations
- drawing predialysis sample after starting blood pump

Causes of low K_{dp}:
- same as causes of high *V*

Causes of high K_{dp}:
- same as causes of low *V*

Appendix G

Useful equations

Additional clearance from ultrafiltration

$$K_d = Q_{bo}(C_{in} - C_o)/C_{in} + Q_f \qquad \text{G.1}$$

or

$$K_d = K_{d0} + Q_f(C_o/C_{in}) \qquad \text{G.2}$$

or

$$K_d = K_{d0} + Q_f(1 - K_{d0}/Q_{bi}) \qquad \text{G.3}$$

K_d is dialyzer clearance including ultrafiltration
K_{d0} is dialyzer clearance with no ultrafiltration
Q_{bo} is blood flow at the dialyzer outlet
Q_{bi} is blood flow at the dialyzer inlet
C_{in} is dialyzer inlet BUN
C_o is dialyzer outlet BUN
Q_f is the hydraulic ultrafiltration rate

BUN values are serum urea nitrogen concentrations (any units)
Blood flow and clearances must be blood water values

Effect of hematocrit and plasma water on urea clearance

$$Q_{biw} = Q_{bi}\ [0.86(\text{hct}) + 0.93(1 - \text{hct})] \qquad \text{G.4}$$

Q_{biw} is effective water flow through the dialyzer

Mass transfer coefficient from dialyzer clearance, blood and dialysate flow (counter-current flow)

$$KA = \frac{Q_b \cdot Q_d}{Q_b - Q_d} \ln \left[\frac{1 - K_d/Q_b}{1 - K_d/Q_d}\right] \qquad \text{G.5}$$

KA is the mass transfer area coefficient (ml/min)
Q_b is blood flow (ml/min)
Q_d is dialysate flow (ml/min)
ln is the natural logarithm
K_d is dialyzer urea clearance (ml/min)

KA, K_d, and Q_b can be either whole-blood values or blood water values. All must be of one kind with no mixing.

Dialyzer clearance from mass transfer coefficient, blood and dialysate flow (counter-current flow)

$$z = (KA/Q_b) \cdot (1 - Q_b/Q_d) \qquad \text{G.6}$$

$$K_d = Q_b \frac{(e^z - 1)}{(e^z - Q_b/Q_d)} \qquad \text{G.7}$$

How to measure recirculation of dialyzer venous blood

$$\frac{Q_r}{Q_b} = \frac{C_p - C_b}{C_p - C_v} = F_r \text{ (reflow fraction)} \quad \text{G.8}$$

Q_b is arterial blood inflow (measured at the blood pump)
Q_p is peripheral blood inflow
Q_r is flow of recirculating blood
C_b is arterial BUN
C_p is peripheral BUN
C_v is venous BUN

Effect of venous reflow on effective dialyzer urea clearance

$$K_{de} = \frac{K_d(1 - F_r)}{1 - F_r + K_d \cdot F_r / Q_b} \quad \text{G.9}$$

K_{de} is effective dialyzer urea clearance
K_d is dialyzer clearance in the absence of reflow

Equation 7.20 shows that reflow will diminish the diffusive urea clearance by a factor that is a complex function of the reflow fraction (F_r), dialyzer clearance (K_d), and blood flow in the arterial blood line (Q_b).

Conversely, the reflow fraction can be computed from the difference between expected dialyzer clearance and true patient clearance either measured or modeled:

$$F_r = \frac{K_d - K_{de}}{K_d - K_{de} + K_d \cdot K_{de} / Q_b} \quad \text{G.10}$$

Effect of the patient's residual urea clearance on *Kt/V*

$$\frac{K't}{V} = \frac{K_d \cdot t + K_r \cdot f}{V} \quad \text{G.11}$$

K_d is dialyzer urea clearance (ml/min)
K_r is patient's residual urea clearance (ml/min)
K' is total effective urea clearance (dialyzer + residual)
t is time on dialysis (min)
V is volume of urea distribution (ml)
f is a coefficient that converts residual clearance into units that allow reduction of K_d while maintaining the same time-averaged BUN when $K_r = 0$.

This formula is an approximation that works well for residual clearances in the 0 to 5 ml/min range. The value of f is 4000 minutes for a three dialyses/week schedule. For dialyses twice/week, the value for f is 6500 minutes.

Protein catabolic rate (PCR) from urea generation rate (*G*)

$$\text{PCR} = 9.35 \cdot G + 0.29 \cdot Vl \quad \text{G.12}$$

$$\begin{aligned}\text{PCRn} &= \text{PCR} \cdot 0.58 / Vl \\ &= 5.42 \cdot G/Vl + 0.17\end{aligned} \quad \text{G.13}$$

PCRn is PCR normalized to ideal body weight
Vl is urea distribution volume in liters

Target therapy (ideal TAC)

when PCRn is below 1.1 g/kg/day:

$$ITAC = 0.60(PCRn) - 0.10 \qquad G.14$$

when PCRn is equal to or above 1.1 g/kg/day:

$$ITAC = 0.25(PCRn) + 0.28 \qquad G.15$$

ITAC is ideal time-averaged BUN (mg/ml)
PCRn is protein catabolic rate in mg/kg BWn/day
BWn is normalized body weight defined as:
Vl/.58
Vl is urea distribution volume in liters

Quick estimate of *Kt/V*

$$K \bullet t\ /V = 0.026 \bullet P - 0.49 \qquad G.16$$

P is percent reduction in BUN

If *P* is 57%, *Kt/V* is 1.0. *Kt/V* changes by 0.13 for each 5% change in *P*. This method gives a quick bedside estimate of the adequacy of dialysis if your goal is to maintain *Kt/V* above 1.0.

First-order kinetics (e.g., for drugs)

$$dC\ /dt = k \bullet C \qquad G.17$$

$$C = C_0 \bullet e^{-kt} \qquad G.18$$

Single-compartment urea kinetics

$$d(VC)/dt = G - K \cdot C \qquad G.19$$

Fixed-volume model:

$$C = C_0 \cdot e^{-K \cdot t/V} + \frac{G}{K}(1 - e^{-K \cdot t/V}) \qquad G.20$$

Variable-volume model:

$$C = C_0 \left[\frac{V + B \cdot t}{V}\right]^{-(Kr + Kd + B)/B} + \left[\frac{G}{K_r + K_d + B}\right]\left[1 - \frac{V + B \cdot t}{V}\right]^{-(Kr + Kd + B)/B} \qquad G.21$$

$B = dV/dt$, the rate of change in volume, i.e., fluid removal during dialysis and weight gain between dialyses

Direct quantification method: simplified urea kinetics in the absence of K_r or Q_f

$$V = \frac{\text{amount removed during dialysis}}{\text{change in BUN}} \qquad G.22$$

$$G = \frac{(\text{change in BUN between dialyses}) \cdot V}{t} \qquad G.23$$

Anthropometric estimates of total body water

Watson formula (3)

males:

TBW = 2.447 - 0.09516 • age + 0.1074 • height + 0.3362 • weight

females:

TBW = -2.097 + 0.1069 • height + 0.2466 • weight

Hume formula (4)

males:

$$TBW = 0.194786 \cdot height + 0.296785 \cdot weight - 14.012934$$

females:

$$TBW = 0.344547 \cdot height + 0.183809 \cdot weight - 35.270121$$

Appendix H

Examples of data collection forms

Table H.1 Urea modeling entry form and instructions

Patient: Example, John ID#: 1111111 Date: 07/19/90

HEMODIALYSIS QUALITY ASSURANCE
Urea Kinetics Data Input Form

INSTRUCTIONS: Fill in the appropriate values for today's dialysis. Alternatively, you may attach a copy of the nursing flow sheet if it includes the information requested. Draw two blood samples from the dialyzer arterial line according to the instructions below. This is a work-sheet, not to be included in the patient's chart.

Careful attention to blood sampling technique and timing is critical. Starting and ending times must be recorded accurately (e.g. 0810). Blood flow rates must be recorded accurately on the nursing flow sheet together with the clock time whenever a change in blood flow occurs. Faulty technique or errors in recording data can lead to an inappropriate dialysis prescription, unnecessary investigation of faulty equipment or inappropriate changes in diet.

PRE DIALYSIS BLOOD SAMPLE:
This must be drawn before injecting heparin or saline and prior to starting the blood pump.

POST DIALYSIS BLOOD SAMPLE:
Just before disconnecting the arterial line, decrease the blood pump rate to 25 - 50 ml/min, clamp the venous line, and wait for 30 seconds. Then draw the blood sample from the arterial port (avoid venous port). Do not draw the postdialysis sample while returning blood to the patient. These precautions are designed to avoid mixing venous blood with the arterial sample.

AT THE END OF DIALYSIS, CHECK IF THE FOLLOWING OCCURRED:	YES	NO
replacement fluids given during dialysis	____	____
blood flow reduced transiently during dialysis.	____	____
pre-dialysis sampling obtained after starting blood pump	____	____
post-dialysis blood sampling delayed > 3 min after stopping blood pump	____	____
maximum venous pressure > 250 mm	____	____
blood access device functioning poorly	____	____
patient dialysis schedule has changed within the past week	____	____

DATE OF STUDY ________ DIALYZER ________ RESIDUAL CLEARANCE ________ (ml/min)

WEEKLY SCHEDULE ______ (e.g. MWF) AVERAGE BLOOD FLOW ______ (ml/min) USUAL DIALYSIS TIME ______ (hours)

REUSE # _____ DIALYSATE FLOW ________ (ml/min)

START TIME ____________ START WEIGHT ________ (kg)

END TIME ____________ END WEIGHT ________ (kg)

Start BUN ______ (mg/dl)
End BUN ______ (mg/dl)

Comments: ____

____

____

Nurse/Tech signature ____________________________________

Table H.2 Dialyzer clearance/reflow: data collection and instructions

Patient: Example, John ID #: 1111111 Date: 07/19/90

DIALYZER CLEARANCE & REFLOW STUDY

This procedure will check the performance of the dialyzer during an actual dialysis. Dialyzer urea clearance and recirculation within the access device can be measured simultaneously. Three blood samples are required, one from the dialyzer arterial line, one from the venous line and one from the patient's other arm (peripheral). If only the dialyzer clearance is to be measured, the peripheral sample is not necessary. Blood samples must be drawn simultaneously (within 30 seconds of one another) and should be drawn shortly after starting dialysis (after achieving stable blood flow). For most accurate results, ultrafiltration should be turned off at least 5 minutes before drawing samples.

To avoid a separate venipuncture the peripheral sample can be obtained from the arterial line using the following technique. Immediately after drawing the arterial and venous samples, reduce the blood flow to a minimum (25-50 ml/min), clamp the venous line and wait 30 seconds. Then draw an arterial sample. Unclamp the venous line and resume dialysis at the previous blood flow rate. This technique is designed to eliminate dilution of the arterial sample with venous blood.

Simple formulas for calculating dialyzer urea nitrogen clearance and percent reflow are given below:

$$\underset{(\text{ml/min})}{\text{Clearance}} = 0.9 \cdot Qb \cdot \left[\frac{\text{Art - Ven}}{\text{Art}}\right] = \boxed{\_\_\_\_\_\_\ \text{ml/min}}$$

Qb = whole blood flow rate (ml/min)

$$\%\ \text{Reflow} = 100 \cdot \left[\frac{\text{Periph - Art}}{\text{Periph - Ven}}\right] = \boxed{\_\_\_\_\_\_\ \%}$$

	Time	BUN (from lab)	
Arterial	________	________	
Venous	________	________	
Peripheral	________	________	(for reflow)

Blood flow rate (important) ______________. This must be an accurate assessment of blood flow when the arterial and venous samples are drawn.

Nurse/Tech signature ____________________________

Table H.3 Residual clearance: data collection and instructions

Patient: Example, John | ID #: 111111 | Date: 07/19/90

RESIDUAL CLEARANCE INPUT FORM

Residual or endogenous urea clearance is the small but often significant pathway for urea elimination remaining in the patient's own diseased kidneys. To measure residual clearance, a timed collection of urine is required.

There is no need to measure blood urea concentrations because urea modeling can predict the blood concentrations at the start and end of the urine collection. These two values are averaged for the clearance calculation.

Data:

	Day of week (e.g. Thurs)	Time*
End of last Dialysis	________	________
Start of urine collection	________	________
End of urine collection	________	________

* use 24 hour format for time (e.g. 1:30 P.M. = 1330)

Urine Volume ________ (ml)

Urine Urea Nitrogen ________ (mg/dl)

Results:

Residual urea clearance ________ (ml/min)

Requirements for residual clearance measurements from modeled BUN:

1) Previous urea kinetic analysis
2) Patient must remain in weekly steady-state
3) No changes in dialysis prescription since last urea kinetic study
4) Timed urine collection for urea nitrogen

Nurse/Tech signature ____________________________

APPENDIX I

EXAMPLES OF MODELING REPORTS

Short report, dialysis center

The abbreviated report shown in table I.1 gives vital information about dialyzer settings for the nurse/technician staff. It lists each patient, the dialysis schedule, dialyzer brand, blood flow, dialysate flow, and ideal dialysis duration determined by the last kinetic analysis.

Table I.1 Short report for dialysis center

PATIENT NAME	I.D. #	SCHEDULE	DIALYZER TYPE	Qb (ml/min)	Qd	IDEAL DURATION
TEST1, 1ST	111111	Mwf	FR-F80	400	800	2 hrs 3 min
TEST2, 2ND	222222	mWf	FR-F80	450	800	1 hrs 37 min
TEST3, 3RD	333333	mwF	FR-F80	350	500	1 hrs 43 min
TEST4, 4TH	444444	Mwf	FR-F80	300	500	2 hrs 7 min
TEST5, 5TH	555555	mWf	CF-1511	250	500	3 hrs 36 min
TEST6, 6TH	666666	mwF	CF-1511	200	500	2 hrs 47 min
TEST7, 7TH	777777	Mwf	CF-1211	300	500	3 hrs 41 min
TEST8, 8TH	888888	mWf	CF-1211	300	500	3 hrs 18 min
TEST9, 9TH	999999	mwF	CF-1211	250	500	4 hrs 2 min
TEST10, 10TH	101010	Mwf	CF-1511	300	500	2 hrs 47 min
TEST11, 11TH	110011	mWf	CF-1511	300	500	4 hrs 39 min
TEST12, 12TH	121212	mwF	CF-1211	200	500	3 hrs 33 min

Complete report, patient or center

The more extensive report, shown in table I.2, tabulates all data, both entered and calculated, for a group of patients or for multiple studies in a single patient. As shown in table I.2, the font is compressed to allow tabular output on 8 1/2 inch paper. The heading abbreviations are explained in table 6.1.

Summary report for the dialysis center

Table I.3 shows mean values and standard deviations for another group of patients, each studied more than once. Typically, data from the most recent 5 to 10 kinetic studies are averaged for each patient. The data are extracted from a master file or from individual files kept for each patient. Averaging several studies enhances the value of urea modeling by providing a more accurate value for urea volume. But care must be taken to examine the data from each individual kinetic study, especially the most recent, to detect changes in dialyzer performance or in protein intake that might require a change in the prescription.

Table I.2 Complete Report for the Dialysis Center

TITLE OF REPORT

Patient name	I.D. #	Study Date	Dialzer Type	Sch-dle	Wt — Kg —	dW — Kg —	Tda hrs	Qb ml/min	Qd ml/min	Kr ml/min	Kd ml/min	Kdp ml/min	Pre (mg/dl)	Post (mg/dl)	TAC (mg/dl)	PREa (mg/dl)	Kt/V	PCRn g/kg	V %bw	Vp %bw	— Ideal — TAC	PRE	hrs
TEST1, 1ST	111111	07/02/90	FR-F80	Mwf	70	3.5	2.5	400	800	0.0	308	328	80	28	48	70	1.29	1.13	51	55	56	78	2.0
TEST2, 2ND	222222	07/04/90	FR-F80	mWf	69	2.8	2.0	450	800	2.0	330	329	70	26	51	72	1.15	1.26	50	50	59	79	1.6
TEST3, 3RD	333333	07/06/90	FR-F80	mwF	70	2.2	2.0	350	500	3.3	266	264	66	30	53	71	0.90	1.24	51	50	59	76	1.7
TEST4, 4TH	444444	07/02/90	FR-F80	Mwf	50	0.3	2.0	300	500	0.0	240	246	92	33	57	81	0.99	1.07	58	60	54	78	2.1
TEST5, 5TH	555555	07/04/90	CF-1511	mWf	65	2.4	4.0	250	500	0.0	165	181	80	26	57	84	1.35	1.35	45	50	62	89	3.6
TEST6, 6TH	666666	07/06/90	CF-1511	mwF	53	1.9	4.0	200	500	4.0	146	145	65	26	52	71	1.10	1.42	60	60	63	81	2.8
TEST7, 7TH	777777	07/02/90	CF-1211	Mwf	63	1.1	4.0	300	500	0.0	162	175	75	28	45	65	1.12	0.97	56	60	48	68	3.7
TEST8, 8TH	888888	07/04/90	CF-1211	mWf	57	0.9	4.0	300	500	0.0	162	148	39	13	28	41	1.25	0.70	54	50	32	46	3.3
TEST9, 9TH	999999	07/06/90	CF-1211	mwF	79	2.0	4.0	250	500	0.0	152	188	86	29	66	97	1.31	1.50	36	45	65	96	4.0
TEST10, 10TH	101010	07/09/90	CF-1511	Mwf	56	1.5	4.0	300	500	0.0	177	204	75	21	41	64	1.47	1.09	52	60	55	77	2.8
TEST11, 11TH	110011	07/11/90	CF-1511	mWf	86	4.0	4.0	300	500	0.0	181	180	90	40	68	92	1.00	1.28	50	50	60	84	4.7
TEST12, 12TH	121212	07/13/90	CF-1211	mwF	55	0.6	3.5	200	500	0.0	136	134	40	16	31	44	1.02	*0.68	51	50	31	44	3.5
Mean values					64	1.9	3.3	300	550	0.8	202	210	72	26	50	71	1.16	1.14	51	53	54	75	3.0
SD ($n = 12$)					11	1.1	0.9	74	117	1.5	67	67	17	7	12	17	0.17	0.26	6	5	11	16	1.0
07-19-1990																							

* = excessively high or low PCRn.

Table I.3 Summary Report for the Dialysis Center

HEMODIALYSIS UREA MODELING: SUMMARY REPORT
Means & SD

Patient name	I.D. #	First Date	Last Date	#/ week	Wt — Kg —	dW — Kg —	Tda hrs	Qb ml/min	Qd ml/min	Kd ml/min	Kr ml/min	TAC mg/dl	PREa mg/dl	PCRn g/kg	Kt/V	V %BW	— Ideal — TAC	PRE	hrs
EXAMPLE1, 1ST	413477	03/27/89	01/17/90	3.0	81.8	2.3	3.0	270	500	204	0.0	44	61	0.9	1.0	47.1	42	58	3.3
SD		(5 files)		0.0	2.6	0.9	0.0	27	0	12	0.0	9	13	0.2	0.2	7.7	12	16	0.6
EXAMPLE2, 2ND	325015	08/15/89	01/18/90	3.0	66.2	3.0	3.3	350	500	222	0.0	53	74	1.1	1.2	56.4	53	75	3.1
SD		(5 files)		0.0	2.0	1.6	0.4	100	0	18	0.0	19	23	0.2	0.3	2.3	7	10	0.6
EXAMPLE3, 3RD	1002479	03/22/89	01/17/90	3.0	72.8	3.5	3.0	290	500	215	0.0	57	81	1.2	1.2	46.2	58	82	2.9
SD		(5 files)		0.0	1.4	0.4	0.0	22	0	10	0.0	7	10	0.1	0.1	6.0	3	6	0.3
EXAMPLE4, 4TH	967729	03/22/89	01/29/90	3.0	64.4	3.9	3.0	380	500	246	0.0	47	67	1.1	1.2	58.6	52	72	2.7
SD		(5 files)		0.0	2.8	1.5	0.0	27	0	8	0.0	8	12	0.2	0.1	6.5	10	14	0.2
EXAMPLE5, 5TH	3188542	03/23/89	01/18/90	3.0	70.4	2.6	3.0	250	500	194	0.0	52	68	0.9	0.8	61.4	41	58	3.9
SD		(5 files)		0.0	0.8	0.7	0.0	35	0	18	0.0	12	17	0.2	0.1	6.5	11	16	0.6
EXAMPLE6, 6TH	1269197	07/13/89	01-23-90	3.0	48.6	2.4	3.0	217	500	179	0.0	47	68	1.1	1.2	55.1	51	71	2.8
SD		(3 files)		0.0	1.6	0.6	0.0	29	0	14	0.0	13	18	0.3	0.3	6.3	14	20	0.5
EXAMPLE7, 7TH	867133	03/23/89	01-18-90	3.0	59.8	2.0	3.0	378	500	242	0.0	47	70	1.1	1.3	58.5	52	74	2.6
SD		(5 files)		0.0	0.7	1.2	0.0	35	0	11	0.0	10	17	0.3	0.2	5.7	8	16	0.3
EXAMPLE8, 8TH	835734	03/27/89	02/12/90	3.0	60.2	2.5	3.0	338	500	241	0.0	52	76	1.2	1.3	56.8	58	82	2.6
SD		(5 files)		0.0	1.1	0.8	0.0	41	0	18	0.0	5	8	0.2	0.1	5.6	3	6	0.2
EXAMPLE9, 9TH	731625	01/25/89	10/18/89	3.0	76.0	4.4	3.0	350	500	236	0.0	56	77	1.1	1.0	54.5	53	74	3.2
SD		(5 files)		0.0	0.5	2.0	0.0	71	0	24	0.0	9	13	0.2	0.2	8.0	9	13	0.5
EXAMPL10, 10TH	881076	03/23/89	01-18-90	3.0	54.5	1.8	2.7	370	500	238	0.0	56	82	1.2	1.2	61.3	57	83	2.6
SD		(5 files)		0.0	3.9	1.6	0.3	84	0	25	0.0	8	12	0.2	0.2	8.5	4	7	0.6
EXAMPL11, 11TH	739098	01/25/89	01-17-90	3.0	104.6	5.9	4.0	400	680	286	0.0	60	83	1.2	1.1	59.8	58	81	4.2
SD		(5 files)		0.0	2.4	1.6	0.0	61	164	46	0.0	10	13	0.2	0.2	1.9	5	9	0.6
EXAMPL12, 12TH	842906	01/26/89	07/13/89	3.0	58.1	2.6	2.8	333	500	228	0.0	36	51	0.8	1.2	57.8	38	53	2.6
SD		(3 files)		0.0	3.1	1.0	0.3	58	0	18	0.0	20	29	0.4	0.1	7.9	21	30	0.1
Mean center values for 12 patients				3.0	68.1	3.1	3.1	327	515	228	0.0	51	72	1.0	1.0	56.1	51	72	3.2

Individual patient reports and BUN profile

One compartment

Table I.4 displays all entered data above the solid line and all calculated data below the solid line. Explanations for some of the symbols are provided at the bottom together with space for comments by the reviewer.

The example patient's dry weight is 70.0 kg; 4.5 kg was gained prior to dialysis on Monday. High-flux dialysis for 2 1/2 hours caused the BUN to fall from 80 to 28 mg/dl. According to the model, the patient's urea volume is 35.8 liters, or 51.1% of dry body weight. Net protein catabolism is 69.6 grams/day or 1.13 grams/kg normalized body weight/day (see chapter 3 for further discussion of normalized body weight). Time-averaged BUN is 48 mg/dl and is below the ideal 56 mg/dl. According to this model, dialysis time can be reduced to 2 hours 3 minutes at the prescribed blood flow of 400 ml/min. If blood flow could be increased to 600 ml/min, dialysis time could be cut to 1 hour 39 minutes. In the second column of data below the solid line, the patient's estimated urea volume is 55% of dry body weight. If this is an accurate figure, then the actual average clearance achieved during this modeled dialysis was 328 ml/min, slightly higher but very close to the clearance estimated from blood and dialysate flow. Figure I.1 shows the weekly profile of BUN values predicted by the model, assuming urea nitrogen generation is constant at 6.32 mg/min. Also shown are the time-averaged BUN and the ideal time-averaged BUN as dotted and dashed lines, respectively.

Two compartments

For comparison, a two-compartment model was used to generate data in the same patient for the same modeled dialysis. The data are displayed in table I.5. To exaggerate the two-pool effect, the intercompartment mass transfer coefficient was lowered to 300 ml/min. This causes a marked concentration difference between compartments during and shortly after dialysis as shown in figure I.2. The model also calculates a lower urea nitrogen generation rate, shown as the slope of the line between dialyses. Although urea clearance is the same,

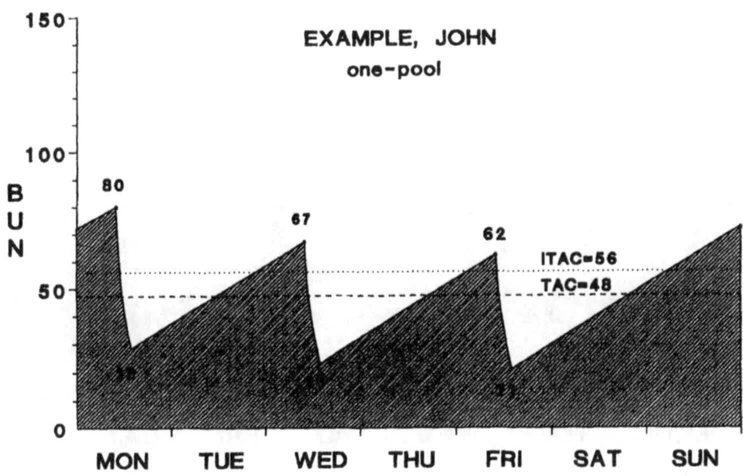

Figure I.1. Weekly BUN profile generated by a single-compartment model. TAC is time-averaged BUN. ITAC is ideal time-averaged BUN.

less urea is removed by the dialyzer, due to lower blood concentrations. This causes higher predialysis BUN levels. Note that a substantial rebound occurs immediately after dialysis. Calculated urea volume is higher because the change in urea concentration at equilibrium is much less than the difference between predialysis and postdialysis BUN assumed by the one-compartment model (see discussion of the rebound effect and error 2 in chapter 5). Ideal time-averaged BUN is lower because the urea generation rate is lower. The duration of dialysis must be extended to 3 hours 12 minutes because of reduced dialyzer efficiency and the lower ideal time-averaged BUN.

These differences are exaggerated in this patient because of the slow diffusion rate between compartments. If we select an average intercompartment mass transfer coefficient closer to normal at 800 ml/min, the differences between the two models are small.

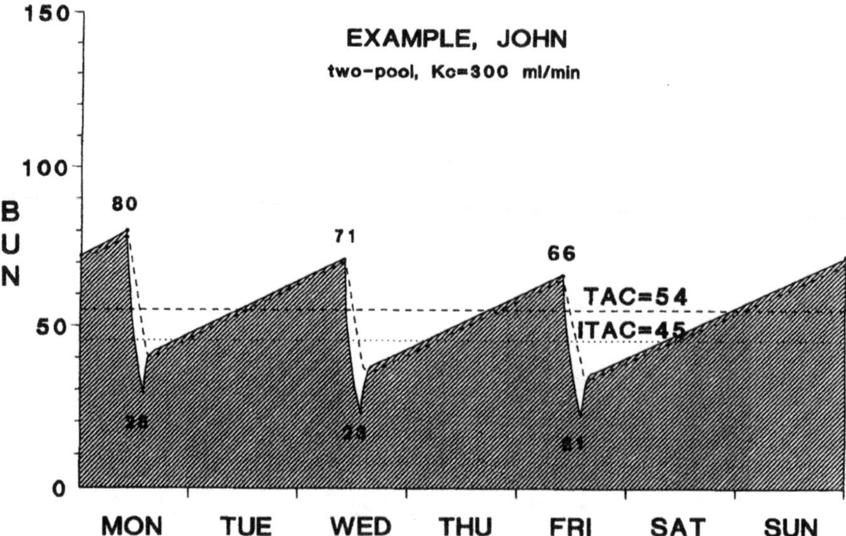

Figure I.2. Two-compartment model profile of urea nitrogen concentrations. Concentrations in the remote compartment are shown as a broken line. TAC is time-averaged BUN. ITAC is ideal time-averaged BUN. The intercompartment mass transfer coefficient (*KC*) is set low at 300 ml/min.

Table I.4 Single patient report, one compartment

EXAMPLE, JOHN			
Date of study		07/02/90	
Patient ID #		111111	
Residual clearance (ml/min)		0.0	
DIALYSIS PRESCRIPTION			
Schedule		MON wed fri	
Duration		2 hr 30 min (this study)	
		2 hr 30 min (usual)	
Dialyzer		FR-F80	
blood flow (ml/min)		400	
dialysate flow (ml/min)		800	
urea clearance (ml/min)		308 (water + *Qf*)	
MEASURED VALUES		BUN (mg/dl)	Wt (kg)
predialysis		80	74.5
postdialysis		28	70.0

MODELED VALUES (1 pool, variable *V*)

		fixed *Kd*	fixed *V*
V	(liters)	35.8	[38.5]
	(% body wt)	51.1%	[55.0%]
G	(mg/min)	6.35	6.74
PCR	(g/day)	69.9	74.3
PCRn	(g/kg/day)**	1.13	1.12
Kd	(ml/min)	[308]	328
TAC	(mg/dl)	48	same
Mean pre	(mg/dl)	70	same
Kd(t)/V	(/dialysis)	1.29	1.28

TARGET VALUES (3 dialyses/week)

TAC	(mg/dl)	56	
Mean pre	(mg/dl)	78	
Kd(t)/V	(/dialysis)	1.06	
IDEAL DURATION		2 hrs	3 min
*Q*b = 300	*K*d = 252	2 "	30 "
400	308	2 "	3 "
500	351	1 "	48 "
600	383	1 "	39 "

COMMENTS

V = volume of urea distribution, *G* = urea generation rate, PCRn = normalized protein catabolic rate
Mean pre = mean predialysis BUN, TAC = time averaged BUN, *K*d = dialyzer clearance including Qf

Table I.5 Single patient report, two compartments

EXAMPLE, JOHN			
Date of study		07/02/90	
Patient ID #		111111	
Residual clearance (ml/min)		0.0	
DIALYSIS PRESCRIPTION			
Schedule		MON wed fri	
Duration		2 hr 30 min (this study)	
		2 hr 30 min (usual)	
Dialyzer		FR-F80	
blood flow (ml/min)		400	
dialysate flow (ml/min)		800	
urea clearance (ml/min)		308 (water + *Qf*)	
MEASURED VALUES		BUN (mg/dl)	Wt (kg)
predialysis		80	74.5
postdialysis		28	70.0

MODELED VALUES (2 pool, variable *V*)

		fixed *Kd*	fixed *V*
V	(liters)	42.4	[38.5]
	(% body wt)	60.5%	[55.0%]
G	(mg/min)	5.87	5.56
PCR	(g/day)	67.3	63.3
PCRn	(g/kg/day)**	0.92	0.95
Kd	(ml/min)	[308]	288
TAC	(mg/dl)	54	same
Mean pre	(mg/dl)	73	same
Kd(*t*)/*V*	(/dialysis)	1.09	1.12
TARGET VALUES (3 dialyses/week)			
TAC	(mg/dl)	45	
Mean pre	(mg/dl)	63	
Kd(*t*)/*V*	(/dialysis)	1.40	
IDEAL DURATION		3 hrs	12 min
*Q*b = 300	*K*d = 251	3 "	56 "
400	307	3 "	13 "
500	348	2 "	50 "
600	379	2 "	36 "

COMMENTS

V = volume of urea distribution, *G* = urea generation rate, PCRn = normalized protein catabolic rate
Mean pre = mean predialysis BUN, TAC = time averaged BUN, *Kd* = dialyzer clearance including Qf
** ideal = 0.8 to 1.4 g/kg (normalized body weight)/day for average adults

Ideal treatment domain map

Figure I.3 shows the ideal range (shaded area) for time-averaged BUN (TAC) at differing protein catabolic rates. Target therapy is the dashed line; Kt/V isopleths are superimposed. Data are shown for the example patients listed in table I.2. Values for Kt/V represent equivalent values for dialysis three times/week and include a component of residual clearance.

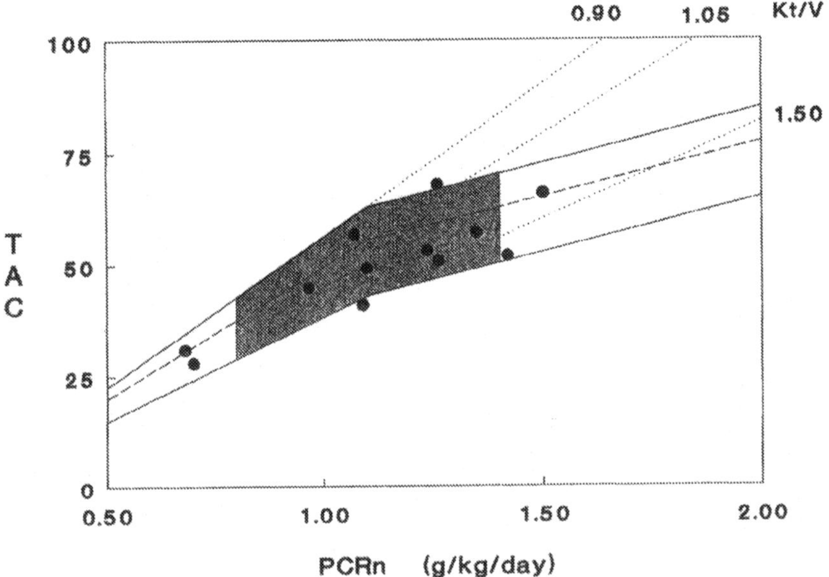

Figure I.3. Patient data superimposed on map of ideal treatment domain. Left axis is time-averaged BUN (TAC) in mg/dl. The horizontal axis is protein catabolic rate, normalized for urea volume.

References

1. Sargent JA, Gotch FA: Principles and biophysics of dialysis, in Replacement of Renal Function by Dialysis (3ed), Maher JF (ed), Dordrecht, Kluwer Academic Publishers, pp 87-143, 1989.
2. Gotch FA: Kinetic modeling in hemodialysis, in Clinical Dialysis (2ed), Nissensen AR, Gentile DE, Fine RN (eds), Norwalk CT, Appleton and Lange, pp 118-146, 1989.
3. Watson PE, Watson ID, Batt RD: Total body water volumes for adult males and females estimated from simple anthropometric measurements. Am J Clin Nutr 33:27-39, 1980.
4. Hume R, Weyers E: Relationship between total body water and surface area in normal and obese subjects. J Clin Pathol 24:234-238, 1971.

INDEX

Printed in the USA
CPSIA information can be obtained
at www.ICGtesting.com
LVHW022135220724
786230LV00003B/35